The Global Chemical Industry in the Age of the Petrochemical Revolution

This book fills a critical gap in modern economic and business history by presenting research by leading scholars to an international audience of academics, business executives, and policy makers. The research is presented in two clusters. The first explores four crosscutting topics, including surveys of the changes in industry structure, corporate strategies, plant technologies, governmental policies, finance, and corporate governance. The second cluster of studies comprises nine country surveys that examine the experiences of representative nations in chemical production and foreign trade. By combining the similar historical cases of a few nations (such as Sweden, Norway, and Finland), the authors are able to deal with eleven chemical-producing nations, including all of the leaders in this area as well as some of the important followers.

Louis Galambos is Professor of History at The Johns Hopkins University in Baltimore, Maryland, and is the editor of *The Papers of Dwight David Eisenhower*. He is the coauthor of *Networks of Innovation* (Cambridge, 1996); *The Fall of the Bell System* (Cambridge, 1987); *Anytime, Anywhere* (Cambridge, 2002); and *Medicine, Science, and Merck* (Cambridge, 2004).

Takashi Hikino is Associate Professor of Industrial and Business Organization at the Graduate School of Management and the Graduate School of Economics at Kyoto University, where he teaches industrial organization, business economics, corporate strategy, and comparative management.

Vera Zamagni has been Visiting Professor of European Economic History at the Bologna Centre of The Johns Hopkins University since 1973. Her published work consists of more than seventy essays, seven volumes, and thirteen edited volumes covering the economic history of Italy from 1860 to the present in the context of European and world economic history of the last two centuries.

COMPARATIVE PERSPECTIVES IN BUSINESS HISTORY

At the dawn of the twenty-first century, the world economy is in the midst of the most profound transformation since the industrial revolution. Firms, communications systems, and markets for products, services, labor, and currencies are all breaking out of national boundaries. Business enterprises today must negotiate a global environment in order to innovate and to compete in ways that will protect or enhance their market shares. At the same time, they are finding it essential to understand the different perspectives growing out of local, regional, and national experiences with business and economic development. This has become a crucial competitive advantage to companies and a vital skill for those who study them. *Comparative Perspectives in Business History* explores these developments in a series of volumes that draw upon the best work of scholars from a variety of nations writing on the history of enterprise, public and private. The series encourages the use of new styles of analysis and seeks to enhance understanding of modern enterprise and its social and political relations, leaders, cultures, economic strategies, accomplishments, and failures.

Series Editors

Franco Amatori, *Bocconi University*
Louis Galambos, *The Johns Hopkins University*

Managing Editor

Jill Friedman

Sponsors

Associazione per gli Studi Storici sull'Impresa (ASSI), Milan
Istituto di Storia Economica, Bocconi University, Milan
The Institute for Applied Economics and Study of Business Enterprise,
The Johns Hopkins University

Others Titles in the Series

Pier Angelo Toninelli, *The Rise and Fall of the State-Owned Enterprise in the Western World* (ISBN 0-521-78081-0)
Franco Amatori and Geoffrey Jones, *Business History Around the World* (ISBN 0-521-82107-X)

The Global Chemical Industry in the Age of the Petrochemical Revolution

Edited by

LOUIS GALAMBOS

The Johns Hopkins University

TAKASHI HIKINO

Kyoto University

VERA ZAMAGNI

University of Bologna

CAMBRIDGE UNIVERSITY PRESS

CAMBRIDGE UNIVERSITY PRESS
Cambridge, New York, Melbourne, Madrid, Cape Town, Singapore, São Paulo

Cambridge University Press
32 Avenue of the Americas, New York, NY 10013-2473, USA

www.cambridge.org
Information on this title: www.cambridge.org/9780521871051

© Cambridge University Press 2007

First published 2007

Printed in the United States of America

A catalog record for this publication is available from the British Library.

Library of Congress Cataloging in Publication Data

The global chemical industry in the age of the petrochemical revolution /
edited by Louis Galambos, Takashi Hikino, Vera Zamagni.
p. cm.
Includes bibliographical references and index.
ISBN 0-521-87105-0 (hardcover)
1. Chemical industry. 2. Petroleum chemicals industry. I. Galambos, Louis.
II. Hikino, Takashi. III. Zamagni, Vera. IV. Title.
HD9650.G576 2007
338.4'766 – dc22 2006016552

ISBN-13 978-0-521-87105-1 hardback
ISBN-10 0-521-87105-0 hardback

Contents

Introduction

TAKASHI HIKINO, VERA ZAMAGNI,
AND LOUIS GALAMBOS

The modern chemical industry emerged during the Second Industrial Revolution of the late nineteenth and early twentieth centuries, alongside other capital intensive industries such as primary metals, electrical machinery, food processing, oil refining, and automobiles. These industries were prime movers in the global economy. All came to be dominated by large international businesses, most of which drew heavily upon scientific knowledge to achieve advances in process and product innovations. None was more geared to research and development than the chemical industry, which led the way in developing industrial laboratories and establishing links with universities and other research centers.

Indeed, the evolution of the modern chemical industry provides many insights into the central strengths of the capitalist system in the twentieth and twenty-first centuries. These include its ability to respond to changes in its economic, political, and scientific and technological environments; its ability to remain innovative over the long term; and its ability to achieve high levels of operational efficiency in production and distribution of essential goods and services. The chemical industry that spawned our modern age of pharmaceuticals, plastics, synthetic fibers, and building materials stands out as a leading example of what an alliance between the institutions of capitalism and of modern science and technology could accomplish.

1

In chemicals and other Second Industrial Revolution industries, these accomplishments have come at a price, and the modern chemical industry can also provide insights into the problems stemming from industrial capitalism. These include wrenching changes for workers and communities, domestic and international political tensions, and the creation of environmental hazards. In the years before World War II, national governments were the primary agents dealing with these problems and determining the extent to which cartelization and mergers would shape the structure of chemical markets. In those decades, a few national "champions" with high degrees of market power emerged in each of the countries playing a leading role in chemical production.

World War II was a major watershed for the industry, for the national champions, and for the governments that had encouraged their development. During the postwar years, petroleum (and, to a lesser extent, natural-gas-based feedstocks) replaced coal and agricultural inputs, a transition that called for major innovations, as did the development of new environmental controls in country after country. The petrochemical revolution tested the flexibility and innovative capacity of all of the institutions – private corporations, governments, and nonprofit organizations – associated with the chemical industries and their responses provide a central concern of this book. All of these institutions were hierarchical and bureaucratic, and Joseph A. Schumpeter, the great analyst of innovation, contended that bureaucracy was inherently opposed to innovation. Business bureaucracies, he predicted, would coalesce with government bureaucracies and inevitably stifle innovation.[1] To the contrary, said Alfred D. Chandler, Jr., the leading historian of modern business. Large multinational corporations in chemicals (and electronics), he said, "succeeded by following virtuous strategies – that is, they used the profits and learning from each generation of new products to commercialize the next generation, and they defined their strategic boundaries around the capabilities in their integrated learning bases." Chandler gave less attention to the political setting than Schumpeter and focused primarily on the internal aspects, the strategies, structures, and capabilities of the leading firms. These capabilities, he said, "provided the internal dynamic for the continuing growth of the enterprise. . . . Such a process of growth has provided this bureaucratic institution [that is, the corporation] with the internal dynamic that has made it powerful and enabled it

---

[1] Joseph A. Schumpeter, *Capitalism, Socialism and Democracy* (New York: Harper & Brothers Publishers, 1942), especially p. 134.

to maintain its position of dominance as markets and technologies have changed and as world wars and depressions have come and gone."[2]

World War II and the petrochemical revolution certainly tested all of the organizations involved in the chemical industries. Following the war, competition was fierce as American chemical firms challenged the previously undisputed leaders in Germany, where in the 1930s the switch from coal to oil had first taken place. The petrochemical revolution enhanced the capital-using and scale-oriented characteristics of chemical technology. With this, the international chemical industry experienced massive expansion of upstream basic petrochemicals such as ethylene, propylene, and butylene for the production of plastics, synthetic rubber and synthetic fibers. By 1970, as a result, world plastics consumption by weight exceeded that of nonferrous metals, whereas synthetic fibers account today for more than half of the world's fiber consumption.

All major chemical firms and many petroleum companies in the industrialized countries invested massive resources in petrochemical facilities in an effort to exploit economies of scale. Meanwhile, several other nations, spearheaded by Japan, became significant producers in the global chemical industry through the rapid growth of basic petrochemical production. The number of players in the industry increased significantly, as did the competition. As John Kenley Smith writes in his chapter: "The headlong rush into chemicals demonstrated the power of markets to reduce prices, but also contributed to wasteful over-investment and undermined organizational capabilities." In particular, the oil companies, which were very large and had the raw material the industry now needed, were able to make massive investments in petrochemical plants. These organizations often lacked the research and marketing know-how that were indispensable for lasting success, but in the short-term, they transformed the competitive setting in the industry.

Then two politically induced, external events once again transformed the industry. The oil shocks of 1973–74 and 1980–81 drastically altered the economic viability of the chemical industry by increasing costs of production in a substantial way and by slowing the rate of growth of consumption. The firms responded in various ways and with various levels of difficulty to this challenge. One segment of the industry moved into

---

[2] Alfred D. Chandler, Jr., *Shaping the Industrial Century: The Remarkable Story of the Evolution of the Modern Chemical and Pharmaceutical Industries* (Cambridge: Harvard University Press, 2005), p. 309. Alfred D. Chandler, Jr., *Scale and Scope: The Dynamics of Industrial Capitalism* (Cambridge: Harvard University Press, 1990), pp. 8, 17.

downstream fine and specialty chemical production, which was much more R&D-intensive than upstream, basic chemical making. Another segment continued to concentrate on the large commodity-producing upstream facilities, which confronted excess capacity worldwide and had to be drastically reorganized. The chemical industry thus experienced restructuring and downsizing in the 1980s before other major Second Industrial Revolution industries were forced to introduce similar measures.

The pharmaceutical branch of the chemical industry also experienced a decisive transition in the postwar years. Here, too, American firms became significant competitors to a European industry severely damaged by the war. Massive U.S. public investments in basic research and in the training of scientific personnel created an innovation system that became the world leader in the medical sciences by the 1950s and 1960s. When the science base for pharmaceutical innovation began to shift away from organic chemistry and toward biochemistry and enzymology, American firms were able to push ahead of their European rivals in new drug development. So, too, when molecular genetics and biotechnology became the hot medical sciences of the 1980s and 1990s. American pharmaceutical firms and biotechs and the American-based subsidiaries of European firms were best positioned to take advantage of the new science and technology. Japanese pharmaceutical firms also made considerable advances during these years but still lagged their European and American competitors in innovative capacity.

These developments brought pharmaceuticals and the rest of the chemical industries to the edge of the Third Industrial Revolution and another wave of reorganizations and restructurings. The new technology of what has also been called the information age–microwave transmission, the transistor, the integrated circuit, the computer, and then the Internet – had begun very early to impact all of the chemical industries. Computer controls reduced the work forces needed to process chemicals, including pharmaceuticals. Ultimately, this changed the balance of power between management and labor by making it possible for supervisory personnel to run the plants and break strikes.

Other changes in the industry's context also fostered change within the industry. The development in Europe of a new and vigorous antitrust policy made the traditional forms of cartelization difficult to implement. The industry had a long and intricate experience with cartels, an experience that had involved American as well as the leading European firms. Meanwhile, a series of ecological disasters brought a wave of

new government regulations and forced producers to develop sophisticated process and product innovations that would better protect their
employees, customers, and local communities. During these same years,
the General Agreement on Tariffs and Trade (GATT) and the World Trade
Organization (WTO) were reducing national barriers to trade and the
breakdown of the Bretton Woods system of controls was fostering a truly
global, market-oriented financial system. The long era of the national
champions in chemicals was not entirely over, but it appeared to be
drawing to an end. In pharmaceuticals, the new science and technological context created a surge of new drugs and preventive medicines,
driving growth in this branch of the industry and utimately encouraging firms to focus on their core pharmaceutical activities and then to
seek through mergers, acquisitions and strategic alliances to sustain high
growth rates.

Despite all of these challenging changes, the bureaucratic firms of the
global chemical industries have, on balance, remained innovative and
successful. In that regard, Chandler has clearly trumped Schumpeter. As
the essays in this volume indicate, however, Chandler's synthesis needs
to be supplemented with more thoroughgoing analysis of the public, private, and nonprofit institutions that constitute essential elements of the
chemical industry's environment.[3] Together, this combination of institutions has enabled the chemical industry to grow in the ten years from
1992 to 2002, at an average rate of 3.3 percent in the EU, 2 percent in
the United States, and 1.4 percent in Japan, everywhere outperforming
the growth rate for the rest of the manufacturing sector. The distribution
of the world chemical production in 2002 can be seen in Figure I.1: the
EU still produces 29 percent and the United States 26 percent of the
world output of chemical products, whereas Japan makes 10 percent.
From the geographical share of world exports and imports reported in
Figure I.2, it can be seen that the EU still commands 55 percent of world
exports, with a large trade surplus (only 46 percent of imports), whereas
the United States has an almost balanced trade at around 15 percent.

As illustrated in Table I.1, the major enterprises in today's global
chemical industry still include some of the traditional leaders that had
secured their market dominating positions well before World War II, but
they are now joined by the chemical divisions of a number of oil companies. Other new entries include Huntsman of the United States, which

---

[3] Note in particular the analysis of "The Evolution of Networks in the Chemical Industry,"
Chapter 2, by Fabrizio Cesaroni, Alfonso Gambardella, and Myriam Mariani.

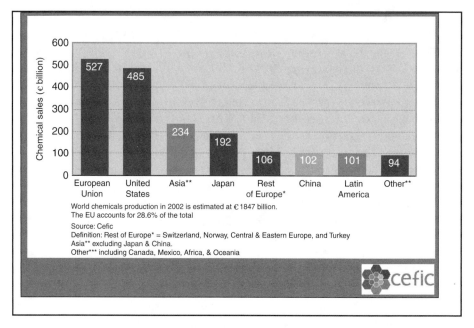

Figure I.1. Geographical breakdown of world chemical sales.

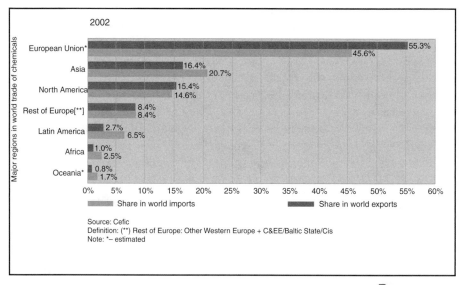

Figure I.2. Regional shares in world trade in chemicals.

Table I.1. Worldwide Sales 2002

Company		In mio euro	In mio USD	Country
1	BASF	32,216	30,441	EU
2	Bayer	29,624	27,992	EU
3	Dow Chemical	29,034	27,434	US
4	DuPont	25,406	24,006	US
5	ExxonMobil	21,494	20,310	US
6	Atofina	19,672	18,588	EU
7	Mitsubishi Chemical	15,967	15,088	Japan
8	Akzo Nobel	14,002	13,231	EU
9	BP	13,236	12,507	EU
10	Shell	12,160	11,490	EU
11	Degussa	11,765	11,117	EU
12	Asahi Kasei	10,097	9,541	Japan
13	ICI	9,740	9,203	EU
14	Sabic	9,604	9,075	Saudi
15	Sumitomo Chemical	9,400	8,882	Japan
16	Takeda	8,849	8,362	Japan
17	Linde	8,726	8,245	EU
18	Sasol	8,678	8,199	South Africa
19	Dainippon Inks & Chemicals (DIC)	8,138	7,690	Japan
20	General Electric	8,071	7,626	US
21	Ashland	7,983	7,543	US
22	Solvay	7,918	7,482	EU
23	Air Liquide	7,900	7,465	EU
24	Merck KGaA	7,473	7,061	EU
25	Huntsman	7,408	7,000	US
26	Sinopec	7,327	6,923	China
27	Sekisui Chemical	6,765	6,393	Japan
28	DSM	6,665	6,298	EU
29	Rhodia	6,617	6,252	EU
30	Basell	6,500	6,142	EU

Sources: Chemical Insight & Cefic-ITC (International Trade and Competitiveness) Analysis.

emerged through active buyouts of commodity chemical operations of other enterprises, and Saudi Basic Industries Corporation (SABIC), which was established by the government of Saudi Arabia as the central force of the nation's industrialization efforts. Most of these companies, however, both old and new, have changed radically during recent years. Even more complex is the picture of pharmaceutical companies reported in Table I.2 because of the extensive processes of mergers that will be discussed in the present collection, and because of the disproportionate presence of

Table I.2. Pharmaceutical Firms in the Top Global 500 in 2002

	Country	Revenues $mil	Employees
Merck	USA	51,790	77,300
Johnson & Johnson	USA	36,298	108,300
Pfizer	USA	35,281	98,000
GlaxoSmithKline	UK	31,874	104,499
Novartis	Switzerland	20,822	72,877
Aventis	France	19,497	78,099
Roche	Switzerland	19,096	69,659
Bristol-Myers-Squibb	USA	18,199	44,000
Astra-Zeneca	UK	18,032	58,700
Abbott Laboratories	USA	17,685	71,819
Pharmacia	USA	16,929	43,000
Wyeth	USA	14,584	52,762
Eli Lilly	USA	11,078	43,700
Shering-Plough	USA	10,180	30,500

Source: Fortune 2003.

the United States. The list comes from the *Fortune* 2003 global 500, and refers to 2002. It reports fourteen companies, but Pharmacia has since been acquired by Pfizer, so that at the time of this writing the companies number thirteen and Pfizer is on top. But new mergers are still being announced, as will be discussed later.

Compared with other knowledge-based, high-technology industries, the chemical industry has also been remarkably stable in terms of national and geographical leadership. This is particularly true because of the resilience of most European nations and their enterprises. In computer manufacturing, by contrast, once significant European countries have all struggled in recent decades. In consumer electronics, both Europe and the United States have lost significant market shares. In both industries, Japan and other East Asian economies have steadily expanded their presence. As set forth in Tables I.1 and I.2 and in the Figures, however, the leading chemical-producing firms in Europe and the United States have for the most part successfully transformed themselves from producers of upstream, basic-chemical commodities into leaders in the manufacturing of downstream, fine, and specialty chemical products and in so doing, have preserved their dominant position in this key global industry. Comparing the two world leaders, the EU is still stronger overall than the United States, but in pharmaceuticals, the United States has rapidly gained a significant position because of the excellence of the nation's research centers and its immense domestic market.[4] The leading European companies have all established operations in the United States to take advantage of a market largely free of price controls and a research establishment that can be tapped for personnel and new ideas.

Japan and the other East Asian emerging economies, which have been so prominent in other major industries, have yet to make deep inroads into the chemical industry, a subject discussed in some detail in Hikino's chapter in this book. In particular, Japanese chemical producers, which have experienced a large quantitative growth in output, have nevertheless continued to concentrate primarily on their domestic market.[5] In this case, the distinctive structure of their firms and interfirm networks have guided them toward the home market rather than exports.

TWO COMPLEMENTARY RESEARCH APPROACHES: THEMATIC AND NATIONAL

In spite of the centrality of the modern international chemical industry to the history of the world economy in the second half of the twentieth century – and its distinguished record of successful adaptation – research on this subject remains embryonic. With the notable exceptions of the general volumes by Haber,[6] Aftalion,[7] Arora-Landau-Rosenberg,[8] and Chandler,[9] there are only few academic studies of chemical industries within the leading industrial nations for the postwar period, and even fewer scholarly works that take a comparative perspective. This gap is especially remarkable when we compare it with the large number of

---

[4] In 1999, the world pharmaceutical market was divided thus: 40.5% United States, 26.2% Europe, 15.2% Japan, and only a meager 18.1% for the rest of the world. But the subdivision of the world market for the new drugs is even more strikingly in favor of the United States. Sales of the new medicines launched during the period 1997–2001 were directed 62% to the United States, 21% to Europe, 7% to Japan and a mere 10% to the rest of the world.

[5] In pharmaceuticals, Europe sells only 43% of what it produces in the internal market, the United States 56%, and Japan 94%.

[6] L.F. Haber, *The Chemical Industry During the Nineteenth Century* (Oxford: Oxford University Press, 1958), and *The Chemical Industry 1900–1930* (Oxford: Clarendon Press, 1971).

[7] Fred Aftalion, *A History of the International Chemical Industry* (Philadelphia: University of Pennsylvania Press, 1991).

[8] Ashish Arora, Ralph Landau, and Nathan Rosenberg, eds., *Chemicals and Long Term Economic Growth: Insights from the Chemical Industry* (New York: John Wiley & Sons, 1998).

[9] Chandler, *Shaping the Industrial Century.*

scholarly studies of other major industries, including automobiles, electronics, and computers. This neglect is especially acute for the years following the second oil shock, a watershed that wrought fundamental transformations in the global chemical industry.

The present volume aims precisely at filling this critical gap in modern economic and business history, by presenting research by leading scholars to an international audience of academics, business executives, and policy makers. This research is presented in two clusters. The first cluster of studies explores five crosscutting topics, including surveys of the changes in industry structure, corporate strategies, plant technologies, governmental policies, finance, and corporate governance. The second cluster of studies comprises nine country surveys that examine the experiences of representative nations in chemical production and foreign trade. By combining the similar historical cases of a few nations (such as Sweden, Norway, and Finland), these authors are able to deal with eleven chemical-producing nations, including all the leading ones and some of the important followers.

The book opens with a chapter by Cesaroni-Gambardella-Mariani sketching the history of the world modern chemical industry since its nineteenth-century origins. The authors' employ a networking analysis, a useful tool in the chemical industry, where networks have played a leading strategic role within the industry (interfirm networking includes collusion, cartelization, mergers); between the industry and the research centers; between the industry and the plants' suppliers; and also between the industry and the users of its products. Given this feature of the chemical industry, countries where networking has proved easier, either because of the natural propensity of the agents to coordinate their efforts or because of the role played by governments, have been the ones registering more lasting success.

Harm Schröter, in the second chapter of this book, provides a thought-provoking analysis of the strategies followed by the fifteen world leader companies across the period of restructuring of the industry after the oil crises. He starts by photographing the situation in the early 1970s and checks changes at intervals of about ten years. In the first period there were no new entries, and only a few of the fifteen companies adjusted their strategies. Although a much more radical change of strategies took place in the second period, up to the early 1990s, the fifteen companies remained the same. However, by now they were not as representative of the world chemical industry as they had been before, because some of the purely pharmaceutical companies originally excluded from the sample on the grounds that they were too narrowly specialized had become larger and more economically significant.

Interestingly, the industry's fundamental strategic reorientation took place during the most recent period, from the early 1990s to the present day. This was a business earthquake that entirely changed the profile of the industry. One of the fifteen companies (Montedison) disappeared, a demise explained in a later chapter by Zamagni. Another company, Union Carbide, was acquired by Dow Chemicals; Rhone Poulenc and Hoechst merged their pharmaceutical divisions into Aventis and sold the rest. This story is analyzed from the point of view of Rhone Poulenc in the chapter by Charue Duboc. Ciba-Geigy merged its pharmaceutical divisions with Sandoz, to form Novartis, while forming specialized companies for other products, a process that is well explained in the chapter by Margrit Müller. Monsanto merged its pharmaceutical division with Pharmacia, Upjohn, and Searle to form the new Pharmacia; Norsk Hydro divested out of the chemical industry and into oil, gas, and aluminum, whereas other oil companies enlarged their chemical divisions (as did Exxon Mobil, BP, Total with Atofina, Shell, and even the Italian ENI, whose chemical division, however, has always remained small).

Even the companies that still appear among the fifteen leaders of the early twenty-first century under their old names, like ICI and Bayer, have profoundly changed. ICI, for instance, in 1992 spun off its pharmaceutical division, called Zeneca (now Astra-Zeneca, after another merger) and slimmed down so much that by 2002 it no longer was included in the global 500 listed by *Fortune*. A panorama of the first fifteen chemical companies of today (2004) includes many new entries, whereas the inclusion in or exclusion from the list of purely pharmaceutical companies makes a great difference: in the global 500 *Fortune* list for 2002, there are twenty-five chemical companies listed, of which only seven are chemical conglomerates and four chemical branches of oil companies. All the remaining fourteen are pharmaceutical firms.

The chapter by Da Rin on finance provides an interesting analysis of the difficulties faced by the chemical corporations in meeting the challenges of restructuring in the 1970s: they had great difficulties in raising equity capital and experienced an enormous increase in their debt to equity ratios. Da Rin shows us that the United States was more efficient in solving these financial problems than Europe, whereas Japan showed the least impressive performance.

The final chapter of the first part deals with the environmental problems that have plagued the chemical industry since its beginnings. Wyn

Grant makes very clear that steps forward in the legislation protecting consumers, workers, and citizens are often incident driven and are highly politicized by the presence of green movements on one side and powerful lobbies of the chemical corporations on the other. A steady improvement in consumer protection, safety in the workplace, and pollution control started only in the 1970s, today reaching high levels of sophistication in the EU with the 4 December 2003 presentation to the European Parliament of the proposal for Regulation, Registration, Evaluation, Authorization and Restrictions of Chemicals (REACH). This new comprehensive regulation aims at proving the safety of all the estimated thirty-thousand-some chemical substances in circulation (not only of the new ones), and shifts the burden of proof to the industry, a very significant development.

Grant's essay focuses primarily on the development of environmental policy in Europe and thus develops a template that can be used to study policies in the United States and the developing nations. America clearly lagged Europe in creating an administrative state with regulatory capabilities, just as it lagged in social welfare protections. But in recent years, U.S. and European patterns of policy development – including the "collapse of the Keynesian project" and the increase in regulation – have tracked more closely without entirely eliminating the gap between the two regions. Because they have played such an important role in the global chemical industry, the EU and U.S. experiences with the environment are of overwhelming importance. The situation in the developing world is more problematical. The Grant template may fit the nations of Latin America and Asia, many of which have experienced environmental problems with chemical production and distribution. But the desire to catch up with the leaders in global economic development is powerful, the friends of the environment in the developing world are weak, and the opportunities for expansion of chemical production so attractive, that lag in this case may undercut both the globalization enterprise that Grant outlines and the existing EU regulatory structure that he so carefully analyzes.

The second part of the book opens with an essay by Ulrich Wengenroth that helps the reader grasp the reasons for the protracted success of the German chemical industry, core of the German economy and pillar of the European industry. In the postwar period, the Germans succeeded in rapidly bridging the gap with the United States that had emerged during the 1930s and World War II. In spite of the loss of around 30% of their capacity (which was left in East Germany), the West German firms quickly accomplished a switch from coal to oil. Above all, they started internationalizing very early, first within Europe; next in the Americas and especially in the United States; and currently in Asia.

Long successful exporting to foreign markets, Germany now employed foreign subsidiaries and made many foreign acquisitions. Overcapacity in petrochemicals was faced without major difficulties, and the industry has remained dynamic and innovative. The dark splotch on the industry's recent history is provided by biotechnology; the "private research foundations and the federal government were aware of the great potential" much before the industry, which is now trying to catch up through acquisition of companies in the field. But in 2000, even a leader like BASF had to sell its pharmaceutical division to Abbott Laboratories, and Hoechst had to merge its division with that of Rhone Poulenc to form Aventis. The problem is all the more serious if we remember that of all the chemical branches, pharmaceuticals is today by far the fastest-growing part of the industry.

The next essay by John Kenley Smith concentrates exclusively on U.S. petrochemicals, where initial American leadership was tempered by an early maturity, from which the author does not see the industry capable of escaping. Pharmaceuticals and biotech have flourished in America, but not the other branches of the industry. In the American context, then, Smith concludes that the industry looks less successful than other Second Industrial Revolution and most Third Industrial Revolution industries. The operative word in this case is "looks." As Smith explains: "It is a large and important sector of the economy; it generates a significant trade surplus; it is technologically sophisticated; it supports a creative R&D extablishment; and because its returns are not much above the cost of capital, it has fallen out of favor with investors." Alfred D. Chandler's recent book advances an even gloomier conclusion: "Thus, early in the twenty-first century," he says, "one of the two technology pioneers of the Second Industrial Revolution, the chemical industry, is no longer a high-tech industry."[10]

Both Chandler and Smith are, we suggest, especially negative because they are evaluating an industry in the trough following a long period of intense innovation. They are, as well, emphasizing the petrochemical branch of the industry and within that branch, the DuPont Company, which has recently struggled to adopt a new strategy that would promise greater returns in the immediate future. The other essays of this volume are thus more optimistic about the U.S. situation, taking note of the fact

[10] Chandler, *Shaping the Industrial Century*, p. 312.

that the country's chemical industry is still one of the two undisputed world leaders. The difference between seeing the glass half full or half empty may be found in two aspects of the authors' perspectives: whether one emphasizes growth potential (as Smith does) or current standing (as do the other authors); and whether one treats only petrochemicals (as does Smith) or scans across the entire industry (as the other authors are inclined to do). An even broader perspective would take into account the manner in which science-based industries like chemicals advance in waves as the underlying science changes and generates new opportunities. Neither we nor our other authors can predict that another wave will follow, but the probabilities that this will occur seem higher than either Smith or Chandler acknowledge.

Outside of Germany and the United States, there are a handful of countries that can boast of a few companies that are successful world competitors. Small nations such as Switzerland and the Nordic countries have taken the approach of strengthening some solid internationalized corporations through mergers. In the chapter by Margrit Müller, the proliferation of chemical companies in Switzerland, particularly in Basle, is described and the process of concentration is followed with great care: Ciba and Geigy merged in 1970 to form Ciba-Geigy, later Ciba. Another merger took place in 1996 when Ciba combined its pharmaceutical division with Sandoz, forming Novartis, a major world player in pharmaceuticals. Recently, Novartis has shown great interest in the possibility of merging with Roche. Müller, like Wengenroth, questions the practice continued in this volume of conducting national analyses in the presence of global firms like these, firms that are less and less dependent on national competitive advantages.

The successful strategies of mergers and internationalization also have been adopted in the Nordic countries dealt with in the chapter by Gunnar Nerheim. He gives a full account of the development of the petrochemical industry in Sweden, Finland, and Norway and carefully considers the role played by the oil companies. Restructuring after the oil crises produced two major companies – Neste in Finland and Statoil in Norway – each of which had a large chemical division that nevertheless was soon considered to be below global scale. So, the two companies organized a new corporation, named Borealis in honor of its geographical origin, which was "the largest polyolefin manufacturer in Europe and one of the largest in the world."

In spite of being a much larger country than Switzerland or the Nordic countries, France's strategy in the end has been similar. As the chapter by

Florence Charue Duboc shows, France has primarily been a producer of bulk chemical products for the domestic market, much like Japan in this regard. To move beyond that strategy and become a world player, France, too, had to resort to mergers. This was the case with Atofina, which emerged in the late 1990s from a long process of mergers in the oil industry and then in their chemical divisions. Sanofi-Synthelabo, as well, was formed in 1999 as a result of a series of mergers among pharmaceutical companies. Most significantly, Aventis, also in 1999, was created by the consolidation of the pharmaceutical division of Rhône Poulenc with the Hoechst health division, thus crossing national borders and forming another major world player in pharmaceuticals. Charue Duboc's essay follows the internal organizational changes that made it possible for Rhône Poulenc to reach this position – a story of continued concentration that is continuing to unfold. At the time of writing this introduction, Aventis was merging with Sanofi-Synthelabo.

The British case is somewhat different, because Britain already had a very large chemical producer and world-class competitor, ICI. Formed in the 1920s, ICI had succeeded after World War II in resisting all attacks on its leadership in the domestic market, as Wyn Grant shows. Although it was one of the best-performing British companies, competition in the global market finally caught up with ICI, which was forced to restructure in the 1990s. The firm sold its chemical division, Zeneca, in 1992, and its industrial chemical division in 2001. Wyn Grant is nevertheless right in concluding that the chemical industry in the United Kingdom is a case of relative success in a country where Second Industrial Revolution industries generally have performed poorly.

This survey of the European countries is not complete, but it enables us to reach a firm conclusion: when we put together all the successful European corporations, the highly competitive status of Europe in the field of global chemical products stands out clearly. Neither companies nor governments opted for the status quo; they have continuously struggled to keep up with the rapid pace of innovation in chemicals. They have crossed national borders not only to establish subsidiaries and make acquisitions, but in the end, also to consumate significant mergers. The EU is still far from having an homogeneous domestic market, but the road is now wide open to reach this goal in a number of industries, among which the chemical industry is far ahead.

In a category of its own must be placed the Japanese case researched by Takashi Hikino, who documents the intense quantitative growth of the chemical industry in Japan. This growth, however, has not produced a world leader, as has been the case in other industrial sectors. Japan remains a supplier only to its domestic market. In spite of the presence of Japanese chemical corporations at the top of international rankings because of their size, Hikino says, "Japan's chemical industry remains a marginal and invisible player in the global industry." To discover the reasons for this unusual performance is not at all easy, and Hikino concludes that "the Japanese chemical industry is still a puzzle." He hypothesizes that the Japanese chemical companies' membership in the different *kigyo shudan* has made it difficult for them to combine into one or more undisputed global leaders. In this case, the export strategy of the Japanese economy has not been as successful as it has been in electronics. Germany, which also has been traditionally an export-oriented economy, gave up this feature in the chemical industry long ago in favor of manufacturing abroad, but Japan has not adopted this strategy. Research also must be a problem, because, as Hikino states, "Japanese chemical companies in general have not yet exhibited their technological competencies in radical product or process innovation."

The two remaining essays in the collection cover other European countries, classified as followers because they do not have major corporations that can be listed among the world's leading competitors. The Italian case reconstructed by Vera Zamagni is of particular interest because it shows that Italy could follow the French pattern of becoming a world player in at least some lines of production. Italy has a strong tradition in chemical production and a major company, Montecatini. But poor decisions at the company and government levels defeated this possibility, to the point of weakening Montecatini's successor, Montedison, so much that it was unable to meet the challenge of radical restructuring in the 1990s and had to exit the field. Also, the Italian oil company ENI, which had successfully developed its chemical division, has failed to internationalize and to make its chemical division large enough to compete in global markets. It too is now trying to exit the field. This left Italian chemical production in the hands of SMEs, some of which have been quite creative about developing niche markets (in the manufacturing style of the country), and subsidiaries of foreign multinational companies.

The final chapter by Nuria Puig documents the intense growth of the chemical industry in Spain since the 1880s, its process of modernization in the years 1960-74, and its subsequent restructuring. The author provides a solid base of quantitative data and traces a full picture of Spain's industry. The story here is that of a rather successful "latecomer." Although the presence of foreign subsidiaries is overwhelming, some local firms have internationalized and become competitive, though their size "puts some question marks on [their] next development."

The appendix by Giannetti and Romei provides a much-needed database (up to the early 1990s) covering the six largest countries included in this book. It provides an excellent guide to the leading roles of the U.S. and German industries, shows the speed with which Japan caught up in quantitative terms, maps the French and British resilience in chemicals, and charts the Italian weakness in this important industry.

In conclusion, it is apparent that most national champions have already teased apart their conglomerate businesses and restructured their operations along more tightly focused lines. Increasingly, they have crossed national borders in one form or another seeking access to resources and national markets and economies of global scale. Some firms have specialized in niche markets, and indeed, the chemical industry is still relatively fragmented, especially in Europe. The future appears to promise more efforts to achieve a global scale of operations, with substantial efforts meanwhile to broaden the firm's reach through vigorous networking. It is the intertwining of national cases with company histories that has enabled the authors of these chapters to improve our understanding of this latest phase in the evolution of the chemical industry, a story that has its roots in national competitive advantages but moves steadily toward a global future that is already less positive than Chandler's vision and still substantially more successful than Schumpeter's prediction about the fate of the bureaucratic industrial corporation.

The authors and editors gratefully thank ASSI, the Italian Association for Business History, for having convened in 1999 at the Bocconi University in Milan a conference from which this book derives, and Franco Amatori for having overcome many difficulties and brought this project to a successful conclusion. Jill Friedman at the Institute for Applied Economics and the Study of Business Enterprise at Johns Hopkins University saw the manuscript through the editorial process. Frank Smith of Cambridge University Press guided it into publication.

PART I

Crosscutting Issues

1

The Evolution of Networks

In the Chemical Industry

FABRIZIO CESARONI, ALFONSO GAMBARDELLA,
AND MYRIAM MARIANI

INTRODUCTION[1]

There are quite a few studies of the chemical industry.[2] None of them, however, examines specifically the rise and the nature of networks among firms, research institutions, and customers. This is the goal of our chapter, which describes the factors that gave rise to different types of networks in the chemical industry and examines the characteristics, the evolution over time, and the role of networks in shaping the structure of the industry. Three types of networks are described: interfirm networks such as the strategic agreements among firms for production, marketing, and research and development (R&D) activities; university-industry networks; and user-producer networks that firms developed in order to be responsive to the needs of their customers. The characteristics of these networks changed over time and influenced the behavior of chemical firms and the evolution of the industry.[3]

[1] Myriam Mariani acknowledges support by a Marie Curie Fellowship of the European Community programme IHP (Improving Human Potential), grant no. HPMF-CT-2000-00694. Fabrizio Cesaroni acknowledges financial support by the Italian National Research Council (CNR), through project no. CNRG00B857.
[2] See, for example, Nathan Rosenberg, Ralph Landau, and David Mowery, eds., *Technology and the Wealth of Nations* (Stanford: Stanford University Press, 1992).
[3] We have not considered the networks within multinational firms, leaving that subject to the chapters on the various national industries. For excellent guides to that subject and the

We also examine the role of national governments in influencing the development of these networks. For example, in the United States, the strict antitrust policy made it difficult and dangerous to form cartels and other collusive agreements among firms. This, in turn, led to the restructuring of the industry by means of market mechanisms such as mergers and acquisitions. In Europe, where the antitrust policy was less severe, cartels were used to reduce costs and to solve the problem of production overcapacity.

Although our focus is on the period after the petrochemical revolution, we briefly address the evolution of networks in the period before this decisive transition. We do so for three reasons. First, the relationships that chemical companies established before the petrochemical revolution created the opportunity to "learn how to interact" both in firm-to-firm alliances and in university-industry relationships. Later, when the specialized engineering firms (SEFs) emerged after World War II, chemical companies were already accustomed to collaborating with other parties. This led to the establishment of numerous linkages with the SEFs for the exchange of process technologies. Second, although the petrochemical revolution represented a great break in the history of the industry, the top companies in terms of technological competencies and turnover before that transformation were still leading the industry after the breakthrough. This was the case, for example, with the German chemical companies, and the evolution of networks before World War II helps explain this aspect of the industry. Finally, the United States – underdeveloped compared with the European countries before World War II – emerged as a leading country in chemicals only after the petrochemical revolution. An understanding of the evolution of the industry since the nineteenth century helps explain this change in regional leadership.

THE NINETEENTH CENTURY: UNIVERSITY-INDUSTRY RELATIONSHIPS AND NETWORKS WITH USERS

The modern chemical industry started in Great Britain in the first half of the nineteenth century with the establishment of inorganic chemical

literature, see Mira Wilkins, *The History of Foreign Investment in the United States, 1914–1945* (Cambridge: Harvard University Press, 2004) and Geoffrey Jones, *Multinationals and Global Capitalism: From the Nineteenth to the Twenty-first Century* (Oxford; New York: Oxford University Press, 2005).

firms. The industrial revolution had created a large demand for acids and alkalis that were used in the production of textiles, soap, glass, and steel, and British firms started by making such inorganic products as soda, soda ash and bleach. However, the organic sector, and particularly dyestuffs, would soon become the real engine of growth. United Kingdom firms understood that sales, distribution, and the volume of output were key factors for taking control over pricing in the inorganic sector. In the meanwhile, rapid technological change in the organic sector led to the methodical application of scientific discoveries to chemical manufacturing. The competitive advantage of firms shifted from production and marketing activities to research and innovation. This allowed Germany to catch up with Britain in chemicals, and by the eve of World War I, Germany had become the world leader in the synthetic organic sector.

German firms succeeded because they invested systematically in manufacturing, marketing, and research capabilities[4] and because they developed a large number of networks with other firms, customers, scientific research centers, universities, and individual researchers. As far as manufacturing capabilities are concerned, German firms such as Hoechst, Bayer, BASF and AGFA started as technology followers that imitated the dyes produced by the British and French manufacturers. However, they soon understood that the demand for dyestuff products was growing rapidly, and that most dyes were based on similar organic intermediates leading to economies of scope in the production of different products and colors. They also lowered their production costs by means of economies of scale and by increasing the internal demand for organic intermediates. Furthermore, when scientific research in the 1880s led to the discovery that some chemical intermediates for the production of dyes had also therapeutic properties, many German dyes firms diversified into the pharmaceutical sector. In 1883, Hoechst produced the pain-reliever Antipyrin, and in 1899 Bayer patented the pain-relieving, fever-reducing, anti-inflammatory, Aspirin. In 1887, AGFA used the same technological convergence of organic intermediates to diversify into photo chemicals.

These diversification strategies led German firms to set up two different types of networks. On the one hand, they started to market their products together with technical assistance services in order to increase the interaction with the users and to satisfy their diversified demand. On the other hand, to achieve economies of scale and scope, these companies

---

[4] Alfred D. Chandler, Jr., *Scale and Scope* (Cambridge: Harvard University Press, 1990).

integrated backward into coal mining and into the production of intermediates and basic chemicals. For example, when the contract with a
supplier of sulphuric acid, an important intermediate product in the production of dyes, expired in 1892, Bayer tried to reduce the production
cost by integrating backward into the sulphuric, hydrochloric, and nitric
acids.[5] This upstream integration often took place by means of interfirm
agreements like the one between BASF, AGFA, and Bayer to buy the coal
mine Auguste Victoria in Marl.[6]

High production volumes called for large markets, and in order to
reach these markets firms had to invest in marketing capabilities and in
distribution networks. In other words, economies of scale in production
activities needed to be combined with economies of scale in marketing.[7]
Again, German firms invested in marketing capabilities and by the end
of the nineteenth century they were world leaders in the manufacturing
of organic chemicals. Their distribution channels spread worldwide and
reached important foreign markets such as those in the United States,
Britain, China, France, Russia, and India.

The formation of networks was a key factor in the development of
marketing capabilities. First, firms established complex interactions with
the users of their products for technical reasons (i.e., dyes had to be used
according to specific procedures depending on the color of the dye, its
chemical structure, and the kind of cloth to be colored). German firms
began to offer technical assistance while selling their products. German
chemical experts in dyes were dispatched to the textile factories, and
technical people from the textile factories were invited by the chemical
firms to attend training courses in Germany. This gave German dyestuff
firms a significant advantage over their competitors.[8] Interaction with
users allowed dye firms to better understand their customers' needs and
to produce additional innovations to satisfy a diversified demand.

As the chemical firms quickly discovered, the improvements in their
distribution and sales networks needed to reach a large and diversified

---

[5] Hans-Joachim Flechtner, *Carl Duisberg: From a Chemist to an Industrial Leader* (Dusseldorf: ECON Verlag GMBH, 1959).
[6] Erik Verg, Gottfried Plumpe and Heinz Schultheis, *Meilensteine* (Leverkusen: Bayer; Köln:
Vertrieb, Informedia, 1988).
[7] Chandler, *Scale and Scope*.
[8] Johann Peter Murmann and Ralph Landau, "On the Making of Competitive Advantage: The
Development of the Chemical Industries in Britain and Germany Since 1850," in Ashish
Arora, Ralph Landau, and Nathan Rosenberg, eds., *Chemicals and Long-term Economic
Growth* (New York: John Wiley & Sons, 1998).

demand were extremely expensive. Moreover, by the end of the nineteenth century, German companies understood that their different sales and distribution networks often overlapped, and that they could reduce the distribution costs by developing interfirm commercial alliances. These factors led to the formation of numerous firm-to-firm networks that allowed companies to reduce their distribution costs while increasing economies of scale and scope.[9] The largest firms combined into two alliances: the *Dreibund* (Union of Three), among Bayer, BASF and AGFA; and the *Dreiverban* (Association of Three), among Hoechst, Casella and Kalle.

During the nineteenth century, the dye market and, more generally, the organic chemical market were growing rapidly, and firms were realizing high profit margins. This profitability, however, induced new firms to enter the market and, as a consequence, prices and profit margins declined again. To prevent the erosion of profits, the largest chemical companies built up their own research and development laboratories and increased product innovation and differentiation. Also, their linkages with the universities and other scientific research centers became more important because the invention of new products was strictly related to the advances taking place in the scientific understanding of the chemical structure of new molecules. Chemical companies started to recruit academic researchers and to develop research collaborations with academia in order to invent new products and apply for joint patents.

This was the case for Hoechst, which established a link with the University of Erlangen (with its researcher Ludwig Knorr), and this joint venture produced Hoechst's first drug, Antipyrin.[10] German dyes firms also tried to promote alliances with German universities to set up special research institutes. Between 1911 and 1914, three new chemical research institutes, like the Kaiser Wilhelm Institutes for Chemistry and Physical Chemistry in Berlin, were formed and largely financed by corporate funds.[11] These research institutes gave rise to a particular brand of "co-operative capitalism" based on both competitive and cooperative ties among chemical firms in Germany.[12]

---

[9] L. F. Haber, *The Chemical Industry, 1900–1930: International Growth and Technological Change* (Oxford: Clarendon Press, 1971).

[10] Murmann and Landau, "On the Making of Competitive Advantage."

[11] Jeffrey Allan Johnson, *The Kaiser's Chemists: Science and Modernization in Imperial Germany* (Chapel Hill: University of North Carolina Press, 1990).

[12] Chandler, *Scale and Scope.*

THE ROLE OF GOVERNMENTS, THE RISE OF THE U.S. DURING 1914–1940, AND THE FORMATION OF COLLUSIVE NETWORKS

World War I produced big changes in the structure of the international chemical industry. Chemical firms produced explosives, drugs, and fertilizers for the war needs, and the fact that dyes and explosives shared a common scientific and technological base led many dye producers to enter the explosives market. The influence of governments on firms' strategies was also strong, although there are differences among countries. Whereas in Germany, the presence of chemical trade associations made it easier to create a link between the government and the individual firms, in Britain the absence of such associations led to a deep public intervention.[13] The British government took control over a large part of the economy and reorganized the chemical industry to supply chemical compounds for war needs. In both nations government-led coordination forced the large chemical firms to become better acquainted with each other and led to the creation of new associations and alliances among firms. British firms founded the Association of Chemical Manufacturers, while in Germany the eight largest dyes producers (the six firms already involved in the *Dreibund* and *Dreiverban*, and the two independent Chemische Fabrik Griesheim-Elektron and Weiler-ter-Meer) formed a "quasi-cartel," called an *Interessengemeinshaft* (Community of Interests). This increasing interaction among firms changed the structure of the chemical industry.

Two important macroeconomic changes shaped the industry in this period. First, while the demand for explosives, drugs, and fertilizers during the war allowed chemical firms to utilize fully their production capabilities, the end of the war and the reconstruction phase threw the industry into a crisis caused by the sharp decline in the demand for chemical products in all countries. Almost all of the governments responded with protectionist policies to solve their political and industrial problems, rather than attempting to deal with the real economic reasons for distress. They imposed import restrictions and tariff walls and limited international trade. In Britain, the Dyestuffs Import Regulation Act of 1921 prohibited the import of major dyestuffs for several years, and the Import Duties Act of 1932 imposed tariffs on all imported goods.[14] For countries like Germany, whose firms were strongly export-oriented, protectionism was a major problem. In general, firms in all countries

---

[13] Haber, *The Chemical Industry, 1900–1930.*
[14] Murmann and Landau, "On the Making of Competitive Advantage."

experienced overcapacity problems during the interwar period, which led to a restructuring phase of the industry via mergers, acquisitions, and cartels.

Mergers and Acquisitions

Most chemical firms used mergers and acquisitions in an effort to "rationalize" their operations and deal with the economic depression after World War I. IG Farben in Germany and ICI in Britain were a result of this process. In Germany, the *Interessengemeinshaft* quasi-cartel could no longer control the industry. Paradoxically, protectionist policies had actually increased foreign competition. By the early 1920s, a full merger of all the largest dye companies became the only apparent solution to these problems. In October 1925, the eight firms that were already members of the *Interessengemeinshaft* merged into a new single entity called *I.G. Farbenindustrie Aktiengesellschaft* (IG Farben). IG Farben became the dominant German chemical firm in terms of capital endowment, research investments, exports, and sales, and it played a decisive role in a variety of chemical and metal product markets.

The companies that formed the conglomerate, however, did not integrate completely, and their lines of business remained fairly distinct. (This turned out to be useful after World War II when IG Farben was broken up.) As a reaction to the formation of IG Farben, British firms undertook a similar process of industry consolidation, and in January 1927, four of the largest chemical firms (Brunner, Mond & Co. Ltd., Nobel Industries Ltd., the United Alkali Company, and the British Dyestuffs Corporation Ltd.) merged into the Imperial Chemical Industries (ICI). Like its German counterpart, ICI became the dominant chemical corporation in its home country.

In an effort to relax competition further, these companies engaged in a large number of alliances with other firms. During this period, ICI established more than eight hundred agreements covering all aspects of chemical production, and in 1927, ICI signed a *Patents and Process Agreement* with DuPont in order to reduce the technological gap between ICI and firms like IG Farben and DuPont. IG Farben joined in more than 2,000 agreements with other firms. These contracts relaxed competition and reduced the probability of new firms entering the market. It is worth noting that, while the industry structure was thus becoming static, dominated by a small number of large companies and the cartels, the process of technological innovation remained dynamic: a large number of new products and processes were introduced and diffused beyond the industry's sectoral boundaries.[15]

Cartels

Cartels played a significant role in the restructuring of the chemical industry. These types of alliances were not new to the industry, and many had been formed before World War I. Germany had a less severe antitrust policy than Britain, but there were many cartels in both countries. In the 1890s, United Alkali and Brunner Mond in Britain developed an agreement to avoid direct competition. In addition to the *Dreibund* and the *Dreiverband*, the Cynamides Convention in Germany (which lasted until the 1930s) attempted to control markets. Some cartels were also formed between German and British firms. The Nobel Trust, which was set up in 1886 among explosives manufacturers from both countries, lasted until 1914 when the war made it impossible to maintain the alliance.

The macroeconomic and sociopolitical conditions of the interwar period led to the formation of a large number of cartels. The world economy simply could not absorb the output of the large chemical plants that were created during the war. Moreover, the Great War produced a close sense of community between the governments and the chemical firms, as well as among individual firms, making it easier to set up new cartels. In Germany, the decline of international demand and the ensuing overcapacity problems, as well as the fact that cartels were considered a legal political tool under Nazi party law[16] led to the formation of additional agreements. In 1905, 13 cartels existed in the heavy chemical industry. In 1923, the number of cartels in the whole industry had increased to 93.[17] The less favorable political and legal climate in Great Britain produced a smaller number of cartels, and the creation of ICI reduced the number of players in the chemical industry and led to a more stable industry structure.

Two international cartels were founded. The first one was in the dyestuffs sector. To prevent entry by firms from other countries, the European dyes producers set up a cartel agreement whose members shared their technological knowledge. The agreement was signed in 1926 among the German (IG Farben), Swiss (Swiss IG), and French dyes

---

[15] Murmann and Landau, "On the Making of Competitive Advantage."

[16] Ibid.

[17] L. F. Haber, *The Chemical Industry During the Nineteenth Century* (Oxford: Oxford University Press, 1958).

producers. In 1932, ICI also joined the agreement, under which each member received a specific sales quota. The whole dyes cartel accounted for about 62% of the world output.[18] The second important combination was the Nitrogen Cartel (*Convention International de l'Azote*). Nitrogen is a key compound in the production of fertilizers, and during the Great War, both ICI and IG Farben had increased their capacity to supply farmers after international trade in nitrogen was forbidden. So when the war ended, both firms faced serious problems of overcapacity. In an effort to keep prices high, they set up an international cartel agreement that was signed in 1930. This agreement was less effective than the dyes cartel because it could not incorporate all the international nitrogen producers. Nevertheless, it allowed ICI and IG Farben to reduce overproduction without important consequences for the prices of their products.[19]

The U.S. Chemical Industry

During the First World War, the United States became an important new player in the international chemical industry. The growth of the U.S. chemical industry mirrors that of the whole U.S. economy. In the nineteenth century, the abundance of nonreproducible natural resources in the United States was the most important factor shaping the economic growth of the country. By the end of the century, the United Stated had become the world leader in the production of minerals such as coal, copper, iron, ore, zinc, phosphate, tungsten, molybdenum, lead, and others.[20] This led the industry to produce inorganic chemicals such as fertilizers and explosives. Soda ash and caustic soda were imported in large quantities from Britain. The organic sector was significantly less developed than the inorganic one: the production of dyestuffs, plastics, coal-tar products, and nitrogen fertilizers was quite small, and by 1914 it accounted for only about one-quarter of the total chemical industry production.[21]

The size of the internal market – shaped by rapid U.S. economic growth and geographical integration – gave the U.S. industry another important advantage. The large national market created high demand

[18] Harm G. Schröter, "The International Dyestuff Cartel, 1927–39, with Special Reference to the Developing Areas of Europe and Japan," in Akira Kudo and Terushi Hara, eds., *International Cartels in Business History* (Tokyo: University of Tokyo Press, 1992).

[19] Schröter, "The International Dyestuff Cartel."

[20] Ashish Arora and Nathan Rosenberg, "Chemicals: A U.S. Success Story," in Arora et al., *Chemicals and Long-Term Economic Growth*.

[21] Williams Haynes, *American Chemical Industry*, vol. 2 (New York: Van Nostrand, 1945–1954).

for chemicals and allowed firms to exploit economies of scale at the plant level, particularly in the production of explosives, fertilizers, and sulphuric acid. The U.S. producers, however, were still importing techno-logical knowledge from Germany and Britain, and it was only after World War I that U.S. companies started to invest successfully in the devel-opment of internal technological capabilities through company-owned research laboratories. At the end of the Great War, the U.S. chemical industry, and the U.S. economy as a whole experienced a sharp recession, mainly because of monetary problems and imports of organic products from German (that had been temporarily suspended during the war). The U.S. chemical industry started to grow again during the 1920s, and despite the Great Depression of the 1930s, it continued to develop thanks to the introduction of new products.

The interwar period was a period of consolidation for industry, and the U.S. experience was somewhat different than that of the European producers. The severe U.S. antitrust policy made cartel formation diffi-cult. As a result, mergers were used to achieve similar results. About five hundred mergers took place during the 1920s,[22] such as the one between Allied Chemical and Dye Corporation in 1920, and that between Union Carbide and Carbon Corporation in 1917. Both were based on comple-mentary activities. Other firms like American Cynamid, DuPont, Ameri-can Home Products, Mathieson Alkali, Monsanto, and Hercules acquired less efficient companies.

Not all were aimed at reducing competition. Different and mutually reinforcing motivations explain the use of mergers and acquisitions for restructuring the industry.[23] The search for economies of scale and scope through large-scale plants was one reason to merge. Nevertheless, the economic integration of the U.S. regional markets and the rising product standardization increased price competition and also encouraged merg-ers and acquisitions. Firms tried to limit the decline of chemical prices by differentiating their products in terms of technological and marketing characteristics such as services, delivery, and brand names. This led firms to invest in downstream marketing and distribution capabilities, and the high fixed cost to enter the market under these conditions forced smaller or less efficient firms to be acquired or to exit the industry.

Moreover, the possibility of reaping economies of scope by using a common knowledge base for the development of a wide range of

---

[22] Williams Haynes, *American Chemical Industry*, vol. IV.
[23] Arora and Rosenberg, "Chemicals: A U.S. Success Story."

different organic products created an incentive to product differentiation. DuPont, for example, used its technological knowledge in explosives to move into cellulose products; Dow Chemical diversified in chlorine and petrochemicals; Union Carbide in petrochemicals and air separation gases; Air Products in air separation gases and catalysts. To increase innovation capabilities many firms set up internal R&D laboratories and increased their size by using the retained profits. Once again the situation called for larger firm size to achieve efficiency and led to a new process of consolidation.

FROM THE 1940S TO 1970S: TOWARD INTERFIRM AND USER-PRODUCER NETWORKS

The Second World War profoundly affected the evolution of the chemical industry. Here we highlight the rise of two types of networks during and after the war (i.e., the networks among firms and the networks with the users of the chemical products), and describe the major factors that contributed to the evolution of the networks and to the changes in the industry's structure: the historical events, two important technological changes (i.e., the rise of polymer chemistry and chemical engineering), and the entry of Japan in the world chemical market.

Industry Dynamics: The Effects of the Second World War

The Second World War was much more devastating than the previous world conflict, particularly for Germany. The war program of the Nazi government forced the German chemical industry into autarky. Enormous resources were devoted to high-pressure technologies for coal hydrogenation and gasification that would be fields of little commercial significance in the future (with the exception of artificial fibers and synthetic rubbers). Much of the German infrastructure and physical plants was destroyed. Also, the German science base suffered from the war. The deportation of many Jewish or "ideologically hostile" scientists and engineers from the leading universities and research centers damaged the quality of German science for several decades because most of these people were unable or unwilling to return to their positions after the end of the war.[24] Moreover, in July 1945, President Truman ordered all the

---

[24] Otto Keck, "The National System for Technical Innovation in Germany," in Richard R. Nelson, ed., *National Innovation Systems: A Comparative Analysis* (New York: Oxford University Press, 1993).

proprietary records of the top four hundred German firms and research institutions to become public. The Allies confiscated German know-how, trademarks, patents, and the total assets of IG Farben without any compensation.

Soon after the war the chemical industry was held highly responsible for making German aggression possible. IG Farben, in particular, was linked to the war crimes associated with gas chambers in the concentration camps. Chemical firms were immediately subjected to Allied control and production halted. Germany was directed to adopt stringent antitrust laws, and cartels were prohibited. IG Farben was broken apart in the early 1950s into three successor companies: BASF, Hoechst, and Bayer, which instead of competing with one another created their own separate markets, as already mentioned. With the end of the collusive networks, the U.S. and European industrial structures began to converge. Many U.S. companies were willing to sell their technologies. This lowered entry barriers and reduced concentration in the European market as many firms entered new businesses.

World War II also had a major impact on the structure of the industries of the winner countries. In the United States, the war opened up important opportunities for interfirm networking. The U.S. government created a massive demand for aviation oil[25] and launched various programs for pulling national companies into co-operative research and production projects. One of them was the Rubber Program that made the four major rubber companies and Standard Oil cooperate in the production of synthetic rubber. The American government invested approximately 700 million dollars in this project.[26] These and other war and postwar programs such as the Marshall Plan fostered the formation of several co-operative networks among the U.S. companies. Firms were forced to exchange information, to coordinate their research efforts, to exchange personnel, and to co-operate in downstream applications. When the conflict ended, and the manufacturing facilities managed under these cooperative networks were sold to private firms (typically to companies operating them during the war), these firms started to co-operate without being forced to do so. They learned that cooperative networks could be used for managing production and research activities more efficiently, leading to long-term relationships that fostered the growth of the U.S. chemical industry after the Second World War.

---

[25] Fred Aftalion, *History of the International Chemical Industry* (Philadelphia: University of Pennsylvania Press, 1989).

[26] Arora and Rosenberg, "Chemicals: A U.S. Success Story."

After the war another type of network arose in the United States: networks with the users, which proved to be extremely important for the commercial exploitation of the new technologies involved in making such products as plastics, synthetic fibers, and drugs. Thanks to excellent management, many American companies established systematic interactions and feedbacks between their laboratories and the downstream users (including, for instance, textile companies) in order to be readily informed about specific needs.

At the end of the Second World War, the United States was the dominant global chemical producer. But only a few years later Britain succeeded in rebuilding its chemical industry, and Germany, which suffered the most from the war, was able to recreate an economic and social environment that encouraged the growth of its postwar chemical industry. In 1952, the Allies removed the major restrictions to the development of the German chemical industry. Because of the upcoming conversion into petrochemical technologies, the German technological losses were minimal, and the three successors of IG Farben entered quickly in this new technology by creating interfirm networks, mainly in the form of joint ventures with British oil companies. They also imported American petrochemical technologies.[27] Britain, in contrast, could not sustain its first-mover advantage in petrochemicals because the postwar ICI did not have strong management capabilities, and the chemical industry as a whole lacked access to cheap capital in order to exploit the big opportunities offered by the petrochemical technologies.

Polymer Chemistry: Networks with the Science Base and with the Users

Polymer chemistry is the science of chemical products. Initiated by Herman Staudinger and other German scientists in the 1920s, polymer chemistry is based on the idea that any material consists of long chains of molecules (i.e., polymers) linked together by chemical bonds. The scientific understanding of the existence and configuration of these long chemical macromolecules led to the principle of "materials by design."[28] According to this principle, there is a relationship between the characteristics of the macromolecular structures and the proprieties of materials. By using different building blocks and by changing the way in which the molecules are assembled, a variety of products can be developed. At that

[27] Murmann and Landau, "On the Making of Competitive Advantage."

[28] Ashish Arora and Alfonso Gambardella, "Evolution of Industry Structure in the Chemical Industry," in Arora, *Chemicals and Long-term Economic Growth*.

time, long experimentation was still needed before obtaining the desired material, but the scientific base made the search for new products more productive. The use of catalysts was a fundamental tool: by controlling the rate and the manner in which monomers were connected, it became possible to obtain the desired length and physical structure of the polymers, leading to new and differentiated materials.

The rise of polymer chemistry dramatically influenced the evolution of the chemical industry in the postwar era. It encouraged the formation of networks between the producers and the users of chemical products. By providing a common technological base for developing applications and product differentiation in five distinct and otherwise unconnected product markets (i.e. plastics, fibers, rubbers and elastomers, surface coatings and paintings, and adhesives) polymer chemistry reduced the amount of time and research needed to develop product innovations. This was a new solution to "how" to innovate. The question, however, shifted to "what" to innovate. In other words, while the process of producing new products was comparatively easy for any chemical firm, the discovery of the "right product" was not, and the competition among firms shifted to the correct anticipation and development of the most suitable users' applications. To innovate successfully firms had to be knowledgeable about the characteristics of different market segments, and to do so, they had to develop extensive linkages with the downstream markets. These networks allowed the producers to collect information about the users' specific needs and helped the firms to train the users to use the new products.

The opportunities created by polymer chemistry were exploited by a large number of companies worldwide that had the size, the scope, and the in-house expertise to exploit them. As Freeman[29] points out, the presence of a large number of firms with comparable capabilities in polymers implied that even "small" information leaks allowed very rapid imitation. Many chemical companies and some oil producers found themselves competing in very similar markets.[30] The increased competition in almost every market segment led to a renewed attention to product differentiation and commercialization strategies as important sources of

[29] Chris Freeman, *The Economics of Industrial Innovation* (London: Francis Pinter, 1982).
[30] For instance, Union Carbide, Goodrich, General Electric, IG Farben, and ICI were all producing and doing research in PVC. Dow, IG Farben, and Monsanto were involved in the polystyrene business. DuPont, ICI, Union Carbide, Monsanto, Kodak, and many other firms invested in other types of polyamides, acrylics, and polyesters (Peter H. Spitz, *Petrochemicals: The Rise of an Industry* (New York: Wiley, 1988); Aftalion, *History of the International Chemical Industry*.

competitive advantage. This encouraged extensive investments in R&D to develop new product variants that were tailored for specific applications. Moreover, since the desired properties of a material varied according to the specific use, the possibility of gathering information about the users increased the probability of designing the proper product. Again, to achieve this goal, it was important to develop systematic networks with the users.

The success of polymer chemistry owes a great deal to the shift from coal to petroleum hydrocarbons. This shift began in the years before the Second World War in the United States, which then had abundant reserves of oil and natural gas. By 1950, petrochemicals covered half of the U.S. organic chemical production. Ten years later this share was 88%.[31] Only the development of a world market for oil, and the international diffusion of the petrochemical technologies led to the early decline of American leadership in petrochemicals and to Western Europe's ability to catch up. The upsurge of chemical engineering helped firms in this process.

Chemical Engineering: University-Industry Linkages and the Development of Vertical Networks

If polymer chemistry is the science of chemical products, chemical engineering is the science of chemical processes. Its main concern is the design, construction, and operation of large-scale chemical processing plants to manufacture and commercialize new products efficiently.

The size of the market led to the early introduction of this discipline in the U.S. to solve the problems of large-scale production of some basic products like chlorine, caustic soda, soda ash, and sulphuric acid. However, what came to distinguish the discipline of chemical engineering was the concept of "unit operation," presented by Arthur D. Little to the Massachusetts Institute for Technology (MIT) in 1915.[32] The "unit operation" refers to the breaking down of chemical processes into a limited

---

[31] Keith Chapman, *The International Petrochemical Industry: Evolution and Location* (Oxford, UK; Cambridge, MA: Basil Blackwell, 1991).

[32] The concept of "unit operation" involves breaking down a complex chemical process into a series of elemental components such as evaporation, filtration, grinding, crushing, etc., that are common to many different chemical contexts (G. Wright, "Can a Nation Learn? American Technology as a Network Phenomenon," in Naomi Lamoreaux, Daniel Raff, and Peter Temin P., eds., *Learning By Doing in Markets, Firms, and Countries* (Chicago: University of Chicago Press, 1999). See Nathan Rosenburg, "Technological Change in Chemicals: The Role of University-Industry Relations," in Arora, *Chemicals and Long-term Economic Growth* for a discussion on the concept of "unit operation," and the role of MIT in the development of the chemical engineering discipline.

number of basic components or distinctive processes that were common to many product lines. This abstract and general concept of the engineering discipline became the "general purpose technology" of the chemical sector, providing the unifying base for more contextualized and problem-solving activities at the plant level.[33]

These features of chemical engineering and the separation between product and process innovation in the chemical industry gave rise to important changes in the sector and brought about new types of networks: the networks with universities and the networks between the chemical companies and the specialized process design and engineering contractors (hereafter, SEFs). The remainder of this section will discuss these two types of networks.

The University-Industry Networks

In the first decades of the twentieth century, many American universities established departments of chemical engineering whose distinctive feature was the strong orientation toward practical industrial utility. At the same time, the shift from coal to petrochemicals and the growth of the petroleum-refining industry led to a higher demand for university-trained engineers and to close university-industry partnerships. The link between the university and the industry, and the partial dependence of the former on industry funding, assured the focus on industrial needs. Moreover, in order to develop many processing technologies and to achieve meaningful results, chemical engineers needed the large-scale operations of the chemical firms, a setting that the university could not supply.

One important example of a university-industry network in the United States is that between New Jersey Standard and MIT at the research facility in Baton Rouge, Louisiana.[34] The PhD degree came to play a role in chemical engineering much earlier than in other engineering disciplines, and the demand for graduate students in chemical engineering grew rapidly. At the time of the presentation of the "unit operation," the consulting company of Arthur D. Little employed a large number of MIT graduates.[35]

[33] Rosenberg, "Technological Change in Chemicals."
[34] Ralph Landau and Nathan Rosenberg, "Successful Commercialisation in the Chemical Process Industries," in Landau et al. *Technology and the Wealth of Nations.*
[35] Rosenberg, "Technological Change in Chemicals" shows that the number of PhD degrees in chemical engineering awarded in the United States between 1905 and 1979 was more than 9,000.

This interaction between profit-seeking institutions and independent or semi-independent professional scientists, a development which influenced the evolution of the discipline, was taking place across a broad front in America during these years.[36] DuPont, for example, interacted extensively with the academic world and pushed toward higher scientific and mathematical bases in chemical engineering. Despite the influence of large companies, the scientists maintained their independence and their professional status.[37] Threatened by the possibility of the university as a potential employment option, firms often adapted their employment conditions to match those typically found in academia. They allowed chemical scientists and engineers to maintain a certain degree of freedom and flexibility and gave them the chance to publish their research achievements.

Germany resisted chemical engineering as an autonomous discipline until the 1960s and drew a clear demarcation between subjects to be studied at the university and those of more immediate usefulness to the industry. In Britain, ICI also had only limited interests in university-trained engineers up to the Second World War; only when Britain entered the refining market did the demand for chemical engineers grow rapidly.

Networks between the Chemical Companies and the SEFs

The rise of chemical engineering, the growing importance of petrochemicals, and the increases taking place in the scale and complexity of plants led to the rise of a market for engineering and process design services for chemical plants. This market was operated by a large number of small, specialized and technology-based firms (the SEFs) that have been an original and persistent feature of the American chemical industry. With a few exceptions, the SEFs did not develop radically new processes. They were good at moving down the learning curve for processes invented by the large oil and chemical companies. Equally important, they acted as independent licensors on the behalf of other firms' technology.

The fact that process technology was made into a commodity that could be traded had a major impact on the structure of the industry, and played an important role in the diffusion of chemical technologies. Strong economies of specialization were achieved at the industry level,

[36] Louis Galambos, with Jane Eliot Sewell, *Networks of Innovation: Vaccine Development at Merck, Sharp & Dohme, and Mulford, 1895-1995* (New York: Cambridge University Press, 1995).

[37] G. Wise, "A New Role for Professional Scientists in Industry," *Technology and Culture*, 21, 1980: 408-429.

and a large number of vertical networks were developed between the chemical companies and the SEFs. These vertical linkages often resolved into partnering relationships of two types: between the SEFs and a number of chemical firms developing new technologies; and between the SEFs and an even larger number of firms buying these technologies.[38] As Freeman[39] points out, between 1960–1966 "(n)early three quarters of the major new plants were 'engineered', procured and constructed by specialist plant contractors," and the SEFs were the source of about 30% of all licenses of chemical processes. The accumulated competencies in process design were the basis of the SEFs' comparative advantage in developing the "market for chemical technologies." They supplied the necessary process technologies, the design, and the engineering know-how of new plants. In so doing, they facilitated the entry of new firms into the chemical industry after the Second World War and allowed other countries such as Germany to catch up quickly in petrochemicals.

The existence of a market for technology in the chemical industry with numerous vertical linkages between the producers and the users of technologies is confirmed by recent data.[40] Arora and Gambardella[41] show that between 1980–1990, the SEFs engineered more than 70 percent of the total number of plants in the world. Although the share of SEFs varied across chemical sectors, it was above 50 percent in all sectors. The rest of the plants were engineered in-house or by other companies, including the chemical-engineering divisions of the chemical companies. This phenomenon was less marked for chemical licenses, even though the SEFs still accounted for about 35 percent of total licensing in the industry.

Tables 1.1 and 1.2 show the country distribution of the market for chemical technologies in 1980–1990. By looking at the nationality of the seller and the acquirer of the technology, the tables highlight the international nature of these vertical networks.

Although the SEFs started as an American phenomenon, only 50 percent of the total value of engineering contracts worldwide between 1960–1966 involved American SEFs.[42] Table 1.1 shows that other

---

[38] For instance, Badger used the fluidized bed catalytic process to develop processes for phthalic anhydride with Sherwin Williams, ethylene dichloride with BF Goodrich, and acrylonitrile with Standard Oil of Ohio. Similarly, University of Pennsylvania had a number of strategic partnerships with Dow, Shell, Ashland, Toray and BP to develop different products.

[39] Chris Freeman, "Chemical Process Plant: Innovation and the World Market," in *National Institute Economic Review*, 45, August 1968: 29–51.

[40] Ashish Arora, Andrea Fosfuri, and Alfonso Gambardella, *Markets for Technology: The Economics of Innovation and Corporate Strategy* (Cambridge: MIT Press, 2001).

[41] Arora and Gambardella, "Evolution of Industry Structure in the Chemical Industry."

[42] Freeman, "Chemical Process Plant: Innovation and the World Market."

Table 1.1. Market Share of SEFs – Engineering Services: 1980–1990 (Share of Plants by Region)

Nationality of SEFs	Regions				
	U.S.	West Europe	Japan	Rest of the World	Share of Total World Market
U.S.	58.8	19.8	3.7	18.9	26.0
West Germany	1.9	18.5	4.6	12.7	11.7
UK	6.9	12.2	2.0	7.3	8.1
Italy	0.3	8.2	0.0	5.8	5.1
France	0.2	2.3	0.3	4.6	3.2
Japan	0.2	0.2	34.0	5.1	4.0

Source: Chemical Age Profile.

countries successfully competed with the United States in this field, particularly in Europe and in the third-world markets. Between 1980 and 1990, Germany, the UK, Italy, France and Japan had 11.7 percent, 8.1 percent, 5.1 percent, 3.2 percent, and 4.0 percent of the total market of SEFs' services, compared with 26% of plants engineered by the American SEFs. These vertical networks are also very regional in nature. Most of the plants in the United States are engineered by U.S. SEFs, those in Europe are done by European SEFs, and the plants in Japan are built by Japanese SEFs. Moreover, when the licensee and the licensor are of different nationality, Table 1.1 shows that the probability of an American SEF serving a foreign chemical company is higher than the probability of a European or Japanese SEF selling technology to the United States. In other words, while the American SEFs have a sizeable share of the European market, the European SEFs have only a small share of the U.S. market. This might be due to the establishment of many American SEFs'

Table 1.2. Market Share of SEFs – Licenses: 1980–1990 (Share of Plants by Region)

Nationality of SEFs	Regions				
	U.S.	West Europe	Japan	Rest of the World	Share of Total World Market
U.S.	18.0	10.3	6.5	16.9	15.1
West Germany	3.1	11.3	1.0	10.2	8.8
UK	1.2	3.0	2.7	1.4	2.4
Italy	0.1	1.4	0.0	2.2	1.6
France	0.1	0.6	0.0	0.9	0.7
Japan	0.1	0.1	1.5	1.1	0.7

Source: Chemical Age Profile.

Table 1.3. Licensing Agreements: 1980–1997 (Shares of Total Licenses by Type of Licensor and Region)

Licensor	Receiving Country				
	Germany	UK	Japan	U.S.	Total
SEFs	8.9	8.3	10.4	23.3	50.9
Top Chem. Firms*	1.7	1.4	2.7	3.7	9.5
Other Chem. Firms*	0.1	0.2	0.2	0.3	0.8
Staff	7.4	5.6	9.5	16.3	38.8
Total	18.1	15.5	22.8	43.6	100.0

* *Top Chemical Firms*: Companies in the top 50 positions in terms of number of plants; *Other Chemical Firms*: Companies with more than 5 plants, excluding the top 50 companies.
Source: Chemintell, 1998.

subsidiaries in Europe after the Second World War, firms that have then become full-fledged "national" companies.

Table 1.2 reports the country shares of SEFs for licensing. The U.S. market share is 15.1 percent; the share of Germany, 8.8 percent. The comparative advantage of the U.S. SEFs in licensing is even larger in terms of the share of U.S. licenses to Europe and to Japan. Table 1.2 suggests that, in the case of licensing, vertical networks typically involve American or German SEFs that license technology to European, Japanese, or third-world firms. Similarly, Table 1.3 looks at the country distribution of 5,442 licensing agreements from the point of view of the licensor type. These data are drawn from the *Chemintell* database, which collects information on about 36,000 plants built worldwide since 1980.[43]

Table 1.3 shows that the SEFs are the most important source of chemical processes technologies in all the developed countries. They control 50.9 percent of the world market for technology. Half of the transactions are in the United States (23.3%). In-house development (16.3%) is the second most important source of technology. The shares of SEFs transactions and in-house technology development conditional upon each receiving country are very similar: about 50 percent of the chemical technologies are supplied by the SEFs, and 40 percent by the companies' staffs. In order to analyse this issue in greater detail, Table 1.4 looks at the type of companies involved in 36,343 licensing agreements since 1980.

[43] The *Chemintell* database reports information for each plant on the type of chemical compounds it produces, its production capacity, the technology used, the owner of the plant, the contractor that provided the engineering services, the licensor, the year of construction, etc.

Table 1.4. Licensing Agreements: 1980–1997 (Shares of Licenses by Type of Licensor and Licensee)

| | Receiving Company | | | |
Licensor	Top Chem. Firms*	Other Chem. Firms*	"Non" Chem. Firms*	Total
SEFs	9.3	39.8	19.1	68.2
Top Chem. Firms*	1.6	6.9	2.7	11.2
Other Chem. Firms*	0.2	0.9	0.4	1.5
Staff	8.6	8.8	1.7	19.1
Total	19.7	56.4	23.9	100.0

* *Top Chemical Firms*: Companies in the top 50 positions in terms of number of plants; *Other Chemical Firms*: Companies with more than 5 plants, excluding the top 50 companies; *"Non" Chemical Firms*: Companies with 5 plants or less.

Source: Chemintell, 1998.

Table 1.4 confirms that the SEFs are the principal suppliers of technologies in the chemical sector. They cover 68.2 percent of the total market for licensing. This is true of all types of companies with at least one chemical plant. The SEFs license almost 50 percent of the technologies used by the top chemical firms, 70 percent of the know-how used by the companies with at least five chemical plants, and 80 percent of the technologies used by the companies with less than five plants. The largest chemical companies use the SEFs less than the other firms probably because they have better in-house capabilities. This is confirmed by the fact that the top chemical companies develop internally almost half of their technological know-how, and they sell technologies also to other chemical companies (Table 1.4). This is suggestive of the role of the SEFs. The existence of the SEFs, whose business is to sell process technologies in order to appropriate rents from innovation, encouraged other chemical and oil firms to license their technologies for making profits out of them.[44] This increased the amount of intra-industry linkages and vertical networking among technology producers and users in the chemical industry.

The Entry of Japan

The Japanese chemical industry developed around the second half of the twentieth century, when the European and the U.S. firms were already dominating the industry both in terms of sales and technological

---

[44] Ashish Arora and Andrea Fosfuri, "Licensing the Market for Technology," *Journal of Economic Behavior and Organization* 52:2 (2003): 277–295.

knowledge. Clearly, Japanese chemical firms existed in the nineteenth century, but the industry was extremely fragmented and the firms were in specialized low-technology sectors. It was only during the 1960s that Japan appeared in the international arena, mainly because of the development of petrochemicals, a move that was often driven by the Japanese government. This late industrialization was characterized by three elements: the use of foreign technologies at the start of the industry (mostly from the United States); strong international competition from Europe and the United States; and extensive trade barriers in Japan.

These factors made the Japanese chemical industry structurally different from the industries in America and Europe.[45] For example, Japanese firms from the downstream sectors such as textiles, detergents, pharmaceuticals, and fertilizers integrated backward to acquire technological capabilities and to produce chemical compounds. This model of "backward integration" was used also after World War II, when many downstream firms integrated into petrochemicals in order to acquire the emerging technologies.

The policies adopted by the Japanese government and the presence of large industrial groups influenced the behavior of companies and the formation of networks.[46] Before World War II, government policy was fragmented and did not promote any structural industrial policy. Only after the war did the Minister of Finance and MITI guide the growth of the petrochemical sector through the definition of specific plans. The first, the Petroleum Industry Development Plan, was formalized in July 1950 with the aim of stabilizing the balance of payments and foreign exchange. This was because the growth of the Japanese economy after the war led many downstream sectors such as textiles, machinery, electronics, and automobiles to increase their imports of chemical feedstock, which negatively affected the balance of payments. This intervention promoted the development of a domestic chemical industry. Some Japanese chemical companies also entered the international market.

Problems arose, however, when the government promoted strong competition among business groups to foster the growth of the industry. Competition led to high investments, company diversification, and overcapacity. During the 1970s, when the world demand for chemical products decreased, prices fell. The government tried to promote the

---

[45] Arora and Gambardella, "Evolution of Industry Structure in the Chemical Industry."

[46] Takashi Hikino, Tsutomo Harada, Yoshio Tokuhisa, and James Yoshida, "The Japanese Puzzle: Rapid Catch-Up and Long Struggle," in Arora, *Chemicals and Long-term Economic Growth*.

restructuring of the industry without leaving it to market forces. The presence of large groups limited the ability of firms to pursue individual strategies, and as a result, a successful stabilization was never achieved.

The existence of large and diversified business groups (the *kigyo shudan* that were an evolution of the existing *zaibatsu* in the prewar period) was another characteristic of the Japanese chemical industry.[47] These groups influenced the relationships among companies and the formation of networks like the three most important chemical groups: Mitsubishi, Mitsui, and Sumitomo. Since the 1950s, the establishment of large-scale petrochemical production facilities required large financial investments that individual firms could not afford. Petroleum companies, chemical firms, downstream chemical users, banks, and trading companies were able to combine their financial resources because of their long-term relationships.[48] Moreover, the working of complex petrochemical facilities required the logistical co-ordination of different processes that were controlled by specialized subsidiaries and joint-ventures. To achieve better coordination, these entities set up structural linkages and merged into business groups that were able to achieve economic stability. Intergroup sales and networks helped reduce transaction costs, particularly when firms exchanged technological knowledge.

However, the members of the groups had limited strategic power. They could not efficiently diversify if other firms belonging to the same group were already active in a specific sector or in a geographical area. This limited the ability to achieve economies of scale and scope at the plant and firm level. The possibility of merging or acquiring other companies was also limited. Firms could not merge among groups because of group rivalry. Mergers and acquisitions within the groups were also limited by noneconomic factors, such as executive and management rivalries. Furthermore, the structure of corporate governance in the Japanese groups gave managers high degrees of autonomy, with little control by stakeholders. Managers oriented their actions toward the growth of sales and market share, while little attention was paid to firm efficiency.

During the 1970s, when macroeconomic conditions required a process of consolidation, the structure of the business groups limited the ability to restructure the industry. MITI created a depression cartel with the objective of reducing production capacity of the major chemical and

---

[47] The terms *kigyo shudan* and *keiretsu* are often used interchangeably, but their historical and economic meaning is different. A *keiretsu* is an array of firms linked by long-term relationships and controlled by the largest entity of the group. *Kigyo shudan* is a broader concept, and a single *kigyo shudan* is usually composed of one or more *keiretsu*.

[48] Hikino et al., "The Japanese Puzzle."

petrochemical companies. The process, however, took many years to complete. For example, the consolidation of Mitsubishi Kasei and Mitsubishi Petrochemical (two petrochemical-related companies belonging to the Mitsubishi group) lasted for more than twenty years. The same problem arose in the Mitsui group. In general, the existence of business groups and public intervention during the process of consolidation of the industry limited the flexibility of the Japanese chemical firms to respond to market forces.

FROM THE 1970S TO THE PRESENT: THE RESTRUCTURING OF THE INDUSTRY AND THE CONTINUING OF INTERFIRM NETWORKING

The rise of the SEFs fostered competition in the chemical sector and led to a substantial increase in the number of chemical firms in most markets. During the 1950s and 1960s, the industry could accommodate this process because the demand for chemical products was growing rapidly. But in the early 1960s, profitability started to decline, a condition that became severe in the 1970s and 1980s because of the entry of competitors from the developing countries. This was also the period of the oil shocks when firms in a large number of chemical markets, especially basic intermediates, experienced substantial excess capacity.

The crisis was felt all over the world, and the adjustment to a new equilibrium was slow and painful. The large chemical producers had sunk large investments in capacity, and many of them were highly integrated, both vertically and horizontally. The first phase of the restructuring of the chemical industry occurred through the rationalization of production capacity. Next, there was a phase of restructuring of the large corporations. These processes were not identical in the United States, Europe, and Japan, even though the problem of excess capacity was common to the three areas. In the United States, the restructuring was market driven. In Europe, it was driven by market and government intervention and lagged the United States by about five years. In Japan, it was entirely co-ordinated by MITI.

With the restructuring of the chemical industry two types of networks emerged: those with users and partnering networks between firms. The reasons for the formation of these networks are different from the reasons for network formation between 1940 and 1970. The linkages with the users developed because of the crisis in the basic and intermediate petrochemical sectors where competition was most severe. A number of companies in the United States and Europe exited from the commodity chemical sectors, and moved downstream. They were replaced by oil companies, many of which took over existing commodity chemical firms. This process led firms to specialize either in commodity chemicals, or in more downstream specialty sectors. Many firms were willing to enter specialty chemicals because these markets were characterized by high product differentiation, moderate price competition, low volumes of production and high margins. Moreover, in these sectors well-developed user-producer relationships were a critical asset enabling them to achieve consumer satisfaction, strengthen their competitive advantage, and gain market share. Many companies invested in building up close linkages with final markets and offered services to their customers in addition to chemical products. Akzo, for example, established special consultancy centers for the purchasers of varnishes. ICI offered a computer-aided optimisation system for explosives.[49] In Japan, where the role of the users was already important, the adjustment toward the customization of high-quality products was quick.

The restructuring process involved also a large number of interfirm networks in production and R&D. The creation of interindustry associations fostered such interfirm agreements. In 1985, 34 large petrochemical firms formed the Association of Petrochemicals Producers in Europe (APPE). APPE did not formally encourage interfirm arrangements, but by collecting and diffusing information on the production capacities in different product markets, the association helped members make decisions about their strategies informed by industry-wide considerations.

Mergers and acquisitions (M&A) and alliances reduced the number of businesses in which chemical companies were active and increased the absolute size and market share of their remaining product lines.[50] American and European firms acted differently in this respect. U.S. firms were more likely to use acquisitions in Europe to consolidate their existing lines of operations in the European market. By contrast, a large number of EU

---

[49] H. Albach, D. B. Audretsch, M. Fleischer, R. Greb, E. Hofs, L. Roller, and I. Schulz, 1996, Innovation in the European Chemical Industry, Research Unit Market Processes and *Corporate Development*, Paper presented at the "International Conference on Innovation Measurement and Policies," Luxenbourg, May 20–21, 1996.

[50] The restructuring of the European PVC market provides an example of interfirm arrangements. BP ceded its PVC operations to ICI in 1981. ICI exited from polyethylene by relinquishing its activities to BP. Polyethylene was concentrated in BP and in PVC in ICI. In 1985, ICI formed a joint venture with Enichem (European Vinyls) to merge the PVC businesses of the two firms. The new company became the largest European PVC producer. Similarly in polypropylene, Statoil and Neste merged their petrochemical operations to form Borealis.

Table 1.5. Interfirm Agreements: 1988–1997 (Shares of Total Number of Agreements by Sector)

Type of Interfirm Agreement	Chemicals & Allied Products	Drugs	Soaps	Rubber	Petroleum	Total
	Chemical Sectors					
Equity Purchase	0.5	2.5	0.1	0.1	0.1	3.3
Funding Agreement	0.1	1.9	0.0	0.0	0.0	2.0
Joint Manufacturing Operations	11.4	6.1	1.0	2.3	1.2	22.0
Joint Marketing Arrangements	3.9	9.8	0.6	0.8	0.4	15.5
Joint Natural Resource Exploration	0.1	0.0	0.0	0.0	0.1	0.2
Joint Research and Development	2.2	11.8	0.1	0.3	0.1	14.5
Joint Venture	10.8	3.8	1.0	2.3	2.0	19.9
Licensing Agreement	2.2	11.6	0.2	0.4	0.1	14.5
Original Equip. Manuf./VAR Agreement	0.0	0.1	0.0	0.0	0.0	0.1
Privatizations	0.1	0.0	0.0	0.0	0.0	0.1
Royalties	0.2	4.2	0.0	0.0	0.0	4.4
Spinout	0.0	0.0	0.0	0.0	0.0	0.0
Supply Agreement	0.4	0.7	0.0	0.1	0.1	1.3
Other	0.5	1.1	0.0	0.1	0.5	2.2
Total	32.4	53.6	3.0	6.4	4.6	100.0

Source: SDC, 1998.

firms acquired U.S. companies in activities that were unrelated to the core business of the acquirer firms with the aim of developing new capabilities, rather than reinforcing their commercial presence in the United States. To put it differently, U.S. firms acquired companies in commodity chemicals to achieve large-scale operations, while EU firms aimed at diversifying in specialty chemicals.[51] Table 1.5 shows the sectoral distribution of 14,818 interfirm agreements in the chemical industry since 1988. These data are drawn from the *Securities Data Companies* (SDC) database.[52]

---

[51] See Arora and Gambardella, "Evolution of Industry Structure in the Chemical Industry" for details.

[52] The SDC data are constructed from SEC filings (10-Qs), financial journals, news wire services, proxies, and quarterly reports. Each record reveals the technological content of

Table 1.6. Trends in Acquisitions in the Chemical Sector (Number of Acquisitions by Acquirer)

Acquirer Nationality*	Years									Total
	1985	1986	1987	1988	1989	1990	1991	1992	1993	
U.S.	210	138	225	206	184	210	252	262	3	1690
	(3%)	(5%)	(5%)	(14%)	(14%)	(15%)	(15%)	(23%)	(6%)	(11.8%)
Europe	78	32	58	79	193	269	242	235	2	1188
	(29%)	(34%)	(34%)	(61%)	(69%)	(61%)	(55%)	(61%)	(100%)	(57.6%)
Japan	4	6	7	12	28	18	24	11	0	110
	(0%)	(0%)	(29%)	(92%)	(79%)	(100%)	(75%)	(73%)	(0%)	(71.8%)

* All deals announced worth more than 1 million dollars, or that involve more than 5% of a firm, or of undisclosed value are covered. Deals that are announced but then cancelled are not included. Entries in each column refer to the nationality of the acquiring company. Fractions of overseas acquisitions are in parentheses. For Europe, systematic coverage begins in 1987. Coverage for Japan does not include possible acquisitions in Japan. Figures for 1993 include some acquisitions announced in 1994 as well.

Source: Calculations from the IDD, Information Service Data Base.

Joint manufacturing (22.0%), joint marketing (15.6%), joint-R&D (14.5%), and licensing agreements (14.4%) are the most frequent type of partnership. Joint ventures in production, marketing, and technology cover 19.9 percent of the agreements. Moreover, joint ventures and joint-manufacturing operations are more often used in the chemical and allied products sector (10.8% and 11.4%), which is a relatively mature branch of the industry. As the sector becomes more research intensive (e.g., drugs) the share of licensing agreements for the exchange of technology (11.6%) and the share of joint-R&D operations (11.8%) increase. Interfirm networks in R&D are a means to overcome the classical problem of market failure in R&D. During the restructuring phase, chemical firms lowered the investment in R&D together with their product portfolio. Interfirm agreements in R&D, university-industry partnerships and industry-wide research projects became a means to sustain R&D efforts, particularly in research-intensive sectors. Tables 1.6 and 1.7 examine the time and country distribution of mergers and acquisitions.

Table 1.6 reports the share of acquisitions in the chemical sector between 1985 and 1993, and classifies them by the nationality of the acquirer company. In the United States, there were 1,690 acquisitions (12% of which were foreign) compared with 1,188 acquisitions in Europe (58% foreign) and 110 in Japan (72% foreign). These data are consistent

the transaction, the names of the granter and recipient of the technology, and the presence of cross-licensing agreements.

with the idea that the United States was the first country in which a market-driven restructuring process took place. In the United States many companies were created in the 1980s because large producers of commodity chemicals sold their plants to other firms while moving downstream. There are many examples of firms focusing on commodity chemicals. Huntsman was founded in 1983 by the acquisition of Shell's polystyrene business. Similarly, Sterling Chemical and Cain Chemical were founded through the acquisition of petrochemical plants of large chemical producers (Monsanto, DuPont, and ICI).

In Europe, the formation of new companies was less pronounced because of the different role of the stock markets in corporate governance. The refocusing process in Europe took place through the reorganization of the existing chemical producers. There are however a few examples of firms moving downstream through acquisitions. Monsanto, for instance, acquired G.D. Searle in 1985 and entered into pharmaceuticals. Also Rhone-Poulenc in France acquired specialty producers while withdrawing from commodity chemicals.[53]

Table 1.7 shows the nationality of the acquirer and target companies of 7,440 mergers and acquisitions in the chemical industry over 1985–1993. Firms from Japan, the UK, West Germany, and the United States were the acquirer and target companies of 82.5 percent of total M&A. The remaining 17.5 percent agreements occurred in other countries. The United States shows the highest share of acquirers and sellers of chemical companies, with about 60 percent of all operations in the "developed" countries, and more than 60 percent of the total world transactions. The United States is also the second major acquirer in each foreign country, second only to the national acquirers. Japanese firms are the less active ones, both as acquirers and sellers of chemical companies. The European Union hosts 27.4% of the acquired companies. Moreover, this "market for firms" is not very international in nature. Most of the target companies bought by American companies are in the United States (91%), and a large share of the companies acquired by West Germany and UK firms are in these same areas (62% and 64%, respectively). Japan is the most internationalized country in terms of acquisitions: only 36 percent of total Japanese acquisitions are in Japan. These trends are consistent with those in Table 1.6 on foreign acquisitions.

---

[53] S. J. Lane, "Corporate Restructuring in the Chemical Industry," in Margaret Blair, ed., *The Deal Decade, What Takeovers and Leveraged Buyouts Mean for Corporate Governance* (Washington, DC: The Brookings Institution, 1993).

Table 1.7. Mergers and Acquisitions (M&A): 1988–1997 (Shares of Total Number of M&A by Acquirer and Target Country)

Nationality of the Target Company	Nationality of the Acquirer Company				
	Japan	UK	U.S.	West Germany	Total
Africa and Middle East	0.0	0.1	0.3	0.1	0.5
Asia	1.5	0.5	1.0	0.4	3.4
Australia and New Zealand	0.1	0.4	0.4	0.1	1.0
European Union	0.9	12.1	8.3	6.1	27.4
Eastern Europe and Not European Countries	0.2	0.9	1.6	0.9	3.6
North and Central America	1.8	4.3	55.3	1.8	63.2
South America	0.0	0.1	0.7	0.1	0.9
Total	4.5	18.4	67.6	9.5	100.0
Japan	1.3	0.1	0.4	0.2	2.0
UK	0.3	9.0	2.9	0.6	12.8
West Germany	0.3	0.9	1.7	3.9	6.8
U.S.	1.7	4.0	53.6	1.6	60.9
Total	3.6	14.0	58.6	6.3	82.5

Source: SDC, 1998.

CONCLUSIONS

Networks have become a common concept in many economic and social studies, and this chapter examined various types of networks that arose in the chemical industry. The 150-year history of this industry provides an opportunity to describe the formation and dynamics of different types of networks, and the way they influenced industry structure and corporate strategies.

We described the factors that gave rise to different types of networks, their characteristics, their evolution over time, and the role of these networks in shaping the structure of the industry in the years before and after the petrochemical revolution. We highlighted the importance of vertical networks developed by chemical firms with the producers of chemical technologies or capital goods. The specialized engineering firms in the chemical processing sector created the possibility of gaining economies of scale at the level of the industry, and led to the rapid diffusion of technologies. Some of these relationships evolved into complex forms of collaborations, including alliances and co-developments of various sorts, while in other cases they were based on arms-length market transactions (i.e., technology licensing). We described also other types of networks such as corporate agreements, collusive relationships as shown by the history of the chemical industry between the two wars, linkages between the chemical firms and their users, and the R&D networks among firms or with the universities and other research institutions that influenced the evolution of chemical technologies and the economic profitability of firms. Networks as we understand them also include mergers and acquisitions. This happened during the restructuring of the chemical industry in the 1980s, a process that took place by means of the "exchange of firms among firms."

An important issue discussed in this chapter concerns the role of governments in the chemical industry, and the role of other noneconomic institutions. As far as governments are concerned, direct policies have favored co-operation among independent agents. Governments also provided indirect support to collaborations when they demanded complex technological services. Finally, the industry-wide associations played a role by co-ordinating the process of industry restructuring and by solving the problem of "market failure" in the investment in fundamental research. It is impossible to understand the history of this industry, one of the central industries of the second industrial revolution, without understanding the many networks that have evolved and shaped chemical production and distribution around the world. These networks take us outside the firm and provide an essential part of our analysis of innovation, of competition, and of the changing industrial structure of the global chemical industries.

REFERENCES

Albach, H., D.B. Audretsch, M. Fleischer, R. Greb, E. Hofs, L. Roller, and I. Schulz. *Innovation in the European Chemical Industry, Research Unit Market Processes and Corporate Development*. Paper presented at the "International Conference on Innovation Measurement and Policies," Luxembourg, May 20–21, 1996.

Arora, Ashish, Ralph Landau, and Nathan Rosenberg, eds. *Chemicals and Long-term Economic Growth: Insights from the Chemical Industry*. New York: John Wiley & Sons, 1998.

Arora, Ashish, and Andrea Fosfuri. "Licensing the Market for Technology," *Journal of Economic Behavior and Organization*, 52:2 (2003): 277–295.

Arora, Ashish, Andrea Fosfuri, and Alfonso Gambardella. *Markets for Technology: The Economics of Innovation and Corporate Strategy*. Cambridge: MIT Press, 2001.

Arora, Ashish, and Alfonso Gambardella. "Evolution of Industry Structure in the Chemical Industry." In Arora Ashish, Ralph Landau, and Nathan Rosenberg, eds. *Chemicals and Long-term Economic Growth: Insights from the Chemical Industry*. New York: John Wiley & Sons, 1998.

Arora, Ashish, Ralph Landau, and Nathan Rosenberg, eds. *Chemicals and Long-term Economic Growth: Insights from the Chemical Industry*. New York: John Wiley & Sons, 1998.

Arora, Ashish, and Nathan Rosenberg. "Chemicals: A U.S. Success Story." In Ashish Arora, Ralph Landau, and Nathan Rosenberg, eds. *Chemicals and Long-term Economic Growth: Insights from the Chemical Industry*. New York: John Wiley & Sons, 1998.

Chandler, Alfred D. *Scale and Scope: The Dynamics of Industrial Capitalism*. Cambridge, MA: Harvard University Press, 1990.

Chapman, Keith. *The International Petrochemical Industry: Evolution and Location*. Oxford, UK; Cambridge, MA: Basil Blackwell, 1991.

Chem-Intell, 1998, Reed Elsevier Ltd., London.

Flechtner, Hans-Joachim. *Carl Duisberg: From a Chemist to an Industrial Leader*. Dusseldorf: ECON Verlag GMBH, 1959.

Freeman, Chris. *The Economics of Industrial Innovation*. London: Francis Pinter, 1982.

Freeman, Chris. "Chemical Process Plant: Innovation and the World Market" *National Institute Economic Review*, 45, August 1968: 29–51.

Galambos, Louis, with Jane Eliot Sewell. *Networks of Innovation: Vaccine Development at Merck, Sharp & Dohme, and Mulford, 1895–1995*. New York: Cambridge University Press, 1995.

Haber, L.F. *The Chemical Industry During the Nineteenth Century*. Oxford: Oxford University Press, 1958.

Haber, L.F. *The Chemical Industry, 1900–1930: International Growth and Technological Change*. Oxford: Clarendon Press, 1971.

Haynes, Williams. *American Chemical Industry*. 6 vols. New York: Van Nostrand, 1945–1954.

Hikino, Takashi, Tsutomo Harada, Yoshio Tokuhisa, and James Yoshida. "The Japanese Puzzle: Rapid Catch-Up and Long Struggle." In Arora Ashish, Ralph Landau, and Nathan Rosenberg, eds. *Chemicals and Long-term Economic Growth: Insights from the Chemical Industry*. New York: John Wiley & Sons, 1998.

Johnson, Jeffrey Allan. *The Kaiser's Chemists: Science and Modernization in Imperial Germany*. Chapel Hill: University of North Carolina Press, 1990.

Jones, Geoffrey. *Multinationals and Global Capitalism: From the Nineteenth to the Twenty-first Century*. Oxford; New York; Oxford University Press, 2005.

Keck, Otto. "The National System for Technical Innovation in Germany." In Richard R. Nelson, ed. *National Innovation Systems: A Comparative Analysis*. New York: Oxford University Press, 1993.

Landau, Ralph and Nathan Rosenberg. "Successful Commercialisation in the Chemical Process Industries." In Nathan Rosenberg, Ralph Landau, and David Mowery, eds. *Technology and the Wealth of Nations*. Stanford, CA: Stanford University Press. 1992.

Lane, S.J. "Corporate Restructuring in the Chemical Industry." In Margaret Blair, ed. *The Deal Decade, What Takeovers and Leveraged Buyouts Mean for Corporate Governance*. Washington, DC: The Brookings Institution, 1993.

Murmann, Johann Peter, and Ralph Landau. "On the Making of Competitive Advantage: The Development of the Chemical Industries in Britain and Germany Since 1850." In Ashish Arora, Ralph Landau, and Nathan Rosenberg, eds. *Chemicals and Long-term Economic Growth: Insights from the Chemical Industry*. New York: John Wiley & Sons, 1998.

Rosenberg, Nathan. "Technological Change in Chemicals: The Role of University-Industry Relations." In Ashish Arora, Ralph Landau, and Nathan Rosenberg, eds. *Chemicals and Long-term Economic Growth: Insights from the Chemical Industry*. New York: John Wiley & Sons, 1998.

Schroter, Harm G. "The International Dyestuff Cartel, 1927–39, with Special Reference to the Developing Areas of Europe and Japan." In Akira Kudo, and Terushi Hara, eds. *International Cartels in Business History*. Tokyo: University of Tokyo Press, 1992.

Securities Data Company (SDC), 1998, Newark NJ, U.S.A.

Spitz, Peter H. *Petrochemicals: The Rise of an Industry*. New York: Wiley, 1988.

Verg, Erik, Gottfried Plumpe, and Heinz Schultheis, *Meilensteine*. Leverkusen: Bayer; Köln: Vertrieb, Informedia, 1988.

Wilkins, Mira. *The History of Foreign Investment in the United States, 1914-1945*. Cambridge, MA: Harvard University Press, 2004.

Wise, G. "A New Role for Professional Scientists in Industry." *Technology and Culture*, 21 (1980): 408–429.

Wright, G. "Can a Nation Learn? American Technology as a Network Phenomenon." In Naomi Lamoreaux, Daniel, Raff, and Peter Temin, eds. *Learning by Doing in Markets, Firms, and Countries*. Chicago: University of Chicago Press, 1999.

2

Competitive Strategy of the World's Largest Chemical Companies, 1970–2000

HARM G. SCHRÖTER

This contribution looks into the competitive strategy of fifteen of the world's largest chemical enterprises from nine countries during the last decades of the twentieth century.[1] These firms are: Akzo-Nobel (Netherlands), Asahi Chemical Industry (Japan), BASF (Germany), Bayer (Germany), Ciba-Geigy (Switzerland), Dow Chemical Company (U.S.), E.I. DuPont de Nemours (U.S.), Hoechst (Germany), ICI (UK), Mitsubishi Chemical Company (Japan), Monsanto (U.S.), Montedison (Italy), Norsk Hydro (Norway), Rhone-Poulenc (France), and Union Carbide (U.S.). Other firms, such as Procter & Gamble, Johnson & Johnson, L'Oreal, Kodak or Fuji Film, and Merck, which at times and from some points of view could be counted among the same group, have been excluded. They strategically concentrated much more on activities such as marketing, rather than on chemical production, processes, and technology. Furthermore, it is only a European point of view that considers the pharmaceutical industry as a chemical one, whereas in North America as well as in Asia it always has been perceived as separate. Thus, our group is

[1] I want to thank all who have given advice or commented on this paper: Ulrich Hemel, Carsten Reinhardt, Philip Scranton, John K. Smith, Michael Wortmann, and all participants in the ASSI conference on the Global Chemical Industry Since the Petrochemical Revolution, in Milan in October 2000.

based on the ISIC code 351 (industrial chemicals), whereas the numbers 352 (other chemical products, incl. 3522: medicines), 354 (oil), 355 (rubber products), and 356 (plastic products) have been excluded for several reasons, including the fact that the relationship between pharmaceutical and chemical industries changed during our period. Although in the 1970s only a handful of large drug firms existed, at the end of the century their number in the top five hundred surpassed considerably that of the chemical industry (10:15)! In 2000, the biggest pharmaceutical firm (Merck) had a larger turnover than its chemical counterpart (BASF). The rise of the drug industry is a special development of its own, which cannot be covered here. Furthermore, as we are looking at *strategy of enterprises*, and not market development, we have excluded parts of certain firms that were active in the market for chemical products (e.g., Shell Chemicals, Exxon Chemicals). Chemical business represented but a small part of such firms, and the focus of these enterprises was not the chemical industry.[2] Our evaluation is based on the annual reports, the companies' presentations on the World Wide Web, and on articles from various international newspapers.[3]

After World War II, important changes took place in the chemical industry: it switched its feedstuff and energy from coal to oil; it started not only to export worldwide but to produce on this scale; but perhaps the most important change was its outstanding growth of production. This was based on traditional product-groups such as industrial inorganics (e.g., sulfuric acid), fertilizer, dyestuff, coatings, and pharmaceuticals, but additionally on organic raw-material such as naphtha and intermediate goods such as ethylene or propylene, synthetic fibers and plastics, resins, rubber, detergents, fungicides, herbicides, electronic intermediate goods (e.g., discs), and, last but not least, consumer products such as cosmetics. Chemical products replaced traditional raw materials such as cotton, hemp, timber, or steel. They were demanded to a large extent by other industries, for example, textile, automobile, machine building, construction, electrical and electronics, and agriculture. As before, the smaller part found its way directly to the consumer. To a large extent, the expansion after World War II was based on a fordistic strategy, characterized by large plants with a high throughput, detailed division of labor,

---

[2] In any case, after they entered the market on a large scale mainly during the 1970s, these players stuck to their strategy on basic commodities.

[3] I want to express my thanks to the Firmendienst of the HWWA in Hamburg, which I again kept busy for quite some time.

hierarchical organization, and so on; in other words on strategies well described by Alfred D. Chandler.[4]

The beginning of our period is marked by the end of the boom, the first oil-price crisis, and the subsequent macroeconomic difficulties. The environment for industrial growth changed abruptly to the negative side, whereas the energy-intensive chemical industry was especially affected. Furthermore, difficulties in this branch were even larger than the average within industry, since with the emergence of ecological issues on a large scale during the 1970s, the chemical industry received exceptional scrutiny. However, a first superficial comparison of the weight of this sector with others shows us a relatively stable picture: in 1980 and in 1990, there were five chemical firms on the list of the fifty largest companies in the world: 1980 Hoechst (rank no. 29), Bayer (30), BASF (31), DuPont (38), and ICI (40), and in 1990: DuPont (22), BASF (32), Hoechst (34), Bayer (39), and ICI (44). But in 2000, not a single chemical enterprise was included anymore! A more comprehensive overview provides the same picture. From Alfred Chandler's lists in *Scale and Scope* (based on assets) and *Fortune*'s lists (based on turnover), the reader gets the impression that during the whole century the weight of the chemical industry compared with other ones did not change significantly – except for the last decade. Around World War I, during the interwar period, and at the beginning of the boom, our branch accounted for about 10 percent of the two hundred largest firms in the United Kingdom, the United States, and Germany. Various *Fortune*'s lists of the five hundred world's largest firms included about forty-plus chemical enterprises, respectively, a little less than 10 percent. However, although the number in 1990 still was a corresponding forty-three, it was more than halved to eighteen in 1996 and was even further reduced down to ten in 2000. Surely there are many questions that would arise if this kind of accountancy were to be used for more than a first impression, but here we are content to raise the hypothesis that there occurred a major change in corporate strategy during the 1990s, a change that was deeper than previous ones, as it affected the industry more than the switch from coal to oil or the oil-price crises. Our impression is underlined by the fact that in the last list we do not find any of the old established major firms such as

---

[4] Alfred D. Chandler, Jr., *Scale and Scope: The Dynamics of Industrial Capitalism* (Cambridge, MA: Harvard University Press, 1990); Steffen Becker and Thomas Sablowski, "Konzentration und industrielle Organisation. Das Beispiel der Chemie- und Pharmaindustrie," in *PROKLA, Zeitschrift für kritische Sozialwissenschaft 113*, 4 (1988): 616–41.

Ciba-Geigy, Hoechst, Monsanto, Montedison, Rhone-Poulenc, and Union Carbide; that is, six out of fifteen disappeared during this decade. None of our firms has dropped from the list because it became too small to be included in the global five hundred but, rather, for reasons of takeover and mergers. There was only one firm (Henkel) that entered our list, a fact that gives the impression that other branches of industry grew faster during the last decade.

For an explanation we have to look into strategy in order to exclude abrupt shifts in the market or in the political sphere. With the market, we have a steady opening of world markets for competition, which lead from national exports over transnational firms to global enterprises, which themselves underline their global identity in contrast to a national one by pushing, for example, a *made by Bayer* instead of the traditional *made in Germany*. The political sphere was characterized by the trend to liberalization and later deregulation; here, too, it was not state intervention that caused major changes. Thus, the change was not forced on industry from outside, but was a consequence of corporate strategy.

The core of corporate strategy is its competitive component. Although other components will be briefly touched on later, we will concentrate on competitive strategy. In reflection of Michael Porter's writings on competitive strategies, we distinguish between two product- and two market-related categories: *commodity* versus *specialties*, and *home-* versus *world-market* orientation.[5] It has to be stressed that all of our firms were diversified, many even heavily. The focus on commodities as a strategy became widespread during the 1970s. It meant high investment for the construction of plants of optimal size, innovation focused on new processes, reduction and control of costs, exclusion of marginal customers, minimum spending for R&D, as well as for all other expenses. In the chemical industry with its high investments, cost-leadership is related to technical leadership. A high market share usually is an indicator of this strategy. In order to keep this up, industry had to modernize its products constantly, of course. *Specialization* entailed concentration on marketing, product-engineering, focus on quality and trademarks, high spending on technology, general as well as basic R&D, and on service. Strategic aims are less cost reductive but, rather, quality and above-average prices. Such exclusiveness may be in contrast to a large market share. A long tradition within the respective branch usually helps with

---

[5] Michael Porter, *Wettbewerbsstrategie (Competitive Strategy)*. (Frankfurt and New York, 1999).

this strategy. Competition is driven by five principal sources: competition within the industry; pressure from the negotiation-power of suppliers, and of employees; as well as the possible substitution of products or services, and new competitors, both from outside the industry. Although the industry, as a capital-intensive one, was less hard-pressed by negotiation over power with its employees, there was a lot of competition, especially from oil companies entering the market.

It has to be underlined that this analysis does not deal with the multitude of strategies *announced* but only with positions *achieved*. A general problem with our large chemical firms is that they are not limited to only one strategy. They are accustomed to acting in various markets at the same time. Therefore, our allocation of firms and strategies is never without contradictions, but it is worthwhile to work out major characteristics and their change over time.

THE ENVIRONMENT OF COMPETITIVE STRATEGY

After World War II, the chemical industry became a mature and, in the end, a declining industry, or to express it more drastically with the words of Sir Denys Henderson, CEO of ICI in 1993: "The chemical world has gone to hell in a hand-cart. So you're bound to increasingly ask yourself whether you should put it off."[6] As in the cases of other mature and declining industries, for example, textiles, or iron and steel, there still was a lot of innovation, especially with processes and with organization. But the two most important signs of an aging industry clearly were written on the wall: a lack of basic innovation for entirely new product lines and a decline in the weight of the whole industry as compared to others. However, this was, of course, not to be seen during the boom when not only could nearly everything be sold, but the industry came up with basic innovations such as plastics and artificial fibers. Even the first oil crisis, which trebled the price of oil from October 1973 to June 1974, did not cause an immediate slump as in other industries, but a prolongation of the boom. Fearing the collapse of all deliveries, customers bought products made from oil at any price. Some chemical companies could not meet the demand and started to allocate their output to good and established customers. For the chemical industry, the real problems didn't start until 1975.

Corporate strategy comprehends many fields, including, among others, industrial relations, financial questions, relations to share- as well as

---

[6] *The Financial Times* (February 27, 1993).

to stakeholders, and so on. The key issue that we will focus on is competitive strategy. However, we will start with a few initial remarks on the other fields mentioned, before concentrating on competitive strategy.

The chemical industry came under heavy pressure as a result of ecological issues, especially after a couple of accidents. Bhopal in India, which cost about three thousand lives, was the largest one; smaller accidents such as those at Seveso in Italy and at Schweizerhalle in Switzerland, which entailed a devastation of an area of land and a major river, respectively, caused a massive loss of faith both outside and inside the enterprises. For Union Carbide, Bhopal was the beginning of its end as a company. Industrial accidents also led to the demise of Hoechst, which during the early 1990s had experienced a chain of relatively small accidents at its main site. In both cases, the management's stupid attempts to cover up the accidents undermined stake- and shareholders trust in the companies more than did the accidents themselves. In both cases, such organizational and political inability struck a massive blow to the company's standing and later fate. During the last decade, ecological questions were no longer of strategic importance, because the major firms established better coping strategies, and the public was no longer as concerned as during the 1980s.

In contrast, financial questions became increasingly important. Aging industries are characterized by enhanced competition, which made all major firms look for the best means of refinancing. Fresh capital became cheap in the United States, which in the 1990s caused major companies such as the German ones to list themselve on the New York Stock Exchange. In order to do so, they had to change their accounting standards. This was by no means only a technical issue but involved introducing a different culture, a self-understanding that was more open to the demands of the shareholders at the expense of the stakeholders. As with the ecological question, here, too, we have a special case. In 1985, the chairman of Montedison, Mario Schimberni, tried to rid the company of the advice and control of the Bonomi-group of FIAT, and the influence of Mediobanca. Although Mediobanca was relatively small, it was in the key position for Italian industry. It was led by Enrico Cuccia, who until his death in July 2000 "was unchallenged the mightiest man in the Italian financial world, without whose information, consent or help no deal of significance could be concluded in that country."[7] Schimberni, who succeeded with the help of the Ferruzzi-group, was ousted by his own

---

[7] *Neue Züricher Zeitung*, International Edition (June 24/25, 2000): 12.

white knight two years later. His strategy of branching out into financial services was stopped, and Montedison was reshaped with a firm concentration on agroindustry. Since the second half of the 1990s, the company can scarcely be considered a chemical one anymore. In 1997, out of its turnover of 24,997 billion Lire only 6 percent (1,577 billion Lire) came from chemicals, whereas 75 percent (18,870 billion Lire) was generated from agrobusiness. Prospects for future growth were sought in electrical energy, oil, and gas, not in chemicals. In fact, we could assign Montedison to the group of industry dropouts whose number would thus be raised to seven out of a total of fifteen.

Last but not least, industrial relations have to be addressed. In general, cooperative-minded environments such as in Japan or Central and Northern Europe, industrial relations are major issues. Most of our chemical firms in these and other regions, pursued a policy of positively binding personnel to their enterprises by using better pay and offering better working conditions than the industrial average, American firms such as DuPont included. For Dow, this policy was even a major issue in its profile. European and Japanese employees felt "wedded" to their company and were prepared to give their best. As with all serious relationships, any disappointment when occurred is deeply felt and has massive repercussions on the performance of the respective firm. Hoechst, and to a lesser degree Bayer, managed to uproot such positive feeling, especially in their middle management during the 1990s. Confrontation with the workforce was one of the reasons for Hoechst's self-destructive performance during the 1990s.

The boom had created many opportunities for production and our firms responded to all possible opportunities in chemical and related industries. In the early 1970s, nearly all had facilities to produce, in addition to the old organic and inorganic chemicals, intermediate goods for petrochemical industry such as ethylene, aromatics, plastics, artificial fibers, paints and dyes, pharmaceuticals, herbicides and insecticides, rubber, explosives, and so on. Some included metals, films, information systems, and machine building as well as engineering services. There were, of course, differences. Ciba-Geigy, which traditionally focused on pharmaceuticals and dyes, was less diversified than the German firms, ICI or DuPont; Dow was fairly concentrated, as were Akzo on fibers and Hydro on fertilizers. The most diversified were the Japanese firms, which were engaged in construction, housing, immobiles, and so on.

From a global point of view, a concentration on the home market can be seen as a niche-strategy. This is true even when describing the United

States. Although that home market was vast, it still was relatively homogeneous, especially when compared with the various states of Europe
with differing traditions, standards, tastes, and laws, which were better described by the concept of the multinational enterprise. However,
the differentiation of national markets in Europe diminished over time,
with the 1990s showing much more mix and mutual adaptation than
the 1970s. The foremost chemical enterprises reflect the change, with
European companies claiming all of Europe as their home market rather
than any single nation. Additionally, the European Union formed a unified market in 1993. For these reasons, we take the EU as a home market from that year onward. Up to the 1980s, our firms could be quite
clearly divided into two groups: those that sold about two-thirds at home,
for example, Asahi, DuPont, Mitsubishi Ch. Co., Monsanto, Montedison,
Rhone-Poulenc, and Union Carbide; and those that, in contrast, obtained
two-thirds of their turnover outside their state of origin, for example,
Akzo, BASF, Bayer, Ciba-Geigy, Hoechst, Hydro, and ICI. The two Japanese
firms were even more concentrated on their home market; they did not
even own any significant production facilities outside Japan. The extent
to which our companies reflect the overall structure of the respective
economies in which they were rooted, the early export-orientation of the
European firms, especially those from small nations, and the concentration on the home markets of U.S. and Japanese firms, is quite remarkable.

Table 2.1 allocates the world's top fifteen chemical enterprises according to a scheme based on work by Michael Porter. *Leading advantage in
commodities* is the announced strategy if matched by a leading market
share in bulkware, for example, ethylene; *special competitive advantage is* the announced strategy if matched by high-priced commodities
with special advantages, for example, coatings. Usually, companies in
this sector invested more into R&D and distribution than the former
group. Foreign revenue above or below 50 percent is used as an indicator for home- or world-market exposition. Because of better comparabil-
ity, all allocation is based on turnover, not on profits or stock-exchange
capitalization. Thus, the four tables in this text reflect a sketch of the
competitive situation at a certain time, not the announced business
strategies.

A PHASE OF REACTION: FROM THE MID-1970S TO THE MID-1980S

The first oil-price crisis reopened the question of raw material and energy.
All companies had switched from the traditional coal to oil for both

Table 2.1.   Chemical Firms' Competitive Situation in the Early 1970s

	Advantages Mostly in Commodities	Advantages Mostly in Specialties
World market	Akzo BASF ICI Hydro	Bayer Ciba-Geigy Hoechst
Home market	Asahi Dow Mitsubishi Ch. Co. Monsanto Montedison Rhone-Poulenc	DuPont Union Carbide

feedstuff and energy, a reorientation that took about a decade, ending in the mid-1960s. Ten years later, the oil-crisis challenged that step. Indeed, DuPont, ICI, and the three German companies explored the possibilities for a greater use of coal or even a switchback.[8] All came up with the same conclusion; a return to coal was not possible. It was blocked by extremely high costs for leaving petrochemistry, a rising price for coal, an immediate scarcity, and a time horizon that was too long.[9] Other alternatives for energy, such as atomic power, were not realized.[10] So, the usual adjustment was to try and save as much energy and as much feedstuff as possible – and wait for better times. True, some companies opting for more special competitive advantages did cut back capacity for bulk products, like Union Carbide, which sold its European plastics division, or Akzo, which reduced its staple fibers. Some even made heavy cuts. In 1977, Rhone-Poulenc launched its "textile-plan" by which it reduced its workforce from thirteen thousand (1977) to twenty-six hundred in 1982. But this reduced number still stood for about 20 percent of the company's turnover in the early 1980s. Others established a better supply of

[8] Harm G. Schröter, "Strategic R&D as Answer to the Oil Crisis, West and East German Investment into Coal Refinement and the Chemical Industries, 1970–1990," *History and Technology, 16* (Autumn 2000): 383–402.

[9] Viz.: Robert B. Stobaugh, *Innovation and Competition: The Global Management of Petrochemical Products* (Boston: Harvard Business School Press, 1988). It took up to ten years from the decision to start a coal mine until production at full capacity, a time lag the chemical industry could not afford.

[10] BASF had planned an atomic power plant at its main site in Ludwigshafen. It stepped back in 1976 for financial reasons. Viz.: Werner Abelshauser, ed., *Die BASF. Eine Unternehmensgeschichte* (Munich: C.H. Beck, 2002).

oil: Monsanto signed a joint venture with Conoco oil in 1977, and BASF went on with its backward-integration strategy. In contrast, Hydro's entry into oil and gas was not triggered by the crisis but, rather, by political considerations. Although it had no expertise at all in this sector, the discovery of oil on the Norwegian shelf compelled Hydro, as by far the largest company, to act as a vehicle for upholding the national interest in this sector.[11] For Hydro, oil investment was more or less by chance.

The oil price-related crisis reached the chemical industry only in 1975, and two years later sales had already started to improve. Generally, the recession for the chemical industry was not as deep as for other industries. The second oil shock in 1978/79 that again trebled the price, hit the chemical industry much harder. Upstream products that were near to crude were hit harder than downstream, more refined ones. As the price of intermediate goods such as ethylene or propylene soared, bulk products such as plastics or fibers became expensive. Those firms that had concentrated on these segments suffered a hard time, for example, Asahi, which in 1982 generated 44 percent of its turnover from plastics and 36 percent from fibers. It had been settled during the first crisis that there were no substitutes for oil and gas; instead, the industry had to deal with the possibility that there were substitutes for its own products, for example, natural fibers for artificial ones. In Europe alone, the chemical industry suffered a loss of about $1.5 billion in just one division (plastics) in 1981.[12] At the height of the crisis in 1982, BASF's CEO, Matthias Seefelder, commented on the situation, saying: "The problem lies in the overcapacities with fibers, standard-plastics, refineries and in cracking. With plastics the situation is like in heavy industry. We have a structural crisis which is fed from several sources. The whole branch of industry in the western world invested too much since 1975, because we did not want to understand that the growth rates we were used to from the time before the first oil-crisis, have passed."[13] With these words, Seefelder summed up the general situation.[14] Although all firms had to react to

---

[11] It was politically decided to safeguard Norwegian interest by using three firms, fully state owned and newly created Statoil, 51 percent state owned but privately run Hydro, and the privately owned newcomer Saga (Ryggvik, Helge).

[12] Walter Teltschik, *Geschichte der deutschen Großchemie, Entwicklung und Einfluß in Staat und Gesellschaft* (Weinheim: VCH 1992).

[13] Ibid., 263f.

[14] See Ashish Arora and Alfonso Gambardella, "Evolution of Industry Structure in the Chemical Industry," in Ashish Arora, Ralph Landau, and Nathan Rosenberg, eds., *Chemicals and Long-Term Economic Growth: Insights from the Chemical Industry* (New York: John Wiley & Sons, 1998), pp. 379–413, 399.

this crisis, those that focused on cost-leadership were challenged more than the differentiated ones. The firms of our sample did so mainly in three ways: reorientation in raw material, reconstruction of the market, and repositioning of the respective firm.

In search of cost-leadership, the first group of firms, Asahi, BASF, Dow, DuPont, Hydro, and Mitsubishi Ch. Co., went upstream and engaged more heavily in energy (oil, coal), raw materials (naphtha), and intermediate goods (olefins, aromatics). Asahi invested in its main plants, Mizushima and Kawasaki, to step up basic materials such as (poly-) ethylene and (poly-) propylene. BASF had already before the first crisis started its investment into raw materials. Thus, in 1980 half of its naphtha-intake and two-thirds of the required heating oil came from its own sources. The company set up new cracking facilities, especially at its principal plant at Ludwigshafen, and at Antwerp. Dow had embarked on a similar strategy. It had built up facilities during the 1970s, and it announced plans to continue with large new constructions in Yugoslavia, Germany, and a huge joint venture in Saudi Arabia. This last one generally reflected a new issue in competition: new and financially strong competitors based on crude and backed with state capital were entering the market. Already, the old established oil firms had built plants for chemical intermediate goods; now, various Arabian states announced plans to do the same. Dow's joint venture in Saudi Arabia was to be supplemented with other ones, for example, Hoechst signed an agreement for a massive investment together with Kuwait in 1984. Alec Flamm, vice president of Union Carbide stated in 1986, that the petrochemistry was profitable only if based own raw materials.[15] The most significant step was taken by DuPont. In July 1981, it surprised the industry by buying two energy firms, Conoco (oil) and Consolidation Coal. Two month earlier, CEO Ed Jefferson had underlined DuPont's commitment to basic chemicals and instantly ruled out investment into raw material, claiming that such a move would cost "several billions of dollars and would involve a commitment not consonant with our strategy."[16] Conoco supplied not only energy and intermediate goods but also sold its products on the petrol market. It was a major asset of DuPont, which doubled its turnover by the acquisition and became the world's largest chemical enterprise again. For Hydro, too, oil became the most important earner during the first half of the 1980s. The company ploughed parts of its oil-generated profits into its mainstream business:

[15] Interview by Industriemagazin, *Industriemagazin* (No. 1, 1986): 12.
[16] *Financial Times* (May 6, 1981).

fertilizers. By internal growth and acquisitions of, among others, the Dutch Windmill Group and the French Cofaz, both in 1986, Hydro established itself as the largest supplier of artificial fertilizer in Europe, and thus realized its strategy of cost-leadership in this segment. Mitsubishi Ch. Co., too, concentrated on its traditional strength, which was coal. In 1980, the firm made 46 percent of its turnover from petrochemicals and 29 percent from coke and coal-related products. When the crude price went up and the petrochemical division ran into troubles, the obvious way out was coal. Mitsubishi Ch. Co. for the second half of the 1980s announced the construction of a really large liquefaction plant, with a capacity of one hundred thousand barrels a day, based on Australian lignite. However, like similar plans in the United States and Germany, this was not carried out because the oil price came down in 1985. For other companies too, the reduced oil price was the main reason for repealing the announced large-scale plants in the Arabian Gulf as well.

A second group of firms reacted to the oil price with an opposite strategy: they tried to reduce their commitment to bulkware and concentrated on the upper end of the market. This was not in all cases a deliberate choice. Monsanto was compelled to give up its secure source of supply when its direct competitor, DuPont, bought the partner of its joint venture, Conoco. Rhone-Poulenc simply had to sell its oil investment to Elf-Aquitaine in order to survive. During the first half of the 1980s, all firms announced an intention to invest in more sophisticated products. Some did so relatively swiftly, whereas others needed more time. Dow was very strong in basic chemicals and was one of the world market leaders in polyethylene and polystyrene. But it, too, feared the Arabian competition and announced a new strategy. It pulled out of its announced massive investments into large petrochemical complexes, terminated its joint venture on organic intermediate goods with Asahi, and sold off its gas and oil investments. It announced a concentration on specialties and pharmaceuticals, but because of its heavy concentration, more than 60 percent of turnover was still in basic chemicals.

Two special cases of swift movers were Rhone-Poulenc and Montedison. Both had reason to move rapidly because of heavy losses. In 1981, Rhone-Poulenc was nationalized. It had already run into difficulties during the 1970s, but the second oil crisis caused a real threat. The French state injected three billion Francs in fresh capital and new loans and thus helped its reconstruction – "better red than dead" commented *The Economist* in May 1984. Indeed, it was the French state that saved the company, or at least its size. Montedison, which had focused on bulk production, was reconstructed by CEO Mario Schimberni during the first half of the 1980s. Its workforce was halved to thirty-five thousand and the costs were huge – $2 billion – but the enterprise was out of the red in 1985. By that year, only half of its turnover was generated from chemicals, the other half from services, such as petrol stations. This alternative sector was enlarged when Montedison bought the large insurance company, La Fondiaria, in 1986.

Other companies pursued the same strategy, moving into higher-margin products, but with less spectacular steps. Bayer, Hoechst, and Ciba-Geigy were not as heavily engaged in basic chemicals as other enterprises. Still, they, too, adjusted by concentrating on the upper end of the market, while relying on the old European traditions of cooperation and, by applying them successfully, influencing the market through cartelization. Both cartels were approved by the European Community as temporary and beneficial ones.

One of the leading actors in the reconstruction of the fibers market was Akzo. In 1982, the ten largest fibers producers in Europe signed a pact by which they agreed to cut textile and carpet fibers by five hundred thousand tons over a three-year period. The Japanese and American firms knew about this step but did not formally take part in the cartel. However, DuPont closed down fibers facilities as well. At the same time, in contrast to the Europeans, DuPont's fibers business started to make money from the last quarter of 1980 onward: "Prospects for fibers in the 1980s look much improved with specialized fibers leading the way," the chairman, Ed Jefferson, stated in May 1981.[17] However, although this U.S. firm already had invested into such special and technical fibers, others still had concentrated on staple ware. The cartelized reconstruction of the European market was successful, for Hoechst fibers were one of the most important "pillars of profits" in 1983.

Except for Union Carbide, which used its new process to heavily cut cost in the production of LDPE, bulk plastics were the next field to cause headaches. Hoechst's CEO, Rolf Sammet, suggested setting up a bureau in Brussels on world capacities of plastic production.[18] The idea was that the bureau should not monitor from outside but collect incoming announcements about their capacity from the enterprises themselves. This reflected another cooperative approach, which led to a solution similar to that on fibers. During the following years, all major producers

[17] Ibid.
[18] *Frankfurter Allgemeine Zeitung* (May 4, 1982).

Table 2.2. Chemical Firms' Competitive Situation in the Early 1980s

	Advantages Mostly in Commodities	Advantages Mostly in Specialties
World market	Akzo BASF Dow ICI Hydro	Bayer Ciba-Geigy Hoechst
Home market	Asah DuPont Mitsubishi Ch. Co. Monsanto Montedison Rhone-Poulenc	Union Carbide

in Europe curtailed their capacity between 20 and 50 percent. The effect was that, between 1986 and 1988, plastics were back again and represented the biggest earners in the industry.

All three fields, refineries and cracking facilities for intermediate goods, staple fibers, and bulk plastics, had again demonstrated how dependent they were on the swings of the economy. During the first half of the 1980s, the chemical industry managed to cope with the challenges from the two oil-price shocks, although it took the companies nearly to ten years to do so. In the end, both products and firms had become more differentiated, but the announced investment into consumer goods, pharmaceuticals, and so on had not yet had sufficient impact to give all the companies a new profile. Large companies move relatively sluggishly, and our firms were the fifteen largest ones. It is therefore no surprise that the tables for the early 1970s and 1980s are similar to each other.

A PHASE OF ACTION: FROM THE MID-1980S TO THE EARLY 1990S

The situation in the industry improved after 1983 and a couple of good years made it possible not only to react but also to actively look for new strategic opportunities. But the situation contrasted to the boom years. During that time, the chemical industry was powered by three new accelerators: new processes based on oil, new basic products such as fibers and plastic, and the emerging mass markets. Thus, for the chemical industry the boom period offered even more opportunities for growth than for other branches. The 1980s showed none of these possibilities for expansion, for example, the sales of plastics grew between 1970 and

1990, but the composition of the whole group, which was based on innovation up to 1970 changed significantly, remaining relatively stable.[19]
Consequently, only three possibilities existed for further growth: geographical expansion, concentration on distinguished fields of strength
with a substantial turnover, and special products with high earnings but
low turnover.

From the mid-1980s on, the traditional export oriented firms thought
it necessary not only to sell but to produce in all regions of the triad, in
the United States, in Europe, and in Japan. Many basic chemicals could be
produced more cheaply in other countries with lower costs for personnel
and environment protection. For more sophisticated products, manufacturing near consumption became more important, when other branches
of industry being main customers set up production regimes such as
"just in time." Although this was a trend underlined by virtually all firms
in their official statements, there remained substantial differences. These
largely reflected traditional strategies; those firms that had been foremost
exporters became more global than those that had concentrated on their
home market. BASF, Bayer, Ciba-Geigy, and Hoechst not only established
large facilities for production abroad but even their own R&D facilities
in all of the triad. The idea behind this was not so much to save costs but
to enhance creativity by carrying out R&D in different cultural settings.
The CEO of Hoechst, Wolfgang Hilger, stated in the 1980s, that the best
Japanese or American scholars would not come to Europe, but the company had to go to their country, or in other words, the best personnel
could be tapped only at the spot. Generally, this commitment to foreign
markets was felt most strongly in Europe. Although nearly all companies
tried to communicate an international profile, among the non-Europeans,
only Dow sold more abroad than at home.

During the 1980s, it became increasingly clear that even the largest
chemical firms would cease being generalists offering all products. At
the same time, nearly all the firms announced strategies to stop offering
cyclical products and protect themselves from economic swings. However, the group consisting of Akzo, Asahi, and Dow did not succeed
extremely well in their efforts, whereas the other group (Bayer, Ciba-
Geigy, Hoechst, ICI, Mitsubishi Ch. Co., Montedison, Monsanto, Rhone-
Poulenc, and Union Carbide) did better. One major reason was the heavy
investment of the former group into commodities. BASF and DuPont still

---

[19] Only HDPE (high-density-poly-ethylene) could double its share up to 11 percent (Teltschik,
Geschichte der deutschen Großchemie, Table 13, 299).

pursued the traditional approach of combining both strategies, whereas
Hydro never intended to shift from its traditional focus on commodities.[20]
    Akzo concentrated on technical and high-end fibers, coatings, and
pharmaceuticals. Still, its largest division was basic organic and inorganic
chemicals. It started to sell off bulk fibers, as in the case of American
Enka in 1985, and peripheral activities such as its textile machinery firm,
Bamag in 1989. Asahi competed mainly on its low costs advantages. Its
investment into R&D with about 3 percent of turnover, was one of the
lowest, undercut only by Mitsubishi Ch. Co. That firm had already in the
first half of the 1980s established information systems and pharmaceuticals as fields of preferred growth. Although in 1992, it enjoyed a world
market share of 30 percent in floppy disks, and established itself after
3M and Sony as No. 3 in special tapes, petrochemicals, anorganics, coke,
carbons, and fertilizer stood for more than three-quarters of turnover.
BASF branched out into pharmaceuticals, coatings, and agrochemicals,
but kept its old strategy of being a low-cost producer on international markets. This was quite similar to Dow, which slowly expanded its divisions
for pharmaceuticals, consumer products, and pesticides. These doubled
their share of Dow's turnover during the 1980s, but still one-half was
generated with plastics and a quarter with chemicals and metals. Since
its acquisition of Conoco and Consolidation Coal, DuPont's turnover was
between one-third and one-half in energy. Fibers and plastics stood for
15 percent each, leaving about a quarter for special products such as
pharmaceuticals, electronics, and so on. Hydro's performance was not
only determined by the swings of chemical markets but also, to some
extent, by the price of crude. It successfully tried to ride the waves of
both. Based on its energy advantages, it established itself as the largest
producer of fertilizer and as number two in aluminum and magnesium
production in Europe.
    The main strategy of the other group of firms was to focus on more
sophisticated products, which obtained higher margins.[21] These were tailored plastics, fireproof fibers, optical or electrical conductive polymers,
ceramics, prepregs (a blend of resins and fibers), pharmaceuticals,

---

[20] DuPont announced a strategy of concentration but still stayed very diversified. Alfred D.
   Chandler, Jr., Takashi Hikino, and David C Mowery, "The Evolution of Corporate Capability
   and Corporate Strategy and Structure within the World's Largest Chemical Firms," in Arora
   et al., pp. 415–58, 427ff.
[21] Again we would like to underline that no firm completely fitted into the scheme; for
   example, DuPont invested in high-tech fibers, and so on. We try to concentrate on main
   characteristics.

biotechnological products, and so on. At the same time, many firms divested themselves of their facilities for bulk production. In 1986, Bayer sold Metzeler, which made tires and special rubber products, to Pirelli, while at the same time investing in high-performance rubber. Ciba-Geigy pulled out of its joint venture with Bayer for basic chemicals, Schelde-Chemie, in 1985. An example of its search for a new strategy was its acquisition of Mettler, a Swiss firm producing highly sophisticated scales (1987). Afterward, it concentrated on pharmaceuticals, agrobusiness (fungicides, herbicides, animal health, and seeds). In 1993, it announced its strategy of concentrating on three sectors: industrial special products, pharmaceuticals, and agrobusiness. For Hoechst, the most significant step was its purchase of Celanese in 1987 for nearly $3 billion, one of the biggest acquisitions in the industry up to that time. Although analysts thought it too high a price because Celanese used to be one of the largest producers of bulk fibers, they overlooked the fact that Celanese had invested heavily into R&D. With its special fibers, it became one of Hoechst's cash cows. Instantly, the German company strengthened its global diversification by establishing itself as the fifth biggest chemical enterprise in the United States. In pharmaceuticals, where it had grown to number two after Merck, Hoechst lost ground in the early 1990s. Mitsubishi Ch. Co. made no spectacular movements but developed its chosen strategy of continuing with established businesses and concentrating on electronic systems and biotechnology. Two major events exhibit the steadiness of strategy: in 1985 it sold its aluminum production after it had generated losses for many years; four years later it founded a joint venture with Hoechst for paints, which made it a leading player on the world scale and number one in Japan.

ICI entered the 1980s with an unfavorable structure. It generated two-thirds of its turnover from bulk chemicals and sold 40 percent in the United Kingdom. It, too, focused its attention on special products and on geographical diversity. In 1990, it had succeeded in reducing bulk chemicals to 40 percent, whereas coatings were up from 8 to 13 percent, agro-related business from 5 to 11 percent, and pharmaceuticals from 6 to 11 percent of turnover. The latter division stood for 47 percent of earnings. At the same time, ICI had reconstructed its fiber branch by exchanging its nylon business with DuPonts acrylics and swapping its European polypropylenes for BASF's acrylics. With these exchanges, all companies added to their specific fields of strength. ICI was still under reconstruction when in 1991 Lord Hanson tried to carry out a hostile takeover for the enterprise combined with the suggestion it should be sold off in parts. The takeover bid was fended off but promoted more radical thinking about ICI's future in the 1990s.

Monsanto and Rhone-Poulenc followed a similar strategy, both concentrating on life-science products, pharmaceuticals, and pesticides. Rhone-Poulenc sold its fertilizer division in 1984 and acquired pesticides from Union Carbide three years later. After it had bought the U.S. pharmaceutical firm Rorer in 1991, the *Financial Times* labeled the firm as "a jewel in France's portfolio." It was reprivatized in 1993. Rhone-Poulenc announced its aim to be sixth in the world list of companies in 1986 and again in 1989. Monsanto, too, moved away from its traditional business into biotechnology. Between 1985 and 1993, it sold its divisions for coatings, films, and others not related to life-science.

In reaction to the second oil crisis, Montedison divested heavily and sold its division of nylon fibers, PVC, polyethylene, pigments, dyestuff, rubber, medical instruments, and so on. In 1985, it had established itself already as a conglomerate, stretching itself far over the borders of chemical industry. Its chemical division stood only for 35 percent of turnover, whereas energy and services each generated about 20 percent. In its chemical business, it concentrated on positions of strength, for example, in polypropylene, where its affiliate Himont captured 20 percent of the world market in the late 1980s. It seems that Montedison had more to fear from its own top management and its shareholders than from competition. CEO Schimberni had departed from the traditional financial controlling groups around Fiat and Mediobanca; instead, control lay with the agricultural concern, Feruzzi. When in 1988 Feruzzi faced financial difficulties, Mediobanca, the traditional key player in the construction of the Italian industrial sector, returned. Schimberni was sacked and the service sector of Montedison reallocated to the mother company. The following years were unstable for Montedison, as it underwent several reorganizations. It became a holding company for a large agricultural sector (sugar, vegetable oil, etc.) and a chemical branch for which the old name Montecatini was chosen.

Union Carbide was a special case: it was already very well positioned in 1984, with a substantial part of its turnover coming from growth sectors such as consumer goods (21 percent), technology and services (24 percent), and industrial gases (15 percent). The company was the first to aim at quitting the aging industry. CEO Alec Flamm announced in 1984 "within ten years we are a *technological* concern."[22] Two month later,

---

[22] *Finanz und Wirtschaft* (October 5, 1984).

the most disastrous accident of the chemical industry occurred at its plant in Bhopal, India. About three thousand people died, thirty thousand were seriously injured, and two hundred thousand were hurt. What followed was sad and embarrassing. Because of the accident, the share prices fell about one-third, which caused GAF, a much smaller chemical firm, to launch a hostile takeover bid.[23] Carbide fought vigorously for its independence, buying back 56 percent of its own stock and thus fending off the bid, but was saddled with a debt of $4.5 billion. Additionally, it sold its consumer division and, from the money obtained, paid a bonus to shareholders who had not sold to GAF. In 1986, it sold its agrobusiness to Rhone-Poulenc to reduce debt. Indeed, "they sold off the future of the company" commented the analyst Michael Eckstut of Booz Allen & Hamilton.[24] At that time, employment was down to nineteen thousand from ninety-eight thousand in 1984.

THE FUNDAMENTAL STRATEGIC REORIENTATION DURING THE 1990S

Unlike the strategic reconstruction and optimization on the micro-level, the situation for the industry did not change generally on the macro-level. No basic innovations offering new possibilities of expansion were at hand. Instead, new competitors entered the industry: firms that were situated in fast-growing markets such as East Asia or in the Arabian Gulf, and firms that specialized in raw materials and intermediate goods. In both cases, they enjoyed not only low costs of labor but also state backing. The strategies of the 1980s proved not to have been wrong, but they were not good enough for the next decade. The options were mostly the same: geographic extension, concentration on the upper or the lower end, more special expertise, a larger share of the market, and so on.[25] However, one strategy was added: exiting the traditional chemical industry and focusing on life science. Exit strategies related to core competencies at arms length lay in pharmaceuticals, biotechnical and consumer products, or, at the other end, in energy.

During the 1990s, the change within the industry accelerated. Many new joint ventures were signed (Table 2.3). Although joint ventures were

---

[23] GAF in the end doubled its value by its attempt (Chandler, et al., "Evolution of Corporate Capability," p. 430).

[24] *The Financial Times* (December 22, 1988).

[25] A concentration had been going on in bulk-chemicals since 1972 (see Arora and Gambardella, "Evolution of Industry Structure," p. 401).

Table 2.3. Chemical Firms' Competitive Situation in the
Early 1990s

	Advantages Mostly in Commodities	Advantages Mostly in Specialties
World market	Akzo BASF Dow ICI Hydro Rhone-Poulenc	Bayer Ciba-Geigy Hoechst
Home market	Asahi DuPont Montedison Union Carbide	Mitsubishi Ch. Co. Monsanto

by no means new, they were used to a large extent to open up markets, for example, Bayer joined with Monsanto to form Mobay, and later took it over. The Japanese market was nearly impenetrable without indigenous help. However, during the 1990s, joint ventures aimed in many cases at obtaining a commanding market share of a certain product. Joint ventures of the first kind were market seeking, the latter ones security seeking. Times had become tougher, another indicator of an aging industry. During this decade, several firms quit the industry and moved to branches with higher margins, especially pharmaceuticals. Although during the 1970s, geographical concentration on their home market still gave shelter to half of our firms, these kinds of advantages were substantially diminished in the 1990s. Globalization, pushed by the industry itself and, helped by new definitions of the role of the state (e.g., deregulation), had reduced national differences. The result was best seen in Europe with its relatively small states. Firms such as Bayer played down their German origin and instead stressed their feeling of being a European company. The same line was taken by Egil Myklebust, CEO of Hydro, who in 1998 publicly considered the possibility of moving the company's headquarters out of Norway into the United Kingdom. National responsibility was – at least within Europe – to a large extent reduced. In this respect, there was hardly any geographical niche strategy left at the turn of the century. Asahi and Mitsubishi Ch. Co. in contrast focused on their home market. When, in 1999, Mitsubishi Ch. Co. published its new strategy, all items mentioned were focused on Japan only, although the firm owned plants abroad.[26] But, as the press release showed, these were not part of the core of the enterprise.

After ICI had fought back the hostile takeover bid, it did exactly what the raider initially had suggested: it split. In 1992, Zeneca was spun off, including those fields that were characterized by high investment in R&D and marketing: pharmaceuticals, agrochemicals, dyestuffs, and intermediate goods for pharmaceuticals.[27] This part had a turnover of $7.5 billion and generated 70 percent of all profits (1991). In contrast, the rump-ICI had a turnover of 70 percent of the old ICI. It included capital intensive products with a high volume of production but has shifted in recent years to special products in paints, industrial chemicals, and materials.

During the 1990s, Monsanto owned an established business in life science, and within this sector it concentrated on gene-manipulated and related goods. For example, in 1991 it started to sell genetically engineered insect-protected potatoes, and introduced its delayed-ripening tomato in 1995. Although such products were fairly well received in the United States, the public in Europe became extremely critical of gene-manipulated food. There were boycotts of even such food in which, for example, gene-manipulated maize was a minor component. Consequently, big enterprises such as Unilever announced their intention not to buy and sell any such foodstuff, a step that entailed repercussions by farmers in the United States who had used Monsanto's products. Monsanto had developed too far from its customers. Culturally, it was too American to act as a global company because its management did not understand the different value systems in Europe and Asia. Faced with this surprise, and impressed by the success ecologists had in other branches, too,[28] Monsanto preferred to look for a partner. The suggested merger with American Home Products in 1998 did not develop, but in April 2000 together with Pharmacia, Upjohn, and Searle, it formed the new Pharmacia.

Montedison formed a joint venture with the leading Italian oil firm, ENI, to form Enimont, which in 1992 was sold to the Italian state. There was a bribing scandal in connection with this transaction, and the Feruzzi holding company had to be reconstructed from scratch to

---

[26] Mitsubishi Ch. Co. home page, http://www.m-kagaku.co.jp/index_en.htm 06/28/00.

[27] ICI, "ICI board to recommend Zeneca de-merger," *ICI Press Release* (March 1993).

[28] The fact that a tiny group, such as Greenpeace, beat a mighty multinational enterprise, such as Shell, on the question of what to do with the Brent Spa oil platform made an impression in other industries, too.

form the Compart-group. This group merged with Montedison, reducing the chemical part of the conglomerate to only 6 percent of the whole turnover. It could hardly be called a chemical company any more.

Union Carbide did not succeed in finding its way back to its path. In 1988, and again in 1991, its CEO, Robert D. Kennedy, suggested demerging it into three units, as he could not detect any synergies in the traditional type of a diversified company.[29] In fact, he preferred to sell Carbide's gas business in order to reduce debt. In the 1990s, the company concentrated on plastics, the only division that was left. When it was taken over by Dow in 1999, it had been reduced to a medium-sized company with an employment of 11,600. Up until the accident, Union Carbide had been a traditional, conservative, and, for U.S. conditions, an unusually reliable employer. One of the main reasons for its long agony was that its upper management was unable to reestablish the old confidence among the rank and file. Middle management especially was shocked by four issues: the accident, the takeover bid, the selling of one division after the other, and the long and dishonorable fight against compensation claims.[30] These claims were settled in 1989 by a payment of $470 million, which was about one-tenth of what the fight for Carbide's independence had cost. By comparison, eight years later Dow offered $3.7 billion to victims of its silicon breast implants.

Rhone-Poulenc's aim of becoming sixth in the chemical world behind DuPont, BASF, Bayer, Hoechst, and ICI was nearly reached in 1998, except that Dow had grown larger than ICI. But the accelerated focus on life-science had burdened the company with debt. Profit was only about 3 percent of the turnover, and in 1997 the company had to take a big loss. Consequently, it had to sell assets. It spun off Rhodia, which took the fibers and plastic divisions, and finally, in 1998, it merged with Hoechst.

During the early 1990s, Hoechst's divisions for polymers and coatings came under strategic pressure. They still made money but were not positioned among the top five in the world. At the beginning, Hoechst tried to correct this, by forming, for example, a joint venture with Courtaulds for its acrylic and viscose fibers (1994), and another one for polypropylene

---

[29] "I think the business should be free to focus on their external markets, not to compete internally for resources and attention" (*The Financial Times*, December 22, 1988). "Carbide has never worked as a conglomerate" (*The Wall Street Journal*, December 17, 1991).

[30] After a considerable lapse of time and after Union Carbide was confronted with extremely high compensation claims, the company suggested the accident was caused by sabotage, a move that worsened its reputation.

film in 1996. In 1993, together with Schering, another German chemical firm, it founded a joint venture for pesticides called AgrEvo, which became market leader in its field. A similar solution was found with DyStar, a joint venture with Bayer (each 50 percent) for dyestuff with a turnover of $1.5 billion, set up in 1996. Hoechst was a market leader in highly specialized technical plastics, having captured 11 percent of the world market in 1995, and in certain promising fields it held a larger share, up to three-fourths.[31] Although Hoechst was under reconstruction, a chain of small but nasty accidents at its main site near Frankfurt undermined its reputation inside and outside the industry. Union Carbide had experienced the same thing, with similar results. In 1994, the old CEO was replaced by Jürgen Dormann, who initially continued with the strategy of alliances but after two years embarked on a course of a ruthless reshaping. He was one of the first German managers to proclaim shareholder value in Germany and promised a return on investment of 15 percent. In 1996, Hoechst acquired Marion Merrell Dow, the pharma-branch of Dow, and Dormann announced that his company would be number one in pharmaceuticals by the year 2000. For this, he was highly praised by analysts, owners, and even by the workforce, all of whom would heavily criticize him two years later. Unfortunately for Hoechst, other companies had spotted pharmaceuticals as their field of concentration as well. In order to reach its aim, Hoechst sold out first peripheral parts such as Uhde (construction of plants) and its fine chemicals division. In spite of increased sales and rising earnings, Dormann came under pressure, as neither analysts nor the employees could understand the implementation of his strategy. The number of employees tumbled as did the shares. It was very remarkable when exactly the group of people created by the firm's strategy, one hundred managers from the pharmaceutical division, in a palace revolt wrote a letter to the controlling board asking "to stop as soon as possible the erratic action of Mr. Dormann, which gives the company a bad name . . . "[32] Reconstruction did cost Hoechst about $20 billion, which left the company loaded with debt. In 1998, Hoechst became truncated when the divisions for plastics, coatings, and

---

[31] GE was by far the largest in technical plastics with 22 percent, followed by DuPont 12 percent, Bayer 9 percent, BASF 5 percent, Mitsubishi Ch. Co. 4 percent, Dow 3 percent, Asahi 3 percent, Toray 3 percent, Teijin 2 percent, and others 25 percent. Hoechst obtained market shares of 75 percent in LCP (liquid crystaline polymeres), 49 percent in PE-UHMW (ultra-high molecular polyethylene), and 42 percent in POM (polyoxymethylene) (*Plastik- und Kautschukzeitung 10* (October 1995).

[32] *Wirtschaftswoche* (March 12, 1998).

gases were sold and Celanese spun off. The rest (45 percent) merged on equal terms with Rhone-Poulenc to form the pharmaceutical firm of Aventis.[33] That was the end of Hoechst, a company that was said to have sought a friendly takeover of Rhone-Poulenc during the 1970s and 1980s when it was two times as large. Hoechst changed too much at one time and therefore went deeply into the red. Consequently, it had to sell what Dormann in his annual report 1994 had labeled as "key-fields" of its future activities: advanced chemicals and plastics, coatings, and technical gases. The announced target – Hoechst as the world's number one in pharmaceuticals in 2000 – was missed to an embarrassing extent. Even the merged Aventis had a turnover of about only a third of that of Merck, the world's number one in pharmaceuticals. Aventis scored no higher than eleventh!

While Hoechst shrank, Bayer made steps in the same direction but with more care concerning speed, not rushing forward at any costs. It, too, had focused on life-science but grew more internally than externally. It entered the 1990s with each of six principal divisions providing largely the same share of turnover, pharmaceuticals, agroproducts, polymers, "organica," industrial products, and information (Agfa). Whereas in 1980 all divisions had nearly the same amount in R&D, pharmaceuticals in 1991 received one-third, and in 1998 more than half. Bayer grew especially in this field. It spun off its Agfa division in 1999 and merged its organica and industrial ones. In the first quarter of 2000, turnover came from polymers (38 percent), pharmaceuticals (33 percent), agrobusiness, and chemistry (the rest) (15 percent). Internal growth was not a dogma; in 1999 Bayer bought Chiron (diagnostics) and with the acquisition of Lyondell became the first to offer the whole variety of polyurethanes. Bayer therefore made similar moves as Hoechst. Although the moves were slower and more balanced, the results were only marginally better. Bayer's attempt to enter the top league of pharmaceutical firms failed and it consequently shrank to a medium-sized global player. DuPont, too, restructured its business. It focused on life science and those chemicals and fibers where it could capture the largest or second-largest market share. This was the case with nylon, one of its traditional strengths. It acquired the respective branches of ICI and of Rhone-Poulenc. Together with Dow, it formed a joint venture for elastomers, and thus became one of the biggest producers. When

---

[33] Aventis fell victim to a hostile takeover by the French pharmaceutical firm, Sanofi. Swiss Novartis was blocked by French political intervention from acting as "White Knight."

DuPont bought the coatings division of Hoechst in 1998, it commanded a market share of 30 percent in automotive coatings. In life science, it pressed forward even harder, founding a joint venture with Merck for pharmaceuticals and another one for seed with Pioneer Hi-Bred, both of which it took over in 1999. Other branches were sold, first the coal business and, in 1999, Conoco. Still, DuPont was quite diversified in 2000 with divisions in pharmaceuticals, pesticides, seeds, nutrition, chemicals, polymers, and fibers.

Even more than DuPont, BASF, now the world's largest chemical enterprise, pursued the strategy of combining bulk and sophisticated products. It, too, made joint ventures in order to obtain a leading share of the world market; for example, it entered DyeStar, the combined dye-divisions of Bayer and Hoechst, merged its PVC-activities with Solvay, bought the polypropylene division of Hoechst and Celanese and started a large-scale joint venture with Shell in polyolefins. BASF's high investment during the 1990s into production and distribution of gas may in the long run open a door for exiting the industry. There is one feature in which BASF distinguished itself from the other enterprises: the *"Verbund."* BASF believed very much in its sophisticated production-techniques of extremely large and interrelated plants. By bundling various series of production chains, materials and energy can be exchanged, and thus use is made of all by-products and heat or cold that would otherwise be wasted. BASF runs the largest industrial site in the world at Ludwigshafen, and has constructed similar ones in Antwerp, Belgium, and in China. When correctly implemented, this approach offers substantial organizational, ecological, and other advantages. First, it makes manufacturing inflexible, as one product cannot easily be taken out of such a system just because its does not sell well. But for BASF, this systematic approach is only part of its strategy. The *Verbund*, as it is called even in the English, represents "BASF's process to progress. *Verbund* is one of BASF's most important strengths and its cornerstone of the company's strategy."[34] For BASF, the *Verbund* is a key for cost-leadership. The *Verbund* includes not only the integration of production sites, but it relies on: "Cooperation with partners, relations with employees, relations with local communities, environment protection, and a worldwide

---

[34] BASF *Home-page,* http://www.corporate.basf.com/en/?id=V00-QS2M78oepbcp0x- (June 27, 2000), see *Annual Report* 1999.

Table 2.4. Chemical Firms' Competitive Situation at the Turn of the Century

	Advantages Mostly in Commodities	Advantages Mostly in Specialties
World market	Akzo Dow Hydro	Bayer DuPont ICI (Sanofi-Aventis, now drugs only) [35] (Novartis, now drugs only) [36]
Home market	Asahi BASF Mitsubishi Ch. Co.	Henkel (new)

knowledge Verbund."[37] In other words, *Verbund* and BASF's strength lay in integrated production and stakeholder value. In that sense, the company's strategy was quite different from most of the other ones, which underlined shareholder value as the core of their strategy. During the entire time, only one chemical company joined the list of the world's five hundred largest firms, and only at the end of our period: the German firm Henkel, a specialist in detergents, adhesives, and so on.

After the turn of the century the change continued at the same pace (Table 2.4). Within only a few years, until 2002, three major changes occurred, compared to only one during the last forty years. Two more companies opted to leave the industry to a large extent: Bayer and Hydro. Hydro dropped out of our list because the company invested heavily in oil, gas, and aluminum, reducing its chemical side to only about one-third of the enterprise. Bayer announced plans to spin off a chemical firm, called NewCo and concentrate on pharmaceuticals and perhaps a few more activities that require heavy R&D. Degussa (Germany) entered the league with a turnover of $12.4 billion, and claimed to be the largest producer in special chemicals worldwide. All this indicates that the chemical industry has matured. The companies foresee no major breakthroughs in the field of chemistry. Consequently, strategies for growth are sought in other fields or in optimizing known processes.

This chapter was written from the point of view of the chemical industry. However, if seen from the market side, focusing on where to buy chemicals, the picture would become a bit different. The heavy

[35] Aventis was created by a merger of Hoechst and Rhone-Poulenc. Chemicals were sold.
[36] Novartis was created by a merger of Ciba-Geigy and Sandoz. Chemicals were sold.
[37] Ibid.

Table 2.5. Competitive Situation on the Market for Chemicals
in 2002

	Advantages Mostly in Commodities	Advantages Mostly in Specialties
World market	Akzo-Nobel Dow Exxon Mobil Chemical	Bayer DuPont Degussa ICI
Home market	BP Chemicals BASF Mitsubishi Ch. Co. Atofina Shell Chemicals Asahi	Henkel

involvement of oil companies in basic chemicals changes the picture (Atofina is the chemical branch of the French oil concern Total).

According to turnover ($ billion in 2002), the four largest suppliers were still pure chemical firms (BASF 34, Bayer 31, Dow 28, DuPont 24), while the next two were oil firms (Atofina 20, Exxon Mobil Chemical 16)[38] (Table 2.5).

CONCLUSION

Our evaluation of results of strategy (not announced strategy) shows an industry that during the 1970s and 1980s sped up its development and started to change more radically in the last decade. After the boom period, the chemical industry, which relied heavily on oil as a feedstuff and for its energy intensive processes, ran into trouble. The major firms adapted through a better concentration on basic competitive strategies, either becoming more cost effective, or growing in fields of special advantages, and by a better coverage of world markets. Still much remained unclear, including the geographical positioning; all firms kept an outspoken national character, even the export-oriented ones. The 1980s brought a lot of change. Firms focused more on their known advantages. Cost leadership was sought through growth; increasing in size was thought to be

---

[38] Others: Mitsubishi Ch. Co.16, Shell Chemical 15, Akzo 14, BP Chemicals 13, Degussa 12, Henkel 10, Asahi 10, ICI 10 (information from the various homepages and *Die Zeit*, December 11, 2003).

both safe and profitable; the strategy applied was mainly internal growth; major takeovers remained the exception. It was with this background that firms such as Rhone-Poulenc (1986, 1989) or Mitsubishi Ch. Co. (1992), respectively, announced their aims of becoming number five or number ten in the world. All firms tried to combine both basic strategies by cost-, market-, or technology leadership in sub- and even sub-subdivisions of their special fields. Although the difficulties during the 1970s were perceived optimistically as temporary ones, it became clear in the late 1980s that the industry was aging and, compared to others, in relative decline. All firms suggested a strategy of a better global positioning, but not all were able to carry this out.

During the 1990s and well into the next millennium, strategies became not only tougher but much more radical, even including top mergers, the breakup of the firm, and exiting the industry. These two latter options had not previously been used and represented a totally new element. Even those firms that did not consider exiting the industry questioned their established structures more radically than before. Size was no longer an aim but, instead, a return on investment. Whereas, in the past, changes in structure had been comparatively slow, because internal growth was preferred, this changed during the 1990s. Quick change was achieved by selling, buying, or spinning off parts of firms. Old structures and established clusters of competence (Michael Porter) were broken up. Such clusters bridging the borders of branches of the industry existed not only in relation to oil or pharmaceuticals, but in the combination of the chemical industry with machine building and the IT-industry as well. In spite of the fact that the latter activities were profitable and represented a potential for growth, they were removed. Thus, Akzo sold Bamag, Hoechst Uhde, and Bayer Agfa; in contrast Asahi pointed to IT as one of its fields of preferred growth. The reason given for doing so was always the same: the respective branch was or was not counted in the defined sphere of core competence. Because the concept of mergers and acquisitions was used more than ever before, repositioning was much quicker when compared to previous phases characterized by strategies of internal growth. However, at least in one case (Hoechst) such movements became so hectic that neither employees nor shareholder could follow – with a disastrous result for the firm.

The instrument of joint venture was used differently over time. In the beginning, joint ventures were mainly applied to open up markets and, in the 1970s, as a reaction to the oil crisis. During the 1980s, the main aim was to obtain larger market shares for single product lines.

The 1990s emphasized this when whole divisions of firms were merged, for example, Bayer, BASF, and Hoechst merged their dye business into DyStar, or DuPont and Dow set up a common rubber division. In contrast to their talk or self-understanding as global companies, mergers were sought on a national basis. Ciba-Geigy merged with the Swiss Sandoz (to form Novartis), Union Carbide with Dow, and Monsanto with other U.S. firms. The first large-scale, over-the-border merger took place when, at the end of our period, Hoechst and Rhone-Poulenc formed Aventis. In 2005, only nine out of our initial fifteen firms still existed in the industry. This represents an indicator of radical change, which occurred in the 1990s. The roots of all of our firms had existed for more than one hundred years. Several enterprises opted to exist in fields related to chemistry but with higher margins, especially the pharmaceutical industry. Although Ciba-Geigy quit the branch quite elegantly, Bayer, Monsanto, and Rhone-Poulenc did so less well but still much better than Hoechst, Montedison, and Union Carbide, which seem to have lost control of the process. Another group consisting of Asahi, BASF, and Hydro stayed in the industry but constructed a possible emergency exit; Asahi into the IT branch, and the other two into energy. Meanwhile, Hydro executed its option. In the end, only Akzo, Dow, DuPont, and Mitsubishi Ch. Co. entered the new century totally committed to industrial chemistry. During the whole period, only two firms, Henkel and Degussa, entered the top league. The extent to which the relative weight of the industry as a branch has faded at the beginning of the new millennium is exemplified in *Fortune*'s Gobal 500 list of 2002, which included only six firms from the chemical but thirteen from the pharmaceutical industry. Times had changed!

3

Financial Systems and Corporate Strategy in the Chemical Industry

MARCO DA RIN

The chemical industry is one of the largest manufacturing industries in virtually all industrialized countries.[1] Chemicals are also one of the most diverse industries, extending from capital intensive, cyclical commodities (industrial chemicals and plastics), to technologically advanced products like agricultural chemicals and paints, to extremely high-tech, research-intensive businesses like pharmaceuticals or new materials. Surprisingly, economists have so far devoted much less attention to chemicals than to such industries as automobiles and electronics.[2] This is true also for the financial aspects of the industry.

Yet, chemicals provide a financially interesting case to study. The industry is both capital and research intensive in many of its segments and thus requires large investments that in some cases have a very long profit horizon. At the same time, the financial aspects of chemical firms vary widely within the industry. Although the high-tech end of pharmaceuticals and specialty products faces very uncertain outcomes for R&D investments, the mature commodity, capital intensive, businesses provide cyclical but

---

[1] I would like to thank Franco Amatori and Takashi Hikino for useful comments and Giampaolo Gagliardi for dedicated research assistance.

[2] For an exception to this rule, see the comprehensive account of the history and of the economics of the chemical industry in Ashish Arora, Ralph Landau, and Nathan Rosenberg, eds., *Chemicals and Long-Term Growth* (New York: Wiley & Sons, 1998).

stable flows of income. The availability of financial resources, their cost, and the way they constrain managerial decisions through corporate governance are therefore extremely relevant for this industry and for our understanding of its evolution.

In this chapter, I highlight some lessons we may learn from a long-term overview of the development of the modern chemical industry. I will argue that a proper understanding of the industry's restless transition through several "revolutions" since the 1970s requires studying its financial aspects. In particular, I will argue that financing is crucial for maintaining leadership at the firm level because it allows investment in long-ranging research and development activities. Financing is closely linked to the nature of national financial systems and to corporate governance, that is, the set of rules that determine how companies are run. I will argue that corporate governance systems help explain the evolution of the industry – market-based governance, for instance, allows faster corporate restructuring than relationship-based governance.[3] This implies that different businesses require different governance structures – commodities are best run by cost-cutting, whereas research-intensive biotech requires longer-range strategies. Finally, I will argue that the increasing globalization of capital markets has been an important determinant of some major changes in corporate strategy and in industry structure.

FINANCE AND CHEMICALS: A LONG-TERM VIEW

The evolution of the industry after the petrochemical revolution is the main focus of this study. In particular, I look at the United States, Germany, Britain, Japan, and Italy, which were the largest producers of chemicals by the end of World War II. I start with the origins of the industry because some of the traits that characterized the petrochemical revolution date back to that period.[4] Only by taking a long-term view can we properly analyze the influence of national financial systems on the financing of chemical firms and their corporate governance over time.[5]

---

[3] As, for example, with family or politically controlled enterprises.

[4] Marco DaRin, "Finance and the Chemical Industry," in Ashish Arora, Ralph Landau, and Nathan Rosenberg, eds., *Chemicals and Long-Term Growth*, pp. 301-19, examines in more detail the financial aspects of the industry in this period.

[5] L. F. Haber, *The Chemical Industry during the Nineteenth Century* (Oxford: Oxford University Press, 1958); and L. F. Haber, *The Chemical Industry, 1900-1930: International Growth and Technological Change* (Oxford: Clarendon Press, 1971) provides a comprehensive overview of the history of the industry until 1930.

Britain

Britain was the first economy to industrialize and also the first to develop a chemical industry.[6] However, British chemical firms soon proved unable to create innovative technologies. The two leading British producers of chemicals were Brunner, Mond & Co. (alkalis) and Nobel Industries (explosives). They adopted technology that had been developed abroad and was licensed from the international cartels, Solvay and Nobel. In consumer chemicals, where technology was relatively stable, firms such as Courtaulds (artificial fibers) or Lever Bros. (detergents) succeeded thanks to good marketing and to the special access they had to large colonial markets. Where technology was crucial, however, Britain lost its preeminence. Although Britain (Perkins in 1856) led the way in synthetic dyes, the first science-based chemical product, the Germans, who had made large investments in basic scientific research in the 1870s, shattered British competitors through a superior command of the science and technology.[7]

This failure was seen in finance as well.[8] Technological implementation required several steps beyond scientific discovery, steps that were financially demanding. British banks were constrained by regulation to remain small, undercapitalized, and local. They lent short-term against trade payables, a practice that accorded well with the simple financial needs of the main industry of the time, textiles. Although the city of London was the world's hub for capital markets, its institutions did not find it profitable to develop the skills required to appraise, value, and market industrial securities. Most chemical firms therefore had to remain private. The strategies of British banks and firms were complementary and squared well with each other. Banks had no incentive to develop

---

[6] William Reader, *Imperial Chemical Industries: A History. Vol. I: The Forerunners* (London: Oxford University Press, 1970); and *Imperial Chemical Industries: A History, Vol. II: The First Quarter Century* (London: Oxford University Press, 1975).

[7] John Beer, *The Emergence of the German Dye Industry* (Urbana: University of Illinois Press, 1959); Johann Peter Murmann, *Knowledge and Competitive Advantage: The Coevolution of Firms, Technology, and National Institutions* (Cambridge/New York: Cambridge University Press, 2003).

[8] Marco Da Rin, "Finance and Technology in Early Industrial Economies: The Role of Economic Integration," *Ricerche Economiche, 51*(3) (1997): 171–200. On the British financial system in this period, see Michael Collins, *Banks and Industrial Finance in Britain, 1800–1939* (London: Macmillan, 1991); Philip Cottrell, *Industrial Finance 1830–1914* (London: Methuen, 1980); Willliam Kennedy, *Industrial Structure, Capital Markets, and the Origins of British Economic Decline* (Cambridge: Cambridge University Press, 1987); and W. A. Thomas, *The Finance of British Industry* (London: Methuen, 1978).

the costly skills necessary to screen and monitor complex technological and market strategies because most firms invested in simple technology and catered to the stable demand of the domestic and colonial markets. Firms, in turn, could achieve satisfactory results without the need to invest in innovative, complex, and risky technology with distant payoffs. In economic terms, it was an instance of "coordination failure," in which no party had the incentive to move away from the status quo.

British chemical technology was thus complementary to British-style finance. Research and capital intensive fields like pharmaceuticals or electrochemicals had hardly developed. British firms, in chemicals but also in other industries, failed to excel in technology and gradually lost ground to innovative competitors.[9] Even Imperial Chemical Industries (ICI), the British champion that was created by the merger of four of Britain's largest chemical firms in 1926, invested a smaller share of its profits in research than its German or American competitors. It lagged in the race to innovation, its only major discovery being polythene in the 1930s. Britain's history shows that even a successful financial system may not be suited to the support of high-tech industries.

United States

The experience of the United States illustrates well the role of market mechanisms and the importance of market size. The American economy developed relatively late, but by 1890 it had become the world's largest. Its huge internal market, much larger than that of the European national economies, was its distinguishing mark.[10] But the effect of the market took time to unfold. In its beginning, the American chemical industry was made up of many small and technologically unsophisticated firms.[11] Fragmentation was reinforced by the nature of the financial system. Interstate banks were prohibited, and the absence (until 1913) of a central bank that could act as "lender of last resort" forced banks to employ

---

[9] Alfred Chandler, *Scale and Scope: The Dynamics of Industrial Capitalism* (Cambridge, MA: Harvard University Press, 1990). The history of the soda business illustrates this point (L. F. Haber, *The Chemical Industry, 1900–1930: International Growth and Technological Change*). British producers failed to switch to the (more complex) Solvay process and instead retained the rudimentary and obsolescent Leblanc process. Collusion rather than innovation seemed the solution, when several of them merged into United Alkali in 1891, in an unsuccessful attempt to consolidate the market.

[10] Alfred Chandler, *The Visible Hand* (Cambridge, MA: Belknap Press, 1977).

[11] William Haynes, *American Chemical Industry* (New York: Van Nostrand, 1954) provides a comprehensive history of the American chemical industry.

short-term loans and to limit exposure to individual borrowers.[12] Nor were stock markets more accessible as industrial securities were considered speculative – by 1900, only three chemical firms were listed on Wall Street,[13] and only DuPont issued large sums of equity.[14]

Chemical firms, then, had to rely on internally generated funds to finance their growth. The pressure to remain financially viable discouraged large investments in risky research and made the United States largely dependent on imports for more sophisticated products. American firms did not have extensive links to research networks along the lines discussed by Gambardella, Cesaroni, and Mariani in this volume. Firms such as Dow or Monsanto became leaders in their markets only after overcoming major financial problems. Once their income stabilized, however, they benefited from the large, growing American market. They could then use their profits to invest large sums in developing new technologies, thus creating new sources of long-term profitability. The post–World War I effort in research was substantial: by 1930, chemical companies employed a third of America's industrial scientists.[15]

A virtuous cycle of profits and investment in technology that resulted in further profits thus helped American firms catch up with their European competitors. The clearest example is that of DuPont, which during the 1920s and 1930s invested nearly $40 million in R&D, paving the way for a stream of lucrative inventions like viscose, cellophane, and above all, nylon. DuPont and the other leading American firms now had extensive links with a rapidly expanding network of research institutions.[16] The profits-R&D cycle also induced consolidation. About five hundred mergers took place in the 1920s,[17] and more chemical firms entered the list of the two hundred largest manufacturing companies than

---

[12] On the development of the American financial system, see Vincent Carosso, *Investment Banking in America* (Cambridge: Cambridge University Press, 1970); Raymond Goldsmith, *Financial Intermediaries in the American Economy Since 1900* (Princeton, NJ: Princeton University Press, 1958); Mark J. Roe, *Strong Managers, Weak Owners* (Princeton, NJ: Princeton University Press, 1994); and Richard Sylla, *The American Capital Market* (New York: Arno Press, 1975).

[13] William Roy, "The Rise of American Industrial Corporations, 1880-1914," computer file, Department of Sociology, University of California at Los Angeles (1990).

[14] Alfred Chandler, and Stephen Salsbury, *Pierre S. Du Pont and the Making of the Modern Corporation* (New York: Harper & Row, 1971).

[15] Chandler, *Scale and Scope.*

[16] David A. Hounshell and John Kenly Smith, Jr., *Science and Corporate Strategy: DuPont R&D, 1902-1980* (Cambridge/New York: Cambridge University Press, 1988).

[17] Haynes, *American Chemical Industry.*

from any other industry.[18] Firms such as Dow, Monsanto, Union Carbide, and Allied Chemical became market leaders largely by acquisitions.

Germany

Germany was another latecomer to industrialization, which it achieved in the latter half of the nineteenth century. German firms, which were accustomed to adopting technology developed abroad, pioneered the systematic use of science in the development of industrial applications. German firms had extensive networks with the nation's leading research institutions in the chemical sciences. By the end of the century, the German chemical industry had reached the technological forefront, and German firms invented and dominated high-tech segments like synthetic dyes, fertilizers, and pharmaceuticals.[19]

The support of the German financial system was instrumental in the first industrialization of Germany. Industrial credit banks nurtured close relationships with their borrowers. Private bankers (Privatbankiers) and joint-stock Kreditbanken invested in the ability to evaluate complex projects on a technological basis. They also were large enough to support the heavy financial requirements of the Second Industrial Revolution.[20] Moreover, the largest Kreditbanken dominated Berlin's stock exchange and (unlike the situation in New York) could thus help firms issue stock and bonds. Firms in innovative industries like chemicals or electrical engineering accounted for a large share of Berlin's capitalization.[21]

---

[18] Chandler, *Scale and Scope.*

[19] Walter Telschik, *Geschichte der Deutschen Großchemie* (Weinheim: VHC, 1994).

[20] See Marco Da Rin, "German Kreditbanken 1850–1914: An Informational Approach," *Financial History Review 3*(2) (1996): 29–47; Wilfred Feldenkirchen, "Banking and Economic Growth: Banks and Industry in Germany in the Nineteenth Century and Their Changing Relationship during Industrialization," in Wang Lee, ed., *German Industry and German Industrialization* (London: Routledge, 1991); Gerd Hardach, "Banking and Industry in Germany in the Interwar Period 1919–39," *Journal of European Economic History, 13*(S) (1984): 203–34; Hans Pohl, "Forms and Phases of Industry Finance up to the Second World War," *German Yearbook on Business History* (1984), 75–95; Jacob Riesser, *The Great German Banks* (Washington, DC: U.S. National Monetary Commission, 1911); and Richard Tilly, "Germany: 1815–70," in Rondo Cameron, ed., *Banking in the Early Stages of Industrialization* (New York: Oxford University Press, 1967) on the development of the German financial systems through World War I.

[21] From 1850 to 1906, German chemical firms raised about 750 million marks on the stock markets. Rolf Grabower, *Die finanzielle Entwicklung der Aktiengesellschaften der deutschen chemischen Industrie* (Leipzig, Duncker & Humblot, 1910) provides a detailed account of the financial development of German chemical firms.

Da Rin shows the existence of a strong complementarity between the size and risk tolerance of Kreditbanken and the industrial firms' dedication to innovation.[22] Banks' ability to discern temporary financial distress from structural weaknesses allowed them to shield clients from the periods of financial trouble that characterize all research and capital intensive businesses. This encouraged firms to invest in research and to develop new technology. The chemical industry exploited this advantage in full. Firms such as BASF, AGFA, Hoechst, and Degussa received crucial support from their bankers during their early years, which made it possible to pursue ambitious investments in technology with long-term payoffs.[23] In the 1870s, the ability to invest for the long run was crucial in helping German chemical firms develop industrial research laboratories that employed scores of scientists. The large German chemical firms invested between 5 percent and 10 percent of their sales in R&D, a large proportion even by today's standards.

Although market size was a primary factor shaping American firms, German firms in the early phase of their development relied on banks to finance the invention of profitable products. In both cases, profits were used to sustain technological leadership: profits from dyes were used to diversify into new businesses like modern drugs (Hoechst, 1883), synthetic rubber (Bayer, 1910), and ammonia (BASF, 1912). Protracted innovation sustained by strong earnings provided a formidable competitive advantage, a factor that has continued to allow these firms to remain, up to now, among the world leaders.

Japan

Japan provides an intriguing example of the influence of finance on corporate strategy. A latecomer to industrialization, Japan's chemical industry was particularly slow to develop and depended on imports of advanced products well into the twentieth century.[24] Whereas the

---

[22] Marco Da Rin, "Finance and Technology in Early Industrial Economies: The Role of Economic Integration," *Ricerche Economiche, 51*(3) (1997): 171–200.

[23] Developing a new synthetic dye could take years (two decades for indigo). The perfection of nitrogen fixation also required two decades and thousands of costly experiments. BASF and Hoechst invested nearly thirty million marks in the race to the discover indigo (see Julia Wrigley, "Technical Education and Industry in the Nineteenth Century," in Bernard Elbaum and William Lazonick, eds., *The Decline of the British Economy* (Oxford: Clarendon Press, 1987).

[24] Barbara Molony, *Technology and Investment: The Prewar Japanese Chemical Industry* (Cambridge, MA: Harvard University Press, 1990).

technological backwardness of the country was an important factor, the financial system also played a role in retarding development.[25] As in the United States, the regulation of commercial banks kept them small.[26] Only in 1890 was national banking allowed but with many restrictions. Securities markets did not develop until World War I, and investment banks appeared even later.

The characteristic trait of financing industrialization in Japan was the central role of industrial groups, Zaibatsu. These were conglomerates that originated in nineteenth-century trading groups.[27] Mitsui, Mitsubishi, Showa, Sumitomo, and the other large conglomerates needed chemical products for their new industrial activities. Agriculture's demand for fertilizers or alkalis, which were very costly to import, also played a strong role in shaping the Japanese industry. Most Japanese chemical firms, then, developed as divisions of the conglomerates, which provided finance for the acquisition of technology and for the construction of plants. This kept their focus on the domestic market, as Hikino's essay in the volume indicates.

Chemical firms within Zaibatsu were different from independent firms. First, they had a relatively small but stable captive customer base, which came from the other firms within the Zaibatsu. Small and stable demand induced these firms to invest in product customization rather than in large-scale production – still, today, a hallmark of Japanese chemicals, as Hikino indicates later. Second, Zaibatsu ownership reinforced this type of specialization.[28] Third, since the 1930s, only Zaibatsu firms became rich enough to buy costly foreign technology for the production of organic chemicals like dyes or drugs. Finally, close links with the political establishment enabled Zaibatsu to secure wartime procurement during the Sino-Japanese war. Only after World War I were entrepreneurial firms such as Nichitsu, a producer of electrochemicals, able to capture sufficient market share to expand and prosper and eventually become small Zaibatsu themselves.

Zaibatsu reliance on the "internal capital markets" shaped the industry by bringing into existence many relatively small producers that did

[25] Takashi Hikino, Tsutomo Harada, Yoshio Tokuhisa, and James Yoshida, "The Japanese Puzzle," in Arora et al., *Chemicals and Long-Term Growth*.

[26] L. S. Pressnell, ed., *Money and Banking in Japan* (London: Macmillan, 1973).

[27] Hidemasa Morikawa, *Zaibatsu: The Rise and Fall of Family Enterprise Groups in Japan* (Tokyo: University of Tokyo Press, 1992).

[28] Oliver Hart, *Firms, Contracts and Financial Structure* (Oxford: Oxford University Press, 1995).

not reach an optimal scale, unlike American or German firms. In turn, product customization for stable customers resulted in little investment in basic research, which meant that Japan could not reduce the technological lag with the world leaders. The case of Japan then illustrates how dependence on financing from a corporate investor (the Zaibatsu) matters for the choice of technological specialization. This is a clear instance of how corporate governance crucially influences corporate strategies and the development of competitive advantage.

Italy

The case of Italy is instructive for understanding how the financial system may be a necessary but not sufficient condition for industrial success. Italy industrialized after Germany, at the turn of the nineteenth century. Like German firms a quarter century earlier, Italian chemical firms relied extensively on foreign (mostly German) technology. Unlike German firms, however, they never leapt ahead by investing in the research establishments and links to research networks that would enable them to develop new technologies. Rather, Italian chemical firms were born small and tended to remain such. A common strategy was to license foreign technology and carve out small market niches where competition was weak. Drugs provide a good example: only two firms became relatively large, Carlo Erba and Schiapparelli, whereas the several firms that remained at suboptimal scale employed either inferior domestic technology or foreign licensed technology.[29] The lack of a large domestic market was clearly an important factor in restraining investment in technology, and most of the nation's high-tech, high-value organic chemical products (explosives, dyes, and drugs) were imported.[30]

Italian industrial finance developed along the German model but on a smaller scale.[31] Only two large banks emerged, Banca Commerciale Italiana, in 1894, and Credito Italiano, in 1895.[32] These banks helped

---

[29] For example, Serono (created in 1902), Lepetit (1910), or Zambeletti (1913); Paolo Amat, in Ernst Homburg, Anthony Travis and Harm Schröter, eds., *The Chemical Industry in Europe, 1850-1914* (Dordrecht: Kluwer, 1998).

[30] Vera Zamagni, "L'industria chimica in Italia dalle origini agli anni '50," in Daniela Brignone, ed., *Innovazione tecnologica e industria* (Roma: Bulzoni, 1998).

[31] Laura Bottazzi and Marco Da Rin, "Banks as Catalysts for Industrialization: Evidence from Italy," mimeo, IGIER (2002)

[32] Antonio Confalonieri, *Banca e Industria in Italia, 1894-1906* (Milano: Banca Commerciale Italiana, 1976); and *Banca e Industria in Italia dalla crisi del 1907 all'agosto 1914* (Milano: Banca Commerciale Italiana, 1982).

foster the country's industrialization by coordinating investments across different industries.[33] However, only a relatively small amount was devoted to financing chemical firms in the 1894–1913 period: chemical firms received about 9 percent of the overall loans by Credito Italiano and only 5 percent of those by Banca Commerciale.[34] These sums were substantially lower than those invested in the electrical and mechanical industries. Another sign that chemicals were not considered a promising investment is the fact that only a dozen chemical companies were listed on the Milano Stock Exchange in this period, fewer than other industries.

The small size of the internal market and the lack of large-scale financing meant firms were not able to acquire enough financial resources, internal or external, to engage in long-term research projects.[35] Fragmentation led to the development of "chemical artisan" firms.[36] The only notable exception was SNIA, a private company that produced cellulose-based fibers after World War I, and that was one of the world's four largest producers until its financial collapse in 1929. Bank ownership played a somewhat restraining role. Banca Commerciale, which was owned by some German Kreditbanken, financed many Italian firms that used German technology, such as Italiana Carburo di Calcio (electrochemicals) and Unione Italiana Concimi Chimici (fertilizers). Italy's only world-class chemical firm, Montecatini, was at the time still mainly a mining concern that benefitted from only minor financing from Credito Italiano.

THE PETROCHEMICAL REVOLUTION AND POSTWAR GROWTH

Petrochemicals had been hardly noticeable until the 1930s. As Arora and Rosenberg show,[37] however, oil gradually became the main feedstock for chemicals in the United States where oil-refining technology – pioneered by Union Carbide, Dow, and Standard Oil – developed well before other countries. By 1940, nearly 70 percent of the world oil-refining capacity was in North America.

---

[33] Marco Da Rin, Marco, and Thomas Hellmann, "Banks as Catalysts for Industrialization," *Journal of Financial Intermediation, 11*(4) (2002): 366–97.

[34] Laura Bottazzi and Marco Da Rin, "Banks as Catalysts for Industrialization: Evidence from Italy," mimeo (IGIER 2002).

[35] For instance, in the 1930s, Carlo Erba gave up plans to engage in new research because of the failure of its main shareholder, Mr. Panzarasa. See Amat, "The Italian Chemical Industry from 1861 to 1918."

[36] Zamagni, "L'industria chimica in Italia dalle origini agli anni '50."

[37] Ashish Arora and Nathan Rosenberg, "Chemicals: A U.S. Success Story," in Arora et al., *Chemicals and Long-Term Growth.*

After World War II, the shift from coal to oil as the basic feedstock was relatively rapid: the share of oil-based organic compounds was 50 percent in 1950, but leapt to 90 percent by 1960. The change required huge investments, as economies of scale are of utmost importance in petrochemicals and only large plants are economically viable. Huge outlays of capital constituted powerful barriers to entry: no new major firm has emerged in the chemical industry since the 1920s. The rapid growth in the 1950s and 1960s can be seen in Figure 3.1. The development of plant and equipment capital stock since the end of World War II is apparent for three of the major "all-round" diversified chemical firms – Bayer, Dow, and ICI. Both Dow and Bayer tripled their capital stock in real terms in the twenty years before the oil shocks of the 1970s, and ICI more than doubled it. At Dow and ICI, growth continued at a fast pace through the 1970s, as in the industry as a whole.[38] Looking at these three firms is instructive since each represented a typical case of a German, U.S., or British large, diversified chemical firm. Also, almost half a century of data helps one appreciate the long-term evolution of these firms' balance sheets.

The way growth was financed depended on each company's financial policy. European companies have relied on debt to a much larger extent than American ones, which issued larger amounts of equity. Figure 3.2 illustrates this pattern in the case of Bayer, ICI, and Dow. Figure 3.2 shows that the capital base of chemical firms (measured in real terms) increased until the first oil shock, and then started a slow but steady decline. By the end of the century, it had fallen back to its 1955 level. This figure also highlights the contrast between the stability of the equity base and the volatility of debt levels. Soaring interest rates in the early 1980s prompted sharp decreases in debt levels.

A closer look at one corporate case, that of BASF (Badische Anilin Soda Fabrik) helps shed more light on how postwar reconstruction was financed in Europe. Like the other large German chemical firms, BASF at the end of World War II had lost 30 percent of its plants, and the great majority of what remained was damaged.[39] Vast amounts of assets had to be written off, and the balance sheet shrank from 706 to 399 million marks. Reconstruction was initially financed solely from internal resources (depreciation), since banks were unwilling to lend to a

---

[38] Ashish Arora and Alfonso Gambardella, "Evolution of Industry Structure in the Chemical Industry," in Arora et al., *Chemicals and Long-Term Growth.*

[39] Paul Dratt, "The History of BASF Since 1945 from a Financial Viewpoint," mimeograph (1995).

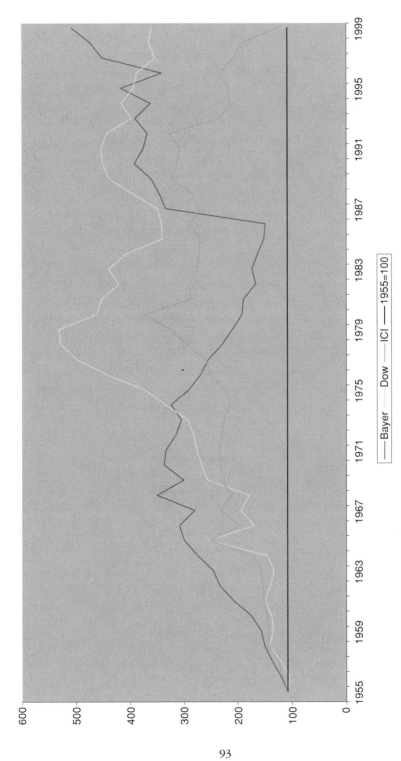

Figure 3.1. Property, plant and equipment. *Note:* Data in constant 1999 dollars, 1955 = 100 for each company. *Source:* Author's calculations based on company balance sheets.

Bayer —— Dow —— ICI —— 1955=100

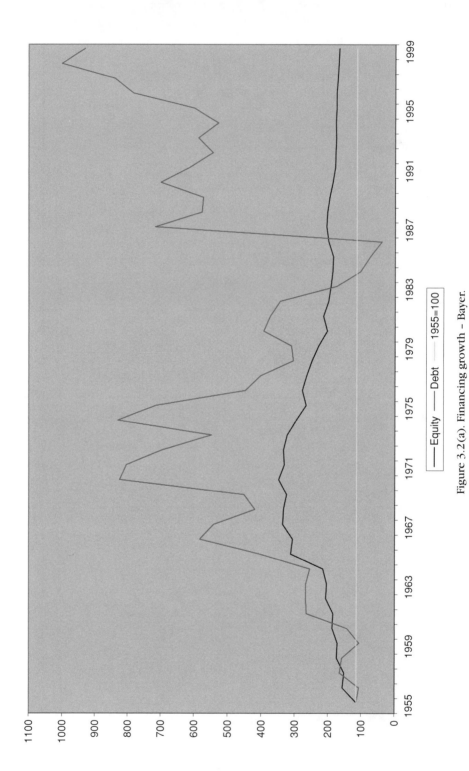

Figure 3.2(a). Financing growth – Bayer.

94

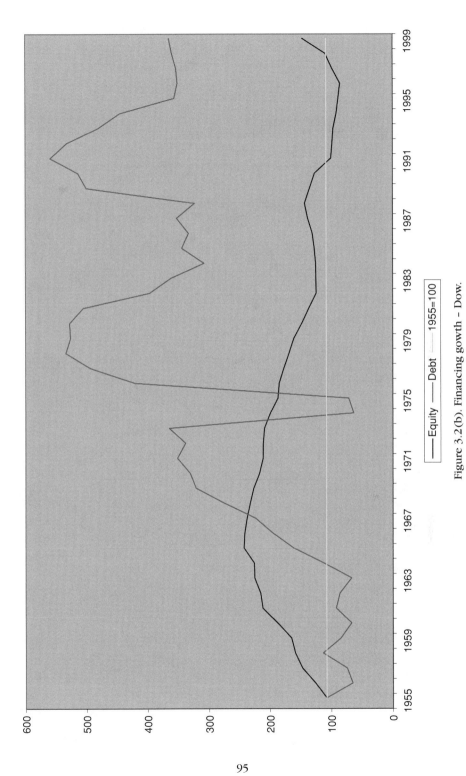

Figure 3.2(b). Financing gowth – Dow.

95

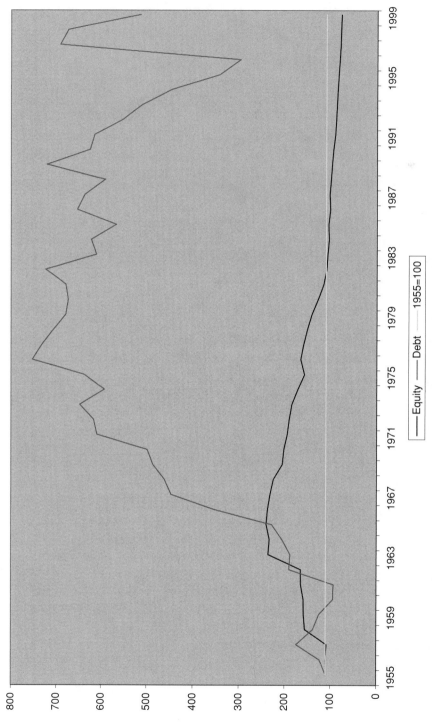

Figure 3.2(c). Financing gowth – ICI. *Note:* Data in constant 1999 dollars, 1955 = 100 for each company. *Source:* Author's calculations based on company balance sheets.

company, that was heavily controlled by the Allied forces and that was formally "in liquidation." Only in 1955 could new equity capital be raised. The large cash flow that was made possible by rapidly increasing sales in the early 1950s was therefore crucial, and financial resources were concentrated on reconstructing the company's capital base.[40] Conversion from coal to oil was undertaken in the 1960s and was financed largely by internal resources and especially bank debt, which grew tenfold over the decade.

In the postwar period, corporate strategies for BASF, Bayer, and Hoechst were driven by the need to exploit the enormous growth potential created by the reconstruction of the European economies. The growth rate throughout the 1950s of these three diversified firms averaged 12 percent, slightly higher than the growth rate of U.S. diversified firms, and almost three times the 4.9 percent growth rate of OCED economies. By comparison, over the 1990s, the industry grew by 3.8 percent in the United States and by 2.9 percent in Europe (American Chemistry Council 2001). Such rapid growth made chemicals a very profitable business, but it had the unfortunate effect of causing the overcapacity that later proved so costly to the industry. The cases of Japan and Italy illustrate well how good financial results induced by fast growth can actually produce bad corporate decisions. It also introduces a subject that turned out to be crucial once the industry had to restructure and cut capacity in the 1980s, namely, corporate governance.

Italy was the first European country to build a petrochemical plant, at Ferrara in 1949. This was followed by two decades of uncoordinated entry by all the Italian major companies – Montecatini, Edison (later merged into Montedison), Anic, and SIR – and even by American firms such as Mobil and Gulf. Montedison became, in fact, one of the European chemical firms with the highest rate of investment in fixed capital. One notable feature of the Italian case was the fragmentation of the industry into many relatively small players: even the earlier-mentioned companies barely reached a viable scale. The small scale meant that the expansion into petrochemicals had to be financed largely from debt, given the stock market's limited ability to provide capital (only ten chemical firms were listed in 1970). This left the firms vulnerable to changes in their economic context. When during the 1970s interest rates increased in response to monetary restrictions, the highly leveraged chemical firms were penalized by a heavy interest-rate burden, which was even larger than in other

---

[40] Dratt, "The History of BASF Since 1945 from a Financial Viewpoint."

industries of the traditionally highly indebted Italian manufacturing firms. The ratio between interest expenses and profit margins reached 0.7 or even 0.8, two or three times the values experienced by other European chemical companies – whose leverage had also increased through the 1960s.

Interest expenses in turn eroded profit margins and further impaired the ability of Italian chemical firms to invest in R&D.[41] R&D intensity, defined as the ratio of R&D expenses to sales, remained between 1 percent and 2 percent, and was reflected in the lack of specialization in fine chemicals and weakness in international trade.[42] The clearest example of the inability to innovate was Montedison's failure to leverage Natta's discovery of polypropylene in the way Du Pont did with nylon. The lack of pressure toward concentration and rationalization was largely a result of the absence of pressure to create value for shareholders, as the industry – a large employer and "national champion" – enjoyed political protection.

Reliance on internal finance also was particularly important in Japan, where securities markets were repressed and conglomerate membership was crucial for obtaining financial resources. Although the Allies dismantled the old Zaibatsu after the war, these soon reappeared under a different guise – the conglomerate organization of Keiretsu. Unlike Zaibatsu, Keiretsu are not based on holding companies, but on a closely knit web of cross-shareholdings of group-affiliated firms.[43] After the war, there was a massive entry into petrochemicals by the Japanese firms, Mitsui, Mitsubishi, Sumitomo, Asahi, and Maruzen Oil.[44] Virtually all the large petrochemical producers were part of a Keiretsu. Affiliation to a Keiretsu provided financing, but at the same time, limited the opportunities for diversification. Mitsubishi Chemicals, Sumitomo Chemicals, Mitsui Toatsu, and Showa Denko are the only diversified Japanese chemical companies. As in Italy, the conglomerate structure has kept Japanese chemical firms smaller than their competitors: in 1995, only three chemical firms ranked among the first 100 industrial companies. As we have

---

[41] Riccardo Azzolini, Giorgio Dimalta, and Roberto Pastore, *L'industria Chimica tra crisi e programmazione* (Roma: Editori Riuniti, 1979).

[42] Ibid.

[43] Masahiko Aoki and Hugh Patrick, eds., *The Japanese Main Bank System* (Oxford: Oxford University Press, 1994), offer a thorough discussion of the financial structure of Japanese industry and of the role of Keiretsu and the associated Main Bank system.

[44] Takashi Hikino et al., "The Japanese Puzzle," in Arora et al., *Chemicals and Long-Term Growth*.

seen earlier, this was a result of the segmented and "self-contained" nature of the conglomerates, and as Hikino points out, this aspect of corporate structure kept the firms focused on the domestic market and prevented them from achieving a global scale of operations.

The postwar high-growth period was therefore marked by intense investments in all countries, profitable growth, and an increase in external financing. In Europe, this resulted in higher leverage (see Figure 3.2), whereas in the United States, issues of new equity were more important. As a result, the close control of the founders became weaker for American chemical companies, as ownership became more dispersed among individual investors. DuPont, still controlled by the family, was an exception rather than the rule. The shift from the concentrated ownership of the entrepreneurial phase to today's oversight by institutional investors in a global capital market therefore was underway in the 1950s and 1960s, a phase of transition in which managers enjoyed an unprecedented degree of freedom. That freedom soon turned out to be costly.

PETROCHEMICALS IN THE AGE OF RESTRUCTURING

The 1970s represented one of the most important decades in the history of the chemical industry, a decade during which the industry was shaken by three major waves of structural change. First, the oil shock exposed the accumulation of petrochemical overcapacity at a global level. Second, the rise of biotechnology severed the longtime synergies between chemicals and the life sciences. Third, the diffusion of technology brought an increase in competition in the form of commoditization to many segments of the industry. As a result of these shocks, the chemical industry has undergone a thorough and painful restructuring.

The oil shock raised the price of the industry's main feedstock and sent industrial economies into a long recession. This in turn brought downward pressure on prices and diminished profit margins, making chemicals less profitable: the return on capital in the U.S. chemical industry had declined from about 15 percent in the mid-1960s to about 8 percent by 1975, and an index of profitability for West European firms also halved between 1965 and 1972.[45] However, it was only in the early 1980s that expectations and strategies changed drastically, as reflected in a declining investment-to-sales ratio. The rise of the biotechnology business had

---

[45] Keith Chapman, *The International Petrochemical Industry: Evolution and Location* (Oxford, UK/Cambridge, MA: Basil Blackwell, 1991), p. 235.

a less visible but more lasting effect, in that it created several new segments of the industry, opening space for new products and processes
that no longer depended on the by-products of oil refining. Biotechnology made the development of new drugs dependent on molecular biology and genetic engineering capabilities, severing the century-old links
between chemicals and pharmaceuticals and destroying economies of
scope between oil refining and sectors like pesticides and herbicides,
pharmaceuticals, and photographic film. Finally, the "commoditization"
of chemicals was driven by the entry of several firms in petrochemicals
and downstream products like fibers and resins, whose technology had
by the 1980s become widely spread. Commoditization lowered margins
and forced the weakest firms to exit, while creating an opportunity to
create value by turning to cost-reduction.

These shocks had a major impact on the dynamics of the industry,
and finance played a large role in the manner in which the industry
responded. First, profitability was dramatically reduced, forcing firms
to find ways to reduce costs and differentiate products; expenditure in
research and development was one of the main casualties, as one can see
in Figure 3.3. Second, restructuring required massive investments. For
instance, life sciences – as drugs and related sectors are called – require
much higher levels of investment in research and development than do
commodities. As a consequence, many firms began to separate their life
sciences operations from their chemical divisions. The demerger within
ICI that in 1994 led to Zeneca, a pharmaceutical firm, and ICI, a chemical
firm, was the first of a series of dramatic separations. Later, this ongoing
process led to deep changes in the corporate nature of entire firms, as in
the cases of Hoechst and Monsanto.

Figure 3.3 shows that R&D, which had been for many decades a
major source of growth in chemicals, dropped sharply in the 1970s.
For instance, at Dow the ratio between R&D and sales reached about 7
percent in the 1930s but was almost halved in the 1960s and 1970s. This
pattern was common to the majority of the large chemical firms.[46] At
DuPont, for instance, the ratio fell from around 8 percent in the 1960s
to around 4 percent in the 1980s.

While the industry was changing its technological and economic traits,
financial markets in all countries – but mostly in the United States – were
going through deep regulatory changes, generally in the direction of

---

[46] Kirkor Bozdogan, "The Transformation of the U.S. Chemical Industry," Working Paper of
the MIT Commission on Industrial Productivity (1989).

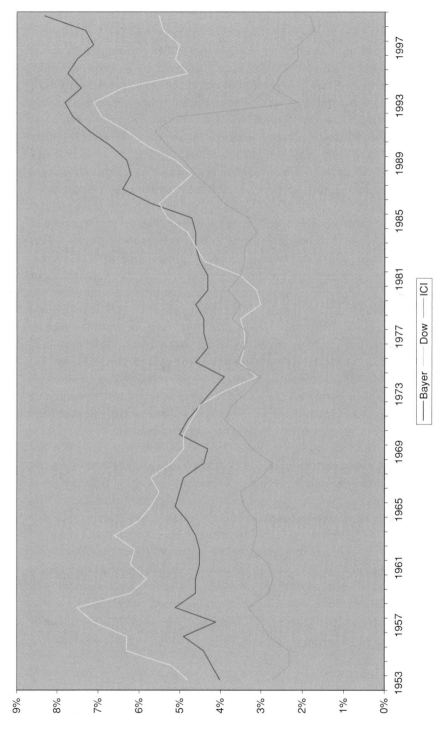

Figure 3.3. R&D Intensity. *Source:* Author's calculations based on company balance sheets.

liberalization and deregulation.[47] This was another important change for the financing of chemical firms. Securities markets grew very rapidly in the 1980s and 1990s, and institutional investors like pension funds, mutual funds, and insurance companies became important shareholders of industrial firms. In the United States, the process was largely completed by the early 1980s, whereas in Europe it took two more decades. Together with the increased freedom to invest across borders, these changes had the effect of increasing shareholders' awareness of the returns on their investments and encouraging them to demand better corporate performance.[48]

A new mode of corporate governance became the norm in the Anglo-Saxon world. "Corporate governance" indicates the set of rules by which companies are run.[49] Variations in corporate governance are large among national financial systems, and in addition to the regulatory environment mentioned already, the differences depend essentially on the development of the equity market and institutional investors. Although equity markets allow firms to raise large sums for investment and therefore to realize their potential, dispersion of ownership also results in more leeway for managers, who gain greater ability to appropriate resources and to buttress their control of the company. Because the cost of monitoring managerial performance is very large, individual investors are unable to effectively ensure discipline at a time when the rise of large institutional investors has improved the efficiency of capital markets.

The wave of restructuring that started in the early 1980s and that has produced sweeping changes in the industry was deeply influenced by financial and corporate governance considerations. First, firms reduced investment, notably capital expenditure for new plants and R&D. Then came the exit from less profitable upstream markets by major industry players, which used the proceeds from the sales of plants to diversify into high value-added sectors such as specialties, pharmaceuticals, and advanced materials. The payoff for quick restructuring was enormous: in the United States, the share of profits in the chemical industry generated by life-science businesses has grown from 21 percent in 1981 to 61.9 percent in 2000 (American Chemistry Council 2001). A number of

[47] Sarkis Khoury, *The Deregulation of the World Financial Markets* (New York: Quorum Books, 1985).

[48] Roe, *Strong Managers, Weak Owners*, describes such evolution in great detail.

[49] Luigi Zingales, "Corporate Governance," in *The New Palgrave Dictionary of Economics and the Law* (New York: Macmillan, 1998).

Table 3.1. Restructuring by Region and Period, 1985–97

Region of Acquirer	Period	1978–82	1983–87	1988–92	1993–97	Total
North America	(1)	344	1,696	2,815	4,087	8,942
	(2)	291	269	125	109	156
Europe	(1)	23	253	1,492	1,719	3,487
	(2)	718	392	191	332	282
UK	(1)	17	266	999	904	2,187
	(2)	658	231	84	98	111
Rest of the World	(1)	45	338	1,298	2,285	3,966
	(2)	601	238	96	62	92
Total Deals		429	2,553	6,604	8,995	18,581

(1) Number of deals.
(2) Average deal amount (millions of 1990 dollars).
Source: Author's calculations based on Securities Data Company data.

new niche players also appeared, among which were several "financial" firms that were created to engineer the purchase of plants in commodity chemicals through highly leveraged transactions.

The speed, mode, and depth of restructuring differed substantially across countries, as documented by Arora and Gambardella.[50] It was American firms that led the pack in shedding overcapacity in commodities.[51] Europe followed almost a decade later and was slower in reshaping the structure of its chemical industry. This trend is illustrated in Table 3.1, which is derived from one of the most comprehensive databases of worldwide merger and acquisition activity (compiled by the Securities Data Company).[52] The table provides a count of the deals by firms that made at least five transactions in the 1985–97 period and supports the view that restructuring in Europe started later. Interestingly, British chemical firms were much more active than were Continental European ones, thus behaving more like their North American competitors. Deals were

[50] Arora Ashish and Alfonso Gambardella, "Evolution of Industry Structure in the Chemical Industry," in Arora et al., *Chemicals and Long-Term Economic Growth.*

[51] Sarah Lane, "Corporate Restructuring in the Chemical Industry," in Margaret Blair, ed., *The Deal Decade* (Washington, DC: Brookings Institution, 1993); Marvin Liebermann, "Exit from Declining Industries: 'Shakeout' or 'Stakeout'?" *Rand Journal*, 21(4), 1990: 538–54.

[52] A limitation of the Security Data Company database is the likely bias toward North American deals, at least until 1985.

Table 3.2. Cross-Border Restructuring, 1985–97

Region of Acquirer		Within-Border Deals	Cross-Border Deals	Total
North America	(1)	7,242	1,700	8,942
	(2)	164	108	156
	(3)	8.8%	16.2%	
Europe	(1)	2,101	1,386	3,487
	(2)	298	267	282
	(3)	53.1%	46.9%	
UK	(1)	1,107	1,079	2,186
	(2)	101	121	111
	(3)	50.7%	49.3%	
Rest of the World	(1)	2,366	1,600	3,966
	(2)	74	120	92
	(3)	59.7%	40.3%	
Total Deals		12,816	5,765	18,581

(1) Number of deals.
(2) Average deal amount (millions of 1990 dollars).
(3) Percentage.
Source: Author's calculations based on Securities Data Company data.

larger in Europe than in the United States.[53] The difference between the United States and Europe also becomes apparent in Table 3.2, which documents the more intense cross-border activity of European firms, that is, their higher propensity to buy and sell plants and businesses outside of Europe. Again, the United Kingdom lies between Europe and the United States, but in this case closer to the latter. Another important difference across the Atlantic was the much higher reliance on asset swaps between European companies, often involving newly created firms like Montell or Borealis.

Restructuring seems to be linked also to research and development, which has constituted the ultimate source of growth for chemicals since the inception of this industry. Restructuring in the 1980s and 1990s has in fact affected the amount and distribution of R&D activities. Arora, Ceccagnoli, and Da Rin[54] show that restructuring since the late 1980s

---

[53] Whereas in the earlier years, this could well be because of a bias in the reporting of smaller European transactions, the trend persists into the 1990s, where such bias is likely to be minimal.

[54] Ashish Arora, Marco Ceccagnoli, and Marco Da Rin, "Corporate Restructuring and R&D: A Panel Data Analysis for the Chemical Industry," in Fabrizio Cesaroni, Alfonso Gambardella,

Table 3.3. Restructuring and R&D Intensity, 1987–97

Industry Segment		Active		Inactive	
		1990–91	1996–97	1990–91	1996–97
Energy	Net acquirers	0.22%	1.65%	2.09%	1.67%
	Net divestors	0.64%	0.89%	1.23%	n.a.
Commodities	Net acquirers	3.31%	3.22%	2.62%	2.11%
	Net divestors	4.39%	3.58%	1.81%	2.30%
Life Sciences	Net acquirers	8.21%	14.07%	16.27%	13.23%
	Net divestors	25.19%	17.85%	32.00%	26.28%
Other Chemical	Net acquirers	1.47%	2.01%	2.98%	2.53%
	Net divestors	1.09%	6.38%	15.82%	10.52%
Other	Net acquirers	2.42%	2.36%	3.20%	2.97%
	Net divestors	4.17%	7.44%	1.65%	2.48%

Note: "Net acquirers" ("net divestors") are firms that buy more (fewer) assets than they sell. "Inactive"/"active" refers to whether a firm had more than the median number of deals in the sample, which is three. Numbers represent R&D intensity (R&D expenditure over sales).

Source: Author's calculations based on Securities Data Company data.

has changed the way R&D is distributed across the main segments of the industry. In particular, restructuring was important in reallocating business portfolios and, with them, the research intensity associated with different activities. For instance, acquisitions of life-science businesses unequivocally increased R&D intensity, whereas acquisitions of commodities reduced it. A point of great interest here is the fact that firms more active in restructuring changed their R&D intensity in a different fashion than less active firms, as Table 3.3 documents.[55] For instance, active net divestors in commodities (a low R&D intensity business) have a higher research intensity, presumably because they started off as diversified conglomerates and ended up more specialized in life sciences or specialty chemicals; likewise, net acquirers (active or inactive) have experienced marginal changes. In life sciences, active net acquirers have seen research intensity grow, whereas inactive net acquirers experienced a small decline.

and Walter Garcia-Fontes, eds., *R&D, Innovation, and Competitiveness in the European Chemical Industry* (Dordrecht: Kluwer, 2004).

[55] Segments are assigned according to firms' primary SIC codes, assigned by Compustat. Energy corresponds to SIC 13, 29, 46; Commodities to SIC 281, 282, 286; Life Sciences to SIC 283; Other Chemical to SIC 284, 285, 287, 289; Others to other manufacturing firms outside SIC 28.

This pattern suggests that the extent to which chemical companies in different countries have been able to refocus their R&D portfolios depends in part on the extent to which financial markets have enabled (or pushed) them to restructure. Operating a commodity business requires skills in costs cutting and capacity management that are quite different from the research and commercialization capabilities necessary to make a life-science business succeed. This means that the ability to quickly reshape business portfolios is one important determinant of corporate success. In fact, one outcome of restructuring has been a reversal in the R&D intensity of formerly diversified firms like BASF or Dow, which have deeply refocused their portfolio of businesses. As Figure 3.3 shows, the late 1990s were a turning point for these firms, with research intensity rebounding and reaching levels around 6 percent.

How was restructuring financed? One important source of finance was restructuring itself; that is, the proceeds from the sale of assets were often used to finance acquisitions. Indeed, the vast majority of companies were active in both selling and acquiring assets. Because the high cost of equity in the 1980s made capital a costly source of finance, debt became a more important source of finance for acquisition in all countries. In the United States, however, the firms used debt much more aggressively than in Europe. Highly leveraged transactions, such as leveraged or management buyouts, were introduced in the early 1980s, thanks to the relaxation of regulations in the United States. This did not take place in Europe. In a typical leveraged buyout, a company newly formed for the purpose of running an existing business issues a large amount of bonds. The productive assets of the firm to be bought provide collateral for the bonds, which carry a high interest rate because of their riskiness – they were indeed dubbed "junk bonds."[56]

Although highly leveraged transactions were most common in traditional industries with stable cash flow.[57] They also played a role in chemicals, notably in commodity products like petrochemicals. Financial companies such as Cain Chemical, GAF, Huntsman, Sterling, Quantum Chemicals, Vista Chemical, Aristech, and others were organized to acquire petrochemical plants, usually at the trough of the cycle, and turn them into profitable businesses by relentlessly cutting costs. These companies were created with the purpose of running a single plant and

---

[56] As the financial community became adjusted to this innovation, they changed the expression to "high-yield securities," a name far more likely to appeal to the general investor.
[57] Blair, *The Deal Decade.*

were often led by financiers with little technical background, like Sam Heyman of GAF. They focused on pressing for cost-cutting, while letting experienced executives run the technical side of the business. Not all of the efforts to restructure were successful. GAF made an unsuccessful attempt to take over Union Carbide; so too with the threat to Allied Chemical. In these cases, as in rumored attempts to take over American Cyanamid, the target firms usually reacted by buying back their shares and divesting several businesses.

The extremely high debt-to-equity ratio that resulted from takeovers and leveraged buyouts forced management to ensure a stable cash flow to repay interest. This was often viewed as providing management with powerful incentives to perform in the interest of shareholders.,[58] while some economists, like Stein[59] take the alternative view that takeovers and leveraged transactions rather induced managerial myopic behavior. Hall[60] evaluates these views with reference to their effect on investment in R&D.

Table 3.4 shows that nearly two hundred high-leverage restructuring deals took place in the United States; of these about a quarter were executed in the 1980s, mostly in commodity chemicals.[61] Overall, nearly six hundred deals with financial acquirers or high leverage took place in the United States, three times as much as in Europe. The largest U.S. petrochemical firms – Monsanto, Dow, Union Carbide, and DuPont – sold many plants. In 1985 alone, Monsanto sold $900 million worth of plants, and Dow divested $1.8 billion. The number and size of the financial transactions in the United States contrasts sharply with those of European deals, which were nearly a third in both number and average size. A likely conjecture is that in Europe two factors have hampered the development of leveraged transactions. One is heavier regulation of securities, which in some cases prevents the use of junk bonds and several other elements of financial engineering. The other is the smaller size of European securities markets, which are much less liquid than those in the United States. As both factors are receding, the use of leveraged transactions in Europe

---

[58] Michael Jensen, "Agency Cost of Free Cash Flow, Corporate Finance and Takeovers," *American Economic Review*, 76(4), 1986: 323–29.

[59] Jeremy Stein, "Takeover Threats and Managerial Myopia," *Journal of Political Economy*, 96, 1988: 215–31.

[60] Bronwyn Hall, "The Impact of Corporate Restructuring on Industrial Research and Development," *Brookings Papers on Economic Activity: Microeconomics* (Washington, DC: Brookings Institution, 1990).

[61] Sarah Lane, "Corporate Restructuring in the Chemical Industry," in Blair, *The Deal Decade*.

Table 3.4. Financial Nonfinancial Restructuring, 1985-97

Region of Acquirer		Financial Acquirer	LBOs and MBOs	Other Financial Deals		Nonfinancial Deals	Total
				Financial Deals			
North America	(1)	205	192	194	591	10,896	11,679
	(2)	218	215	223		213	216
	(3)				6.7%	93.3%	
Europe	(1)	38	67	73	178	3,865	4,110
	(2)	119	84	51		366	349
	(3)				6.0%	94.0%	
UK	(1)	26	83	78	187	2,508	2,778
	(2)	173	46	40		165	155
	(3)				9.7%	90.3%	
Rest of the World	(1)	37	39	56	132	4,642	4,813
	(2)	159	175	241		117	119
	(3)				5.6%	94.4%	
Total Deals		306	381	401		21,911	23,380

(1) Number of deals.
(2) Average deal amount (millions of 1990 dollars).
(3) Percentage of deals.
Source: Author's calculations based on Securities Data Company data.

is growing, but this is happening more than a decade later than in the United States.

Corporate governance has been another important element shaping the different patterns of restructuring. In Europe, equity markets have remained closed, protected from capital outflows, and uncompetitive until the end of the century, when the adoption of a common currency, the euro, and the coming of age of institutional investors have increased companies' willingness to create value for shareholders by targeting stock price growth. Thus, the consolidation of pharmaceuticals in the late 1990s occurred with much less delay in Europe than had the consolidation of petrochemicals. Mergers, which created AstraZeneca, Adventis, Clariant, and Novartis, took place at roughly the same time as similar deals in the United States.

Nevertheless, Europe has been much slower in moving toward effective corporate governance structures than the United States, with a lag of about two decades. With the deregulation of banking and of the stock market in the 1980s, Britain was the first to move in the direction of improved corporate governance. In the United Kingdom, the division of ICI in 1992 into two businesses, one devoted to specialty chemicals and one (Zeneca, now AstraZeneca) devoted to life sciences, came from a threat of a takeover by the Hanson Trust of the ailing chemical conglomerate. Pressure from the stock market brought Zeneca to merge in 1998 with the Scandinavian firm Astra in order to become more innovative and to reach critical mass. ICI has also undergone a major turnaround.

In Germany, the traditional bank-centered system has been resistant to change until very recently. Gorton and Schmidt[62] show that the *hausbank* system, which proved functional in the phase of rapid growth and stable markets in the 1970s, has proved less effective at helping German firms to adapt to the more turbulent and dynamic environment of the globalized economy of the 1980s and 1990s. One major chemical company, BASF, listed on the New York Stock Exchange in 2000, and other companies have announced plans to follow suit. Although only 6 percent of BASF was held outside Germany in the 1960s, now it is close to 25 percent. Still, as late as the turn of the century, an environment supportive of takeovers is yet to develop in Germany. This is true even though for chemical firms the capital market of reference is gradually becoming global. In this sense, national borders are becoming less important for global players. In fact, the growing percentage of shareholdings by foreign institutional investors, which adopt a much less forgiving attitude than German banks or insurance companies, has been a factor in pressing the large German firms, BASF, Bayer and Hoechst, to split their life-science and chemical businesses.

Japan and Italy provide an even less thrilling picture, as government intervention further blurred the strategic perspectives in industries already characterized by inefficient national corporate governance systems. In Italy, restructuring came at first through the 1981 state "national chemical plan." Some asset swaps occurred between (state-owned) ENI and (private) Montedison in petrochemicals in order to specialize the former in commodities and the latter in specialties. But these swaps did not change much, and as technology raised the minimum efficient scale of plants, Italian firms failed to restructure and consolidate (see the chapter by Zamagni in this volume). In Japan, the ownership structure based on Keiretsu has influenced not only industrial structure but also restructuring. The integrated nature of the Keiretsu continued to make

---

[62] Gary Gorton and Frank Schmid, "Universal Banking and the Performance of German Firms" *Journal of Financial Economics*, 58(1–2), 2000: 29–80.

it extremely difficult to shed capacity. Each conglomerate had its own self-contained "chemical industry," which it did not want to lose. Shedding capacity in one product would mean beginning to depend on the market for supplies. The market, in turn, was made up of competitors and was also much less willing than Keiretsu firms to provide the degree of product customization on which conglomerates had always counted.

In Japan, the vertical integration of plants also made it difficult to vary capacity only at some stages of the value chain. During a recession, belonging to a Keiretsu may limit the risk of consolidation or business failure, as intra-Keiretsu mergers of firms are made difficult by intragroup politics. The troubled mergers between Mitsubishi Petrochemicals with Mitsubishi Kasei and of Mitsui Toatsu with Mitsui Petrochemicals after the oil shock are two instances of this sluggishness. Table 3.1, in fact, shows that transactions in the "rest of the world," of which Japan represents a large share, have been even more sluggish than in Europe. Table 3.3 also shows that financial transactions have been of small importance for restructuring. Indeed, hostile takeovers and leveraged buyouts are practically unheard of in Japan. This is in part because Keiretsu provide a very stable ownership structure and thus make a market for corporate control impracticable.[63]

CONCLUSIONS

The influence of finance on corporate strategy and industry structure can be seen with great clarity in the recent history of the chemical industry. Finance matters for many crucial strategic decisions and a long-term look at how it has affected the evolution of an industry as capital- and research-intensive as chemicals is quite instructive. One main point that emerges from this study deserves to be stressed: There is no single successful "recipe" for an effective support of corporate strategy by the society's financial systems.

Different systems may work best when industry conditions are different. In the stable environment of postwar reconstruction, bank-based European systems worked much better than when economic and technological changes required swift and deep changes in the 1980s and 1990s.

---

[63] Although oversight by investors may not be feasible, some argue that the Keiretsu structure itself, and in particular the role played by the Main Bank, may act as a substitute (see, for instance, Masahiko Aoki and Hugh Patrick, *The Japanese Main Bank System*, and Steven Kaplan," Top Executive Rewards and Firm Performance: A Comparison of Japan and the United States," *Journal of Political Economy*, 102(3) 1994: 510–46.

As the structure of the economy, technology, and markets changed, so did the needs of industrial firms. In the most recent transformation of this important industry, the European and Japanese financial systems were in part responsible for the delay in restructuring corporate governance in chemicals – and other industries as well – creating a decade-long lag before the European and Japanese companies were able to refocus their firms and adapt to the intense global competition that now characterizes this industry.

REFERENCES

Amat, Paolo (1998). "The Italian Chemical Industry from 1861 to 1918," in Ernst Homburg, Anthony Travis and Harm Schröter (eds.), *The Chemical Industry in Europe, 1850–1914*. Dordrecht, Kluwer.

American Chemistry Council (2001). Key Industry Data, Arlington.

Aoki, Masahiko, and Hugh Patrick (eds.) (1994). *The Japanese Main Bank System*. Oxford, Oxford University Press.

Arora, Ashish, Marco Ceccagnoli, and Marco Da Rin (2004). "Corporate Restructuring and R&D: A Panel Data Analysis for the Chemical Industry," in Fabrizio Cesaroni, Alfonso Gambardella and Walter Garcia-Fontes (eds.), *R&D, Innovation, and Competitiveness in the European Chemical Industry,* Dordrecht, Kluwer, 2004.

Arora, Ashish, Ralph Landau, and Nathan Rosenberg (eds.) (1998). *Chemicals and Long-term Growth*. New York, Wiley & Sons.

Arora, Ashish, and Alfonso Gambardella (1998). "Evolution of Industry Structure in the Chemical Industry," in Ashish Arora, Ralph Landau, and Nathan Rosenberg (eds.), *Chemicals and Long-Term Growth*. New York, Wiley & Sons.

Arora, Ashish, and Nathan Rosenberg (1998). "Chemicals: A U.S. Success Story," in Ashish Arora, Ralph Landau, and Nathan Rosenberg (eds.), *Chemicals and Long-Term Growth*. New York, Wiley & Sons.

Azzolini, Riccardo, Giorgio Dimalta, and Roberto Pastore (1979). *L'industria Chimica tra crisi e programmazione*. Roma, Editori Riuniti.

Beer, John (1959). *The Emergence of the German Dye Industry*. Urbana, University of Illinois Press.

Blair, Margaret (ed.) (1993). *The Deal Decade*. Washington, DC, Brookings Institution.

Bottazzi, Laura, and Marco Da Rin (2002). "Banks as Catalysts for Industrialization: Evidence from Italy," mimeo, IGIER.

Bozdogan, Kirkor (1989), "The Transformation of the U.S. Chemical Industry." Working Paper of the MIT Commission on Industrial Productivity.

Carosso, Vincent (1970). *Investment Banking in America*. Cambridge, Cambridge University Press.

Chandler, Alfred (1977). *The Visible Hand*. Cambridge, MA, Belknap Press.

Chandler, Alfred (1990). *Scale and Scope: the Dynamics of Industrial Capitalism*. Cambridge, MA, Harvard University Press.

Chandler, Alfred, and Stephen Salsbury (1971). *Pierre S. Du Pont and the Making of the Modern Corporation*. New York, Harper & Row.

Collins, Michael (1991). *Banks and Industrial Finance in Britain, 1800-1939*. London, MacMillan.

Confalonieri, Antonio (1976). *Banca e Industria in Italia, 1894-1906*. Milano, Banca Commerciale Italiana.

Confalonieri, Antonio (1982). *Banca e Industria in Italia dalla crisi del 1907 all'agosto 1914*. Milano, Banca Commerciale Italiana.

Cottrell, Philip (1980). *Industrial Finance 1830-1914*. London, Methuen.

Da Rin, Marco (1996). "German Kreditbanken 1850-1914: An Informational Approach," *Financial History Review 3* (2), 29-47.

Da Rin, Marco (1997). "Finance and Technology in Early Industrial Economies: The Role of Economic Integration," *Ricerche Economiche, 51* (3), 171-200.

Da Rin, Marco, and Thomas Hellmann (2002). "Banks as Catalysts for Industrialization," *Journal of Financial Intermediation, 11*(4), 366-97.

Dratt, Paul (1995). "The History of BASF since 1945 from a Financial Viewpoint," mimeo.

Feldenkirchen, Wilfred (1991). "Banking and Economic Growth: Banks and Industry in Germany in the Nineteenth Century and their Changing Relationship during Industrialization," in Wang Lee (ed.), *German Industry and German Industrialization*. London, Routledge.

Goldsmith, Raymond (1958). *Financial Intermediaries in the American Economy Since 1900*. Princeton, Princeton University Press.

Gorton, Gary, and Frank Schmid (2000). "Universal Banking and the Performance of German Firms." *Journal of Financial Economics, 58* (1-2), 29-80.

Grabower, Rolf (1910). *Die finanzielle Entwicklung der Aktiengesellschaften der deutschen chemischen Industrie*. Leipzig, Duncker & Humblot.

Haber, L. F. (1958). *The Chemical Industry During the Nineteenth Century*. Oxford, Clarendon Press.

Haber, L. F. (1971). *The Chemical Industry, 1900-1930*. Oxford, Clarendon Press.

Hall, Bronwyn (1990). "The Impact of Corporate Restructuring on Industrial Research and Development," Brookings Papers on Economic Activity: Microeconomics, Washington, DC, Brookings Institution.

Hardach, Gerd (1984). "Banking and Industry in Germany in the Interwar Period 1919-39." *Journal of European Economic History, 13* (S), 203-34.

Hart, Oliver (1995). *Firms, Contracts and Financial Structure*. Oxford, Oxford University Press.

Haynes, William (1954). *American Chemical Industry*. New York, Van Nostrand.

Hikino, Takashi, Tsutomo Harada, Yoshio Tokuhisa, and James Yoshida (1998). "The Japanese Puzzle." in Ashish Arora, Ralph Landau, and Nathan Rosenberg (eds.), *Chemicals and Long-term Growth*. New York, Wiley & Sons.

Kaplan, Steven (1994). "Top Executive Rewards and Firm Performance: A Comparison of Japan and the United States," *Journal of Political Economy, 102* (3), 510-46.

Kennedy, William (1987). *Industrial Structure, Capital Markets and the Origins of British Economic Decline*. Cambridge, Cambridge University Press.

Khoury, Sarkis (1985). *The Deregulation of the World Financial Markets*. New York, Quorum Books.

Jensen, Michael (1986). "Agency Cost of Free Cash Flow, Corporate Finance and Takeovers," *American Economic Review, 76* (4), 323-29.

Lane, Sarah (1993). "Corporate Restructuring in the Chemical Industry," in Margaret Blair (ed.). *The Deal Decade*. Washington, DC, Brookings Institution.

Liebermann, Marvin (1990). "Exit from Declining Industries: 'Shakeout' or 'Stake-out'?" *Rand Journal, 21* (4), 538-54.

Molony, Barbara (1990). *Technology and Investment: The Prewar Japanese Chemical Industry*. Cambridge, MA, Harvard East Asian Monographs.

Morikawa, Idemasa (1992). *Zaibatsu*. Tokyo, University of Tokyo Press.

Murmann, Johann Peter (2003). *Knowledge and Competitive Advantage: The Coevolution of Firms, Technology, and National Institutions*. Cambridge; New York: Cambridge University Press.

Pohl, Hans (1984). "Forms and Phases of Industry Finance up to the Second World War," *German Yearbook on Business History - 1984*, 75-95.

Pressnell, L. (ed.) (1973). *Money and Banking in Japan*. London, Macmillan.

Reader, William (1970). *Imperial Chemical Industries: a History. Vol I: The Forerunners*. London, Oxford University Press.

Reader, William (1975). *Imperial Chemical Industries: a History. Vol II: The First Quarter Century*. London, Oxford University Press.

Riesser, Jacob (1911). *The Great German Banks*. Washington, DC, U.S. National Monetary Commission.

Roe, Mark (1994). *Strong Managers, Weak Owners*. Princeton, NJ, Princeton University Press.

Roy, William (1990). "The Rise of American Industrial Corporations, 1880-1914," computer file, Department of Sociology, University of California at Los Angeles.

Stein, Jeremy (1988) "Takeover Threats and Managerial Myopia," *Journal of Political Economy, 96* (1), 215-31.

Sylla, Richard (1975). *The American Capital Market*. New York, Arno Press.

Telschik, Walter (1994). *Geschichte der Deutschen Großchemie*. Weinheim, VHC.

Thomas, W. A. (1978). *The Finance of British Industy*. London, Methuen.

Tilly, Richard (1967). "Germany: 1815-70," in Rondo Cameron (ed.), *Banking in the Early Stages of Industrialization*. New York, Oxford University Press.

Wrigley, Julia (1987). "Technical Education and Industry in the Nineteenth Century," in Bernard Elbaum and William Lazonick (eds.), *The Decline of the British Economy*. Oxford, Clarendon Press, pp. 162-88.

Zamagni, Vera (1998). "L'industria chimica in Italia dalle origini agli anni '50," in Daniela Brignone (ed.), *Innovazione tecnologica e industria*. Roma, Bulzoni.

Zingales, Luigi (1998). "Corporate Governance," in *The New Palgrave Dictionary of Economics and the Law*. New York, Macmillan.

4

Government Environmental Policy
and the Chemical Industry

WYN GRANT

Before analyzing the development of government environmental policy in relation to the chemical industry, it is necessary to make three general points. One relates to the character of the industry (its internationalization); one relates to the changing form of the state (the emergence of the regulatory state); and one relates to the changing nature of the political process (the replacement of a politics of production by a politics of collective consumption).

The chemical industry has a long history of internationalization. One only has to recall the effective pre–World War II division of the world into three spheres of commercial influence: DuPont in the Americas; ICI in the British Empire and Commonwealth; and IG Farben in Europe. The emergence of a petrochemical industry after World War II brought into play a new set of actors, the oil companies, who were already accustomed to operating at an international level.

The internationalization of the industry is not an accident. It is a product in particular of the research-intensive character of the industry. "Given the cost of research and development activities, and the market lead which can be provided by a breakthrough, there is an incentive to exploit new processes and products on an international scale."[1] Even though the original driver of internationalization might have been technological, it acquired a momentum of its own through the development of an international management culture.

The implication of the internationalization of the industry is that many forms of regulation have to be at a global or at least at a regional level. International trade might be disrupted if there were too many variations in forms of regulation at a national level. There are also issues that arise from the transport of chemical products across national borders or by sea.

Above all, however, there have been doubts about the ability of national governments to display sufficient independence from their chemical industries. Governments have displayed a high level of dependence on their chemical industries for both security and economic reasons. Even if the industries have not been nationalized, the leading firms in them have been regarded as "chosen instruments" enjoying a special relationship with government. Indeed, one of the distinctive characteristics of the industry has been the extent to which policy communities and networks in the industry have been "firm led" in contrast to the more usual pattern of their being state led.[2]

When demands for effective environmental regulation of the industry began to be made, it was evident that, at the very least, there would have to be a framework of action above the national level. A measure of harmonization of regulation in so far as it affected trade was in the interests of the companies themselves. However, there were also limits to this process. "The momentum toward international policy convergence is hardly overpowering, however."[3] Nevertheless, within the European Union, the framework and content of regulation is largely drawn up at the European level.

The immediate postwar period saw the establishment in most Western countries of what is often referred to as a Keynesian welfare state. This was not, of course, a uniform process. Some countries had already established comprehensive welfare systems before World War II,

---

[1] Alberto Martinelli and Wyn Grant, "Conclusion," in Alberto Martinelli, ed., *International Markets and Global Firms: A Comparative Study of Organized Business in the Chemical industry* (London/Newbury Park, CA: Sage Publications, 1991), p. 276.

[2] Wyn Grant, William Paterson, and Colin Whitston, *Government and the Chemical Industry: A Comparative Study of Britain and West Germany* (Oxford: Clarendon Press, 1988).

[3] Ronald Brickman, Sheila Jasanoff, and Thomas Ilgen, *Controlling Chemicals: the Politics of Regulation in Europe and the United States* (Ithaca, NY: Cornell University Press, 1985), p. 298.

whereas Keynesian ideas of economic management were not universally accepted.[4] Nevertheless, the main contours of the project were clear. A commitment to maintaining full employment was matched by the provision of "cradle to grave" welfare services.

By the 1970s, the Keynesian welfare state was running into difficulties. There were proximate causes such as the end of the long postwar boom and the two oil shocks that produced the phenomenon known as "stagflation." A more fundamental problem was that of inflation. Keynesian economics had no real answers to the problem of wage driven inflation in a full employment economy, as Keynes himself was the first to admit.[5] The solution pursued by Keynes's disciples was that of incomes policy, which worked more successfully in some European countries than others. Another difficulty was the rising tax burden that increasingly affected those on median wages and made them more susceptible to the appeals of parties promising tax cuts.

The apparent collapse of the Keynesian project created a vacuum that the advocates of neoliberalism and monetarism attempted to fill, most noticeably in Britain. In one sense, this effort was a failure as it led to no reduction in government's share of GDP in the British case. However, this reflected in large part the continuing importance of transfer payments. There was a rolling back of many of the traditional roles of the state in the economic sphere.

THE EMERGENCE OF THE REGULATORY STATE

The question then arose, was there any role for the state beyond the traditional ones of defending life and property externally and internally, providing a medium of exchange and a legal framework for the conduct of commerce? What happened was that the state's role as a regulator increased. In part, this was because changes in its economic role such as privatization created a new need for the supervision of the activities of privatized utilities. In part, it was because of the increasing importance of environmental and health and safety considerations.

From the perspective of those who wished to see the state enjoying a continuing authoritative role (for example, bureaucrats), the assumption

---

[4] Peter A. Hall, *The Political Power of Economic Ideas: Keynesianism across Nations* (Princeton, NJ: Princeton University Press, 1989).

[5] A. Jones, "Inflation as an Industrial Problem," in Robert Skidelsky, ed., *The End of the Keynesian Era* (London: Macmillan, 1977), pp. 50–58.

of the mantle of regulator had a number of advantages. Although regulation created or maintained bureaucratic positions, it did not require large sums of public expenditure, an important consideration in a period when reduction of the tax burden was an objective in many countries. In general, the transaction costs were largely borne by those being regulated.

It was, however, the European Union (EU) that emerged as the regulator state par excellence. Many of the traditional functions of the state, notably those of redistribution through the fiscal system, were not generally available to the EU. The EU did not have a large budget at its disposal and has not generally had a financial impact on the lives of its citizens except through the Common Agricultural Policy. Where the EU could create a comparative advantage was in regulation. The actual implementation of the regulations could be left to the member states. In areas such as environmental policy, the creation of a new framework of regulation could be seen as a potentially effective means of directly impacting the lives of citizens in a positive way.

Part of the story of environmental regulation was then one of the state seeking new roles to replace that of domestic economic management, which lacked credibility in an era of globalization. However, this was not simply a "top down" process driven by state "entrepreneurs" marking out new territory. There also was a considerable "bottom up" dimension in terms of public demand for action on environmental issues.

THE EMERGENCE OF A POLITICS OF COLLECTIVE CONSUMPTION

This may be conceptualized in terms of a shift from a "politics of production" to a "politics of collective consumption."[6] A politics of production is centered on a struggle between management and labor over the distribution of the fruits of the production process. Thus, the politics of production centered around issues such as wages and conditions; attempts by government to influence the outcomes of collective bargaining through incomes policies; the rights of trade unions' industrial relations law; and arrangements for worker participation in decision making. It is a politics in which adjustments at the margin are often possible, for example, a slightly higher wages norm.

The politics of collective consumption is concerned with the outcomes of the production process, rather than what happens within the

---

[6] Wyn Grant, *Pressure Groups and British Politics* (Basingstoke: Macmillan, 2000).

process itself, that is, the externalities of the production process. It is called a politics of collective consumption because at its core it is concerned with collective goods, or at least goods that have some of the characteristics of public goods. Examples that are relevant to the case of the chemical industry would include the air quality or the quality of water in rivers.

It is a politics that is less amenable to elite bargaining that makes policy adjustments at the margin. In part, this is because new actors enter the policy process who are less amenable to accommodative systems of politics. Moreover, many of the core values of production are called into question. In the most extreme cases, the need for a chemical industry at all would be challenged.

The politics of production has not disappeared and the politics of collective consumption will not necessarily predominate. Old productionist coalitions of employers and unions (as in the German case) retain a capability to moderate and channel regulatory demands. As green parties enter government, they have to accept compromises that are based on what is feasible rather than what they or their members might think is desirable.

ENVIRONMENTAL POLICY AND THE CHEMICAL INDUSTRY

Before commencing the discussion of particular environmental measures affecting the chemical industry, it is necessary to think about what is distinctive in relation to environmental policy and the chemical industry. When modern environmental policy was first developed in the 1970s, there was an effort to think in terms of an overall ecological system which had a limited absorptive capacity and in which changes in one part of the system had consequences elsewhere. In practice, however, environmental policy has developed, particularly at the EU level, in terms of a series of distinctive policy sectors and initiatives, for example, air pollution policy, water pollution policy, recycling policy, and so on. Each of these policy sectors then experiences a series of further subdivisions. To take the example of air pollution policy, there are distinct agendas relating to stationary and mobile sources, and these in turn can be subdivided into directives dealing with particular pollutants such as ozone precursors, carbon monoxide, sulphur dioxide, and so on. In large part, this reflects the fact that the expertise on which policy is based is highly specific, so that a group of scientists who know how one can best measure pollution from mobile sources would know nothing about how to measure water pollution. Crosscutting these vertical divisions are politically driven concerns about particular industries such as nuclear power and chemicals. What all this adds up to is an "environmental policy" that is highly complex and segmented.

Nevertheless, it is possible to make five generalizations about the interface between government and the environment in the case of the chemical industry. First, it is *incident driven*. From time to time, chemical plants experience catastrophic incidents that have consequences both inside the plant (and in that sense are seen as health and safety problems) and outside it (and in that sense are seen as environmental problems). Bhopal, Flixborough, Seveso: each of these is a well-known incident that helped to shape public perceptions of the chemical industry and the legislative agenda.

Second, the industry has been *highly politicized* in the sense that it has been a major target of the green movement. This has not been a constant over time and was more characteristic of the 1970s than today. In the 1980s, the emphasis switched to nuclear power issues, and at the turn of the century, the focus was on genetically modified foods and biotechnology. Nevertheless, the debate over the European Union's chemicals policy led to renewed political interest in the chemical industry, an issue revisited later in this chapter.

Third, although environmental policy has in practice been highly segmented, the activities of the chemical industry affect a number of these different areas of concern. Thus, discussing plastics, Mol states: "The environmental problems associated with the production-consumption cycle of plastics relate . . . to the depletion of non-renewable natural resources, the emission of toxic and environmental harmful substances during production, the use of toxic additives in products, and the release of production and post-consumption waste."[7] This is a substantial list, and it suggests that in a vertically segmented environmental policy, the chemical industry displays *horizontal characteristics* in terms of its impact.

The first three political characteristics of the industry have created the need for it to develop *sophisticated forms of political organization.* Hence, both in Europe and the United States, the industry has been in the forefront of the development of government relations or public affairs divisions at firm level. It has developed well-organized industry associations in both the United States (the Chemical Manufacturers' Association)

---

[7] Arthur P. J. Mol, *The Refinement of Production: Ecological Modernization Theory and the Chemical Industry* (Utrecht: van Arkel, 1995), p. 218.

and in European countries. In June 2000, the Chemical Manufacturers'
Association was renamed the American Chemistry Council to reflect significant business changes occurring in the industry and the desire to
create a more positive reputation. Within Europe, the European Chemical Industry (CEFIC) is widely recognized as one of the most effective
sector-level associations operating at a European level. It has developed
its organizational structure in such a way as to ensure input from individual firms, national associations, and product-level bodies. Environmental
questions have been one of its central priorities.

Given the internationalization of the industry discussed earlier, the
industry is also one of the most effectively organized at a global level.
This is achieved through the International Council of Chemical Associations (ICCA), which brings together trade associations from all the continents of the world. It presents an international chemical industry view to
intergovernmental organizations such as the World Trade Organization
and the Organization for Economic Cooperation and Development. It is
interesting that the first set of "Policy issues of international significance
to the chemical industry [forming] the agenda of the ICCA" listed on its
Web site are health, safety, and the environment.[8]

The political sophistication of the industry has led to suspicions that it
is *politically privileged*. On the basis of a review of chemical regulation in
Britain and the United States, Vogel felt compelled to ask, "Do the advantages that the industry enjoys because of its close ties to the scientific
community and the limited ability of nonindustry interests to challenge
regulatory policies publicly expose British workers and consumers to
unnecessary hazards?"[9] Similarly, in Germany, "the dominance of the
three large firms in West German society occasionally caused many alter-
natively inclined West Germans to resent and challenge these great pillars
of the postwar economic establishment."[10] It would, however, be misleading to see the industry as simply resisting demands for more environmentally friendly forms of production. It showed a considerable capacity
to adjust to the new conditions it faced. Writing as green activists, Porritt
and Winner comment: "the larger, more respectable chemical companies positively welcome a coherent regulatory framework, for without it
their less conscientious competitors and the out-and-out cowboys of the

[8] http://www.icca-chem.org, June 28, 2000.
[9] David Vogel, *National Styles of Regulation: Environmental Policy in Great Britain and the United States* (Ithaca, NY: Cornell University Press, 1986), p. 211.
[10] C. S. Allen, "Political Consequences of Change: The Chemical Industry," in Peter J. Katzenstein, ed., *Industry and Politics in West Germany: toward the Third Republic* (Ithaca, NY: Cornell University Press, 1989), p. 171.

chemical industry can undercut them simply by disregarding their environmental responsibilities."[11] Working within a framework derived from ecological modernization theory, Mol's basic thesis is that the chemical industry underwent a process of ecological transformation and restructuring. He concludes:

. . . it has become clear that the environment has moved from the periphery to the centre of processes of continuity and transformation in the chemical industry. As the ecological modernization theory proclaims, the environment is gaining importance in the process of constant restructuring in the chemical industry . . . environmental considerations have become institutionalized, at least as far as the institutions governing social practices in the chemical industry are concerned.[12]

ENVIRONMENTAL REGULATION BEFORE 1970

Nineteen hundered seventy has been taken as an approximate date for the start of the new wave of environmental regulation of the chemical industry for reasons discussed more fully later. From the time of the formation of the chemical industry, there were attempts at environmental regulation. They were, however, ad hoc in character and limited in their effect. "These early regulatory regimes bore comparatively lightly on the chemical industry."[13] Complaints from particular groups in society were relatively easily disposed of, and any institutionalization that occurred was of a primitive nature. The predominant discourses were those of technological modernization and the need to build national chemical industries for defense related reasons. The relationship between the state and the industry was a largely asymmetric one and focused on economic and security considerations. In Germany, the state was highly dependent on the chemical industry during World War I, while "IG Farben was central to the political economy of the Third Reich. This dependence by the state on the chemical industry was also evident in Britain, where a great deal of official effort was invested in trying to respond to the German lead in chemicals."[14] Particular problems were caused by

---

[11] Jonathon Porritt and David Winner, *The Coming of the Greens* (London: Fontana, 1988), p. 137;

[12] Mol, *The Refinement of Production*, pp. 390–91.

[13] Grant, Paterson, and Whitston, *Government and the Chemical Industry*, p. 272.

[14] William Paterson, "Self-Regulation under Pressure: Environmental Protection Policy and the Industry's Response," in Alberto Martinelli, ed., *International Markets and Global Firms: A Comparative Study of Organized Business in the Chemical Industry* (London/Newbury Park, CA: Sage Publications, 1991), p. 229.

large-scale emissions of hydrogen chloride by the Leblanc process used
for soda production in the nineteenth century that led to the devastation
of large tracts of countryside. Landowners had sufficient political influence during that period to secure the establishment of what has been
called the first pollution control agency in the world, the Alkalai Inspectorate, established in Britain in 1863. However, "Attempts by environmentally conscious citizens who formed groups such as the 'Alkali Acts
Extension Association' to tighten controls further were a failure."[15] In the
Netherlands, "in a number of cases, the environmental consequences of
chemical production during the eighteenth and early nineteenth centuries resulted in governmental action. Initially, relocation of industrial
plants to selected areas inside or outside the cities and the increase in the
height of chimneys were the main actions taken."[16] Both in the Netherlands and elsewhere, the relocation of chemical plants to "green field"
sites diminished what public concern there was. Technological improvements (such as the displacement of the Leblanc by the Solvay process)
also removed some of the worst problems. Above all, however, in Britain
and Germany (the leading chemical producers in Europe):

The undisputed technical brilliance of the chemical industry and the states'
dependence on the chemical industry ensured that early measures such as
the Alkali Acts did not develop into a tight regulatory regime for the chemicals
sector. The pattern that developed was one of consultation between the state
and industry with the overwhelming emphasis on self-regulation.[17]

The American chemical industry was a later starter than those in
Britain and Germany, and for much of the first half of the twentieth
century was engaged in "catching up." It really took off with the development of a modern petrochemical industry, although at first the oil
companies were not particularly interested in chemicals, while the older
companies such as DuPont argued: "that it could produce any chemical
from coal."[18] Although the history may have been different, the outcome
in term of government-business relations was similar. "In the 1950s and
1960s, Washington trusted industry to manage its own affairs, a decision
reinforced by a legacy of remarkable growth and expansion."[19] Thus,

[15] Ibid.
[16] Mol, *The Refinement of Production*, p. 124.
[17] Paterson, "Self-Regulation under Pressure," p. 229.
[18] Peter H. Spitz, *Petrochemicals: The Rise of an Industry* (New York: Wiley, 1988),
p. 106.
[19] Brickman, Jasanoff, and Ilgen, *Controlling Chemicals*, p. 239.

"before the sixties, environmental issues had only a very marginal impact on the development of chemical industry in Western industrialized societies."[20] Insofar as the state was engaged in a regulatory role, it was in relation to economic and antitrust issues, particularly in the United States. At the end of the 1960s, it would have been possible to write an acceptable account of the chemical industry without mentioning environmental issues. Ten years later, this would have been impossible. How did this change come about?

THE ESTABLISHMENT OF AN ENVIRONMENTAL AGENDA

In 1962, Rachel Carson published her book *Silent Spring,* which was concerned with the environmental consequences of pesticides. Hitherto, products such as DDT had been seen as a beneficial result of technological process. What had been accepted unquestionably as the benefits of progress now began to be questioned by citizens who were, in any case, benefiting from the postwar expansion of education, particularly higher education.

It is not the purpose of this chapter to explore in any detail why the environmental movement emerged when it did either in terms of proximate factors or deeper underlying causes. What is clear is that a number of key developments took place from 1970 onward:

1970 UK Department of the Environment founded
1970 Environmental Protection Agency formed in United States
1971 Friends of the Earth formed in the UK
1972 UN sponsored Stockholm conference on the environment
1972 European Community Paris Summit calls on Commission to draw up environmental policy
1972 EC establishes Environment and Consumer Protection Service
1972 Club of Rome *Limits to Growth* report

There was, thus, a significant institutionalization of environmental concerns and agendas. With the development of specialized institutions within government, flanked by nongovernmental organizations and, later, green parties, administrative and organizational capacity building for policy development and implementation occurred at a rapid pace. Environmental policy had become incorporated into the standard political agenda, even if it was still viewed with some suspicion by other

[20] Mol, *The Refinement of Production,* p. 125.

agencies. A similarly rapid growth occurred in the case of agricultural policy institutions in the 1930s, which, once established, could not be displaced, despite changing economic and social conditions. The chemical industry thus faced a permanently changed political environment.

THE DEVELOPMENT OF ENVIRONMENTAL POLICY FOR CHEMICALS IN THE EU AND THE UNITED STATES

Environmental policy in relation to chemicals developed through a complex series of legislative measures and regulations at both the national and international level. "For example, a review of UK legislation on the Control of Chemicals (excluding pharmaceuticals and poisons) lists 25 relevant Acts of Parliament, which were overseen by 7 government departments and augmented by over 50 sets of regulations, a pattern of response which is similar to many EU countries."[21] The strategy followed in this part of the chapter will be first to highlight the major pieces of legislation in the EU and the United States, and then to present a case study of chlorine.

In 1967, the EU passed a Council Directive (67/548) on classification and labeling of chemicals. This was a framework directive that provided for a uniform system of listing classification, packaging, and labeling of hazardous substances. It was in itself a rather weak measure, but it provided a basis for subsequent legislation, notably the 6th Amendment (79/831) and the 7th Amendment (92/32). In the United States, the landmark legislation as far as the chemical industry was concerned was the Toxic Substances Control Act of 1976 (TSCA). Originating in an influential report from the Council on Environmental Quality, the measure gave the Environmental Protection Agency (EPA) the ability to track seventy-five thousand industrial chemicals produced or imported into the United States. EPA repeatedly screens these chemicals and can require reporting or testing of those that may pose an environmental or human health hazard. It also can ban the manufacture and import of those chemicals that pose an unreasonable risk.

When he appeared before a House committee in 1975, the then Deputy Administrator of the EPA, John R. Quarles, argued that legislation to prevent the proliferation of dangerous chemicals throughout the environment "is one of our most urgently needed environmental

---

[21] European Environmental Agency, *Europe's Environment: The Second Assessment* (Luxembourg: Office for Official Publications of the European Communities, 1998), p. 124.

laws. . . . Existing federal laws fail to deal evenly and comprehensively with toxic substances problems."[22] Existing federal legislation on air and water pollution was designed to prevent harmful exposure only after the substances have been introduced into production. The toxic substances were dealt with at the point at which they become emissions or effluents that could make them difficult to control. The EPA's position was that only by requiring premarketing notification for all chemicals and testing for selected chemicals would it be able to assess adequately the risks that new chemicals pose to human health and the environment.

The passage of the TSCA gave the EPA a comprehensive mandate for dealing with chemicals issues:

The United States has by far the most fully developed legal framework for discovering and controlling existing chemical hazards. TSCA, for example, establishes a framework for dealing with both new and existing industrial chemicals. . . . EPA can demand information from industry on the production, distribution and use of existing chemicals, as well as data on exposure and health and information effects.[23]

After many years of benign neglect by government, the passage of the TCSA was something of a shock for the representatives of the chemical industry. "TCSA was the first American law aimed at the chemical industry as a whole."[24] The industry found itself on a steep learning curve:

In the five years of debate prior to the law's enactment in 1976, the industry, relatively naive in the ways of Washington, received an intensive education in regulatory politics. Much of that learning took place after 1985, when the industry resigned itself to the act's passage and organized to ensure that intervention would be manageable.[25]

The industry found the TSCA as passed "tough but acceptable."[26] Concerns about confidentiality, research and development, and mixtures of substances known to be safe were met. Nevertheless, the regulatory burden imposed by TSCA and subsequent measures was substantial. By 1986, "the chemical industry spent about $624 million on capital expenditures

---

[22] EPA press release, July 10, 1975.
[23] Brickman, Jasanoff, and Ilgen, *Controlling Chemicals*, p. 36.
[24] Ibid., p. 242.
[25] Ibid.
[26] Ibid., p. 243.

for pollution abatement; annualized operating expenditures for pollution abatement accounted for another $2.7 billion."[27]

The conversion of the industry into a regulated one did, however, have the effect of improving its political organization. By the time this author studied the U.S. chemical industry toward the end of the 1980s, there was an effective integration of the work of government relations officers representing individual companies and that of the CMA. "There is a closely knit community of chemical industry government relations officers in Washington who make up the government relations committee of the [CMA]."[28]

The passage of the TSCA gave a higher priority in Europe to updating the 1967 directive. It changed the calculus of costs and benefits for European firms who had originally been opposed to state regulation of the introduction of new chemical products on the market:

This opposition to the idea of an official regulation of market entry of chemicals was significantly weakened by the [TSCA]. The large European chemical companies, especially the three German giants, were very alarmed by this and the potential burdens it would impose on their chemical exports to the US. They were keen to come to some accommodation with the US, and this could only be done on the basis of arriving at a common EC-wide position and then negotiating with the US.[29]

The Commission published a draft directive in 1976. The negotiations on the draft directive led to tensions between Britain and Germany. The British, in accordance with their regulatory tradition, preferred a more flexible approach based on negotiation between the affected parties. The Germans wanted legally prescribed formal standards with as little flexibility as possible, a stance reinforced by the growth of environmental consciousness in Germany. By 1979, a proposed French law that would have created technical barriers to trade within the common market, and the imminent implementation of the TSCA, led to a compromise that incorporated elements of both positions.

The 6th Amendment was basically a notification procedure that regulated the entry of new chemical substances on the market. New substances were to be notified to national regulatory agencies and tested before being brought on to the market. In accordance with British wishes,

---

[27] United States Department of Commerce, *U.S. Industrial Outlook 1990* (Washington, DC: Department of Commerce, 1990), pp. 12-14.

[28] Wyn Grant, *Government and Industry* (Aldershot: Edward Elgar, 1989), p. 214.

[29] Grant, Paterson, and Whitston, *Government and the Chemical Industry*, p. 299.

the procedure was waived if the chemical was to be marketed in quantities of less than one ton. The German Chemicals Law of 1980 followed the directive very closely and went into effect in 1982. In Britain, the regulations came into force in 1983 and "were somewhat looser than the German Chemicals Law."[30] The next two major pieces of legislation in the United States and Europe were, at least in part, incident driven. The notorious Love Canal incident that became public in 1978 related to a canal in upstate New York, which was used as a chemical dump and then filled in. In the 1970s, the contents began to leach out, leading to a series of medical problems in the neighborhood, including birth defects and miscarriages. The affected houses had to be evacuated and eventually demolished with a concrete cap being placed over a forty-acre site.

The Love Canal incident was a major influence in the political context that led to the passage by Congress in 1980 of the Comprehensive Environmental Response, Compensation and Liability Act (CERCLA), commonly known as Superfund. This law created a tax on the chemical and petroleum industries and provided broad federal authority to respond directly to releases or threatened releases of hazardous substances that may endanger public health or the environment. Over five years, a sum of $1.6 billion was collected. The tax went to a trust fund for cleaning up abandoned or uncontrolled hazardous waste sites where no responsible party could be identified, a common problem given the passage of years since material was dumped and the extent to which sites changed hands.

In 1976, a dioxin release at the Hoffman la Roche chemical plant in Seveso in Northern Italy led to hundreds of casualties. Politically, it resulted in the 1982 directive (EC82/501) on the prevention of major accidents, commonly referred to as the Seveso Directive. This directive is designed to prevent accidents at plants that produce, store, or handle hazardous substances as defined by the 1967 directive on dangerous substances. Such plants are required to evaluate the risk of an accident and to take adequate measures to minimize the risk of an incident occurring. More stringent requirements and reporting obligations are imposed where specified inventories of certain substances are being used. In relation to those substances, the directive also makes provisions about safety equipment, contingency plans, public information, and cooperation with the relevant public authorities.

In 1983, the Green Party entered the German Bundestag. This put pressure on the Social Democrats (SPD) to win back Green voters, leading

---

[30] Ibid., p. 287.

to an adoption of a more stringent attitude toward environmental regulation (reflected, for example, in the chemicals policy of 1986). The
broader significance of this development was that Germany increasingly
assumed something of a leadership role in environmental policy in the
European Union, backed up by other "green" states such as Denmark
and the Netherlands. In part because of the increasing use of referenda,
California was seen as playing a role in the United States "analogous to that
of West Germany in the EC in terms of raising environmental issues."[31]

One area of concern in relation to the chemical industry was water
pollution. In 1976, the EC adopted a Directive on Dangerous Substances
in the Aquatic Environment (76/464). This was a framework directive
aimed at the minimization of water pollution by harmful materials. Using
biodegradability as a criterion, materials were divided into a "black list,"
referring to a few substances (e.g., mercury) that are almost unbiodegrad-
able where the aim was to eliminate pollution, and a "gray list" where
pollution was to be controlled rather than eliminated. Considerable difficulty was experienced in the design and the implementation of daughter
directives such as that on mercury enacted in 1982. The procedure for
producing daughter directives for "black" (List 1) substances "has proved
burdensome and slow."[32]

Because of a political compromise adopted largely as a result of British
pressure, member states were allowed to choose between an emission
limits approach or a limit values approach based on quality values in the
water receiving the pollution. "The emission limit approach is based on
estimates of the maximum levels of reductions in pollution that could
reasonably be expected given the best available techniques not involving excessive costs. The meanings of 'best available techniques' and
'excessive costs' have often proved difficult to ascertain."[33] Although
the 1976 directive has improved surface water quality, it illustrates the
fact that defects in directive design resulting from political compromises
and problems in implementation may mean that environmental measures
may have less impact on the chemical industry than appears likely when
they are first introduced. The limited impact of some earlier measures
has, however, led to an increased focus on "the general *prevention* of
hazardous chemical use and exposure, rather than on detailed *control*

---

[31] Grant, *Government and Industry,* p. 211.
[32] Wyn Grant, Duncan Matthews, and Peter Newell, *The Effectiveness of European Union
Environmental Policy* (Basingstoke: Macmillan, 2000), p. 162.
[33] Ibid., p. 161.

at the point of use and disposal."[34] Incidents, however, continued to lead to unfavorable media coverage of the chemical industry. There were more than two thousand deaths at Bhopal in India in 1984. In November 1986, some of the water used to fight a fire at the Sandoz plant near Basle entered the Rhine. A slick of contaminated water containing mercury compounds and other toxic substances swept downstream, killing an estimated half a million fish and raising fears about the risk of contamination of water supplies. Public concern was intensified by spillages into the Rhine and Main involving all the three leading German chemical firms. These events seriously undermined the position of the German chemical industry that had issued a new set of guidelines on environmental policy earlier in the year. These had received some support from the new environmental minister. However, the spillages, which occurred at the height of a federal election campaign, "seriously weakened the political allies of the industry . . . and severely compromised the chances of maintaining a policy based on self-regulation."[35]

The EC's Fourth Environmental Programme launched in 1987 at a time when environmental policy had been formally incorporated in the treaties through the Single European Act set out to tighten environmental standards, "which had not been sufficiently strict in the past. The chemicals sector was identified as an appropriate target for the introduction of substance-oriented controls."[36] One of the consequences of this approach was the introduction of integrated risk assessment procedures. Council Regulation 793/93 and Commission Regulation 1488/94 dealt with Risk Assessment of Existing Chemicals. However, "Progress with risk assessment and toxicity testing has been understandably slow, given the size and nature of the task."[37]

SELF-REGULATION AND PARTNERSHIP WITH GOVERNMENT

The industry has made new and more proactive efforts at self-regulation. The idea of "Responsible Care" was initiated by the Canadian Chemical Producers' Association in 1984 but has since been taken up by all the major industry associations in an interesting example of cross-national policy learning. It was emphasized in the statement accompanying the

[34] European Environmental Agency, *Europe's Environment*, p. 124.
[35] William Paterson, "Self-Regulation under Pressure," pp. 242–43.
[36] Pamela M. Barnes and Ian G. Barnes, *Environmental Policy in the European Union* (Cheltenham: Edward Elgar, 1999), p. 40.
[37] European Environmental Agency, *Europe's Environment*, p. 124.

launch of the American Chemistry Council as a commitment "to enhance industry product and environmental stewardship [with] a new vision statement of 'Zero accidents, injuries and impact on the environment.'"[38] Indeed, leaving aside train accidents in the United States involving chemicals being transported, there has not been a repetition of the major chemical incidents that occurred in the 1970s and 1980s.

In Europe, Responsible Care has, however, been more enthusiastically adopted in Britain than in Germany.[39] One of the problems was that a company might sign up to the principles but not follow them in practice. "RC was partially successful, as it helped to improve the image of the chemical industry. However, one of its problems was that it lacked credibility."[40] In the United States, the CMA worked with the EPA to create an international high production volume (HPV) testing program that was launched in 1998, based on an ICCA initiative. Companies that manufacture or import HPV chemicals volunteer them for the program. HPV chemicals not volunteered could be subject to a mandatory test rule being developed by the EPA. This suggests a less adversarial relationship between the industry and the EPA than was the case in the 1980s when, of six sectors studied, "the chemical industry had the most difficult relationship, even under the relatively favourable conditions of the Reagan administration."[41]

NEW PRESSURES IN EUROPE

As the EU worked on a new framework directive on chemicals that was originally scheduled to be published in draft form by the end of 2000, the chemical industry found itself facing new pressures. For once, the incidents that damaged public confidence in the regulatory process were outside the chemical industry, but they had a spillover effect. These include controversies about BSE, genetically modified organisms, and dioxins in food in Belgium. "All these scares have adversely affected public opinion. Cumulatively, they create an atmosphere in which regulators feel they must take a tougher precautionary approach to all aspects of consumer

---

[38] Press statement, Arlington, VA, June 12, 2000.
[39] J. F. Franke and F. Wätzold, "Voluntary Initiatives and Public Intervention – The Regulation of Eco-Auditing," in Francois Lévêque, ed., *Environmental Policy in Europe* (Cheltenham: Edward Elgar, 1999), pp. 180–81.
[40] Ibid., p. 175.
[41] Grant, *Government and Industry*, p. 212.

protection, including the need to protect a vulnerable public from the perceived all-pervading hazards of chemicals."[42]

There was a change of atmosphere in the Commission. From an industry perspective, it was claimed that: "Emotive phrases such as 'chemicals have more rights than humans' have been bandied about at the highest echelons of EU governance."[43] In part this change of mood reflected a concern about the adequacy of the patchwork of laws that have grown up since the 1967 directive. The new environment commissioner in the Prodi Commission, Margot Wallström, gave a greater priority to chemicals questions, taking a tougher stance than her predecessors. She stated that the EU's system of assessing the risks of chemicals had "no credibility," adding, "There are 20,000 chemicals out there, and we are looking at four – that is absurd."[44]

The European Environmental Bureau and the consumers' organization, BEUC, joined together in 2000 to call for a complete overhaul of chemicals policy. They argued that current chemicals policy ignored the precautionary principle and the public's right to information. The snails pace at which chemicals were currently tested meant that there were tens of thousands of potentially hazardous substances on the market that had never been assessed. Hazardous chemicals should be phased out completely, and all substances that had not been properly tested by 2005 should be removed from the market. The "burden of proof" should be reversed so that companies would have to prove that a product was safe. Policy should be based firmly on the precautionary principle, which enables policy makers to ban potentially hazardous substances, and on the substitution principle, which seeks to replace dangerous chemicals with safer alternatives.

There was broad agreement that EU chemicals policy had become too complex and cumbersome to be effective. By the end of the twentieth century, it was made up of more than three hundred directives, decisions, regulations, and amendments. The main deficiency of the arrangements existing in 2001 was the failure to provide equivalent information about the hazards of "existing" and "new" substances. A new system of chemicals control introduced in 1981 applied only to products developed after that date, leaving the one hundred thousand or so substances in use before then uncovered.

---

[42] B. S. Gilliatt, "Executive Director's Message" (http://www.chlor.org, 1999).
[43] Ibid.
[44] *European Voice,* May 4–10, 2000.

In February 2001, the European Commission published a White Paper setting out the future strategy for EU policy for chemicals (European Commission, 2001). Sustainable development was set out as the overriding goal of the policy, and there was an emphasis on the fundamental importance of the precautionary principle. One of the key elements in the White Paper was the shift of the burden of proof and responsibility to the chemical industry and to the downstream industries that use its products. The main responsibility was placed on manufacturers, importers, formulators, and industrial users to generate and assess data, prepare risk assessment reports, and give adequate information about safety to the public. However, the emphasis on safeguarding human health in the White Paper was balanced by an expression of the need to ensure the competitiveness of the chemical industry. It was the nature of that balance that was to provide the focus of the subsequent drawn-out and intensive controversy.

CEFIC's strategy initially paid considerable attention to the six working groups on testing, registration, and evaluation set up by the Commission that met between October 2001 and March 2002. CEFIC thus played to its strengths using its specialist resources in a detailed and highly technical discussion of the implications of policy. In particular, CEFIC became involved in an argument with the Commission about the cost of risk assessment. The Commission estimated the cost to the industry at €2.1 billion. CEFIC estimated that testing would cost between €7 and €10 billion. According to CEFIC, the discrepancy in part arose from the Commission's failure to account for industrial chemicals that are not marketed as final products, a substantial proportion of total industry output. In May 2002 the Commission organized a conference to present a report on the cost impact of the EU chemicals policy conducted by a consultant, Risk and Policy Analysis (RPA). CEFIC claimed that: "The RPA study shows that the missing chemical data can be collected at an estimated cost of €7 billion. These figures are in line with those presented by the chemical industry one year ago."[45]

The Commission's proposals for REACH (register, evaluate, and authorize chemicals) put forward in May 2003 were presented as one of most important regulatory reforms ever. Although modified from the original proposals, they were heavily criticized by both chemical and general industry associations. For example, the Federation of German

---

[45] CEFIC, "CEFIC composite statement on EU Commission's White Paper Working Group discussions," Brussels, March 2002, p. 1.

Industries (BDI) claimed that the policies could cost Germany 1.7 million jobs. At the final stage of the decision-making process, opposition was expressed in a joint letter by the political heads of the three leading member states, Britain, France, and Germany. Faced with this level of opposition, the Commission had to make further substantial concessions when it announced its response to the consultation process in October 2003. It was evident that Erkki Liikanen, the enterprise commissioner, was happier with the outcome than Margot Wallström, the environment commissioner, even though the precautionary principle was retained. The Commission estimated that the cost of the revised proposals to the chemical industry would be €2.3 billion over eleven years, an 82 percent reduction from the May proposals. A requirement to provide safety information was softened for some twenty thousand chemicals produced in quantities of less than ten tons a year. Registration requirements were dropped for polymers, one of the industry's demands. A requirement to switch to alternative chemicals (the substitution principle) was made less binding and firms would have the right to keep some information about products confidential. Constraints on imports into the EU would be reduced.

The outcome represented something of a victory for the industry. German Green MEP, Hildrud Beyer, admitted that "industry has been successful in shifting the focus of the debate," with the real coup being getting Blair, Chirac, and Schröeder to insist that REACH should mainly concern itself with competitiveness.[46] Nevertheless, some subsectors of the industry have faced fundamental challenges to their production processes and even their existence as the case study of chlorine that follows shows.

CHLORINE: A CASE STUDY

Chlorine and caustic soda are key building blocks of the chemical industry. These raw materials are made by passing electricity through brine. One-third of the products made using chlorine contain no chlorine, but depend on it for synthesis. Much of chlorine's value is in its co-product, caustic soda. This alkali is used in a wide range of applications such as the manufacture of soaps, detergents, textiles, pulp and paper, aluminum, and glass.

---

[46] *European Voice*, November 13, 2003.

The first environmental crisis to hit the industry arose from the organo-chlorines known as polychlorinated biphenyls (PCBs). These started to be produced in the interwar period with a substantial expansion of production in the 1950s. They were used for such purposes as insulating agents for electrical transformers, lubricants, printing inks and adhesives, and plasticizers. Their advantages included chemical stability, non-flammability, and easy solubility.

PCBs are a classic example of something that was industrially useful but environmentally dangerous. They are chlorinated hydrocarbons that do not occur naturally in biological systems. PCBs have been shown to cause cancer in animals and have as well a number of other serious noncancer health effects. Studies in humans suggest that they have produced carcinogenic and other health effects, including developmental abnormalities. PCBs can enter the body through the lungs, gastrointestinal tract, and skin. They circulate throughout the body and are stored in the fatty tissue. The consumption of PCB-contaminated fish is a major source of exposure.

PCBs are among the most stable organic compounds known, but the very chemical stability that made them useful in industrial processes was a problem from an environmental perspective, as PCBs are persistent, toxic, and tend to bioaccumulate. They are also persistent organic pollutants (POPs), prone to disperse widely. They are found globally in the lower atmosphere and oceans.

Environmental concern about PCBs was aroused by a series of incidents and discoveries in the late 1960s. A Swedish scientist found eggshell thinning among seabirds as a result of bioaccumulation of PCBs, leading to reduced reproductive capacity. In 1968, nearly two thousand people in Yusho, Japan, became ill after eating rice oil heavily contaminated with PCBs. The PCBs had been used as heat exchangers in machinery used to refine the oil and are thought to have entered it accidentally through a leak.

The use of PCBs in open applications such as printing inks and adhesives was banned in the EC in 1976 (Directive 76/403). Their use as a raw material or chemical intermediate has been banned since 1985. In 1996, the EU directive of 1976 was replaced by Directive 96/59/EC that controls the disposal of PCBs and the equipment used in their disposal. In the United States, the TSCA prohibited the manufacturing, processing, and distribution of TCBs, although the first PCB regulations were not published until 1978. In July 1998, this was supplemented by a regulatory Final Rule that introduced major changes in the disposal of

PCBs. PCB-containing equipment is still extensively used. In part, this is because PCBs were inadvertently introduced into a large proportion of the mineral oil – filled equipment in use before 1978, primarily through mixing PCBs and mineral oil during servicing operations. EPA allows these uses to continue, provided the equipment is properly monitored and maintained. This illustrates how long it takes to bring a toxic substance released into the environment under control.

Around two-thirds of West European plants use the mercury cell process to produce chlorine. It is used less extensively in the United States, by about 13 percent of plants (hence the EPA was still drafting a rule involving emissions limits based on control techniques and management practices in the summer of 2000). Mercury is used as the negative electrode or cathode in the production process. It serves the purpose of keeping the highly reactive products apart. However, mercury is a toxic metal, a persistent, bioaccumulative pollutant that affects the nervous system. Chorine producers have made significant progress in reducing mercury emissions. However, they admit that: "Although the process is essentially enclosed, with the mercury recycled, very small unintentional losses do occur."[47] The members of Euro Chlor have agreed voluntarily not to use the mercury process for any new plants being built. The industry is progressively converting plants to other technologies as they come to the end of their economic lives. Greenpeace has called for a phaseout of the production of chlorine, which they argue would enable much of the world's most severe toxic pollution to be stopped. The industry is concerned that even "compulsory phase-out of mercury plants prior to them reaching the end of their economic lives would severely damage the competitiveness of the European chlorine industry as a result of the very high reinvestment costs required with only a marginal benefit to the environment."[48]

The chlorine sector has been very much in the firing line as far as environmentalist critiques of the chemical industry are concerned. Environmentalists would like to see the disappearance of the industry. A more moderate position would be the more rapid phasing out of the mercury cell process, but that option still has implications for the industry's competitiveness.

The industry is not just facing conventional political pressures. "Green consumerism" is becoming an increasing force. Pressure by

---

[47] http://www.eurochlor.org/mercury/br1.htm, 27/6/00.
[48] Ibid.

environmentalists on shareholders have led medical products and hospital management companies in the United States to seek alternatives to PVC medical devices. An international sports goods manufacture decided to stop using PVC, supposedly for environmental reasons. Under these circumstances, Euro Chlor's strategy of seeking a balanced perspective on chlorine chemistry within a rational regulatory environment becomes more difficult to maintain.

CONCLUSIONS

The days when the chemical industry held a benign image as a provider of cutting-edge technology that produced clear benefits for the consumer have long since gone. It is viewed with suspicion, making it a prime target for the regulatory state and the nongovernmental organizations that have arisen from the new politics of collective consumption. In many respects, however, the industry has managed this transformation to new forms of politics and a completely changed interface with government relatively well. This can be seen in the way in which its business interest organizations have reconfigured themselves to make environmental policy their central priority, producing innovative ideas such as "Responsible Care." Despite the costs imposed by regulation, the industry remains competitive and profitable. This would not have been possible had the industry not shown itself capable of going through a process of internal readjustment.

It has been helped in finding a new balance in its relations with government by the fact that other targets appear to distract attention from the chemical industry. For much of the time, the relationship is a routinized one in which the industry can deploy its technical expertise to considerable effect. However, within the EU at least, there have been signs of a new offensive against the chemical industry. The prospects of the industry in the twenty-first century are significantly dependent on its success in satisfying its critics that it can produce its products safely and with minimal environmental impact.

REFERENCES

Allen, C. S., "Political Consequences of Change: The Chemical Industry" in P. J. Katzenstein, ed., *Industry and Politics in West Germany* (Ithaca, NY: Cornell University Press, 1989), 157–84.
Barnes, P., and I. G. Barnes, *Environmental Policy in the European Union* (Cheltenham: Edward Elgar, 1999).

Brickman, R., S. Jasanoff, and T. Ilgen, *Controlling Chemicals: The Politics of Regulation in Europe and the United States* (Ithaca, NY: Cornell University Press, 1985).

CEFIC, "CEFIC composite statement on EU Commission's White Paper Working Group discussions," Brussels, March 2002.

European Commission, *White Paper: Strategy for a Future Chemicals Policy*, COM (2001) 88 final (Brussels: Commission of the European Communities).

European Environmental Agency, *Europe's Environment: The Second Assessment* (Luxembourg: Office for Official Publications of the European Communities, 1998).

Franke, J. F., and F. Wätzold, "Voluntary Initiatives and Public Intervention – The Regulation of Eco-auditing," in *Environmental Policy in Europe*, ed. F. Lévêque (Cheltenham: Edward Elgar, 1999), 175–199.

Gilliatt, B. S., "Executive Director's Message" (http://www.chlor.org, 1999).

Grant, Wyn, *Government and Industry* (Aldershot: Edward Elgar, 1989).

Grant, Wyn, *Pressure Groups and British Politics* (Basingstoke: Macmillan, 2000).

Grant, W., D. Matthews, and P. Newell, *The Effectiveness of European Union Environmental Policy* (Basingstoke: Macmillan, 2000).

Grant, W., W. Paterson, and C. Whitston, *Government and the Chemical Industry* (Oxford: Clarendon Press, 1988).

Hall, P. A., *The Political Power of Economic Ideas: Keynesianism across Nations* (Princeton, NJ: Princeton University Press, 1989).

Jones, A., "Inflation as an Industrial Problem," in *The End of the Keynesian Era*, ed. R. Skidelsky (London: Macmillan, 1977), 50–58.

Martinelli, A., and W. Grant, "Conclusion," in A. Martinelli, ed., *International Markets and Global Firms: a Comparative Study of Organized Business in the Chemical Industry* (London: Sage, 1991), 272–88.

Mol, A. P. J., *The Refinement of Production: Ecological Modernization Theory and the Chemical Industry* (Utrecht: van Arkel, 1995).

Paterson, W., "Self-Regulation under Pressure: Environmental Protection Policy and the Industry's Response," in A. Martinelli, ed., *International Markets and Global Firms: A Comparative Study of Organized Business in the Chemical Industry* (London: Sage, 1991), 228–48.

Porritt, J., and D. Winner, *The Coming of the Greens* (London: Fontana, 1988).

Spitz, P. H., *Petrochemicals: The Rise of an Industry* (New York: John Wiley, 1988).

United States Department of Commerce, *U.S. Industrial Outlook 1990* (Washington, DC: Department of Commerce, 1990).

Vogel, D., *National Styles of Regulation: Environmental Policy in Great Britain and the United States* (Ithaca, NY: Cornell University Press, 1986).

PART II

World Players

Leaders

5

The German Chemical Industry
after World War II

ULRICH WENGENROTH

POINT OF DEPARTURE AFTER WORLD WAR II

The German chemical industry had fallen behind through the war and in the immediate after-war years. "I.G. Farbenindustrie AG" (IGF), the dominating company, founded in 1925, had lost its firm grip on international markets. Germany's share in the world production of chemicals had declined from 22 percent in 1938 to 8 percent in 1951.[1] In 2002, its market share stood at 7.2 percent,[2] which, however, is not a sign of failed reconstruction but of vigorous development of the world's chemical industry in newly industrialized countries and of the internationalization of the German chemical industry itself.

Although the main reason for the decline in the 1940s was the war itself, reconstruction in the immediate postwar years was slowed by limitations inflicted on the industry by the victorious allies. In a kind of déjà vu, German industry had once again lost its patent rights, trademarks,

---

[1] Ashish Arora, Ralph Landau, and Nathan Rosenberg, "Dynamics of Comparative Advantage in the Chemical Industry," in David C. Mowery and Richard R. Nelson (eds.), *Sources of Industrial Leadership. Studies of Seven Industries*, Cambridge: Cambridge University Press, 1999, p. 233.

[2] VCI, Chemiewirtschaft in Zahlen 2003, Frankfurt/Main: VCI, 2003, p. 104.

and subsidiaries in all major foreign markets.[3] In addition, the East-West partition of Germany interrupted a well-established division of production among IGF plants in particular, and the German chemical industry in general. The hydrogenation plant in Leuna, one of the backbones for the supply of carbohydrates, ended up in the Russian zone of occupation. The Allied ban on the production of synthetic rubber, in which the German chemical industry had excelled during the war, idled some of the newest plants. More consequential, however, was the self-inflicted backwardness that was a result of the strategic twins of autarchy policy and technocratic management, both of which had fostered industrial structures and mentalities rooted in market avoidance. The German chemical industry was leading in processes like hydrogenation, which future markets would demand, while trailing behind in promising new fields like petrochemicals and pharmaceuticals. Whatever the shortcomings of management or industrial structure, however, they proved to be of little influence on the breathtaking reconstruction process in the 1950s and, with the exception of shortfalls in developing biotechnology, they were overcome well before they could have turned into a serious liability.

Hydrogenation plants had been destroyed, dismantled, or lay idle, never to be put back in use in West Germany. It was a chapter closed with no adverse aftereffects. The backwardness in some crucial pharmaceuticals was more serious, however. As Achilladelis and Antonakis have shown quantitatively, the German pharmaceutical industry in general and Hoechst and Bayer, whose pharmaceutical divisions had been the first modern pharmaceutical companies, in particular already had fallen behind their American competitors before World War II. They continued to rely on technologies initiated at the turn of the century.[4] Mass-produced American penicillin turned out to be superior to sulfonamides that had been pioneered by IGF. During the war, IGF scientists at the Hoechst plant managed to produce minute quantities of penicillin using a very inefficient process rather than the large fermenters that had come into operation in the United States. Only with substantial help from Merck/Rahway did Hoechst arrive at a position to start penicillin production on an industrial scale in 1950.[5] The Bayer story of penicillin

---

[3] Harm G. Schröter, "Die Auslandsinvestitionen der deutschen chemischen Industrie 1930 bis 1965," in *Zeitschrift für Unternehmensgeschichte 46* (2001), pp. 186–89.

[4] Basil Achilladelis and Nicholas Antonakis, "The Dynamics of Technological Innovation: The Case of the Pharmaceutical Industry, in *Research Policy 30* (2001), pp. 535–88, esp. p. 579.

[5] Ernst Bäumler, *Ein Jahrhundert Chemie,* Düsseldorf: Econ, 1963, pp. 232–34.

was similar. Again, it was the importation of American know-how and technology that initiated the production of marketable penicillin in 1951.[6]

But it was not only penicillin that marked the falling behind of the pharmaceutical branch of the German chemical industry. In fully concentrating on chemical synthesis, the path to success since the late nineteenth century, IGF pharmaceutical research had failed to participate in the development of biotechnological processes for manufacturing antibiotics. Streptomycin, a powerful supplement to penicillin, was another example of a highly successful antibiotic that was pioneered in the United States, and whose production was introduced at Hoechst with American help in the early 1950s.[7] The prewar leaders in antibiotics had become dependent on transfer of both knowledge and technology.

The prohibition on the manufacture of Buna (synthetic rubber) pushed manufacturers into other products. Because Allied restrictions had not come as a surprise, companies had made preparations for postwar production. Chemische Werke Hüls, one of the major producers of Buna and an IGF subsidiary, in 1945 had already moved into developing standard plastics as well as fexibilizers, the production of which would employ existing technology and know-how.[8] The close technological relationship of processes leading to "Buna S" and those leading to standard plastics like PVC, polystyrene, and polyethylene made the transition easy. When the Allied High Commission renewed the ban on producing "Buna S" in March 1950, the chemical industry was well on its way to using its expertise and plant for the production of plastics.[9] In so doing, it was following a policy of change that had its roots in the prewar era. As Peter Morris observed in his study of IGF management: "I.G. Farben's leaders ... wanted to get away from dyes, on the one hand, and the ruinously expensive high-pressure chemistry, on the other. During the Third Reich, ter Meer and Ambros (top managers of IGF, U.W.) also sought to diversify away from any 'political' products such as Buna S toward potentially more profitable consumer products, such

[6] Erik Verg, Gottfried Plumpe, and Heinz Schultheis, Meilensteine. *125 Jahre Bayer 1863–1988*, Leverkusen: Bayer AG, 1988, p. 510.

[7] Bäumler, *Ein Jahrhundert Chemie*, pp. 238–40.

[8] Mechthild Wolf, "Unternehmensstrategien zwischen Wiederaufbau und Globalisierung," in *Nachrichten aus Chemie, Technik und Laboratorium 47* (1999), p. 1043.

[9] Leo Kollek, "Die Bedeutung des Gesetzes Nr. 61 für die Kunststoff-Industrie. Zur Lockerung der Industriekontrolle," in *Kunststoffe 41* (1951), pp. 409–13.

as oil-resistant Bunya N, nylon and polyurethanes."[10] Gottfried Plumpe, in his study of IGF stated that in the interwar years the chemistry of polymers for the production of plastics and synthetic fibers was the most important innovation of chemical technology between the wars.[11]

In plastics, the American lead was less dramatic than in pharmaceuticals. In both fields, however, as in chemical engineering, Joachim Radkau's leitmotif of West German economic and technological history in the 1950s – "learning from the USA" – remained true.[12] In the end, Allied restrictions on German production proved to be short-lived. Buna, like steel, once considered building blocks of potential German rearmament, were taken off the list during the Korean War and used to build and strengthen Western industrial supremacy.

Deconcentration, the nightmare among the IGF management, had proved to be harmless in the three Western zones of occupation. The early cooperation of the Allies in dismantling the fabric of the German chemical industry ended in 1948. The outcome was viable IGF-successors in West Germany, with Bayer and especially Hoechst winning back their independence from BASF's dominating position in IGF.[13] Hoechst had continued to be an important pharmaceutical producer in the IGF era, but its products were sold under the Bayer trademark. This might create the wrong impression, as if Hoechst was of little importance in pharmaceuticals, whereas the company in fact had great experience and a formidable knowledge base that became visible after 1945. Already in 1950, production by the West German chemical industry surpassed the prewar level.[14] Soviet exploitation and socialist planning in isolation from world markets, however, heavily compromised development in the

---

[10] Peter J. T. Morris, "Ambros, Reppe, and the emergence of heavy organic chemicals in Germany, 1925-1945," in Anthony S. Travis et al. (eds.), *Determinants in the Evolution of the European Chemical Industry, 1900-1939: New Technologies, Political Frameworks, Markets, and Companies,* Dordrecht/Boston: Kluwer, 1998, p. 104. (L38Aut / DM-Bib)

[11] Gottfried Plumpe, *Die I.G. Farbenindustrie AG: Wirtschaft, Technik und Politik 1904-1945,* Berlin: Duncker & Humblot, 1990, p. 154.

[12] Joachim Radkau, "'Wirtschaftswunder' ohne technologische Innovation? Technische Modernität in den 50er Jahren," in Axel Schildt and Arnold Sywottek (eds.), *Modernisierung im Wiederaufbau. Die westdeutsche Gesellschaft der 50er Jahre,* Bonn: Dietz, 1993, pp. 134-5.

[13] Raymond G. Stokes, "Von der I.G. Farbenindustrie AG bis zur Neugründung der BASF (1925-1952)," in Werner Abelshauser (ed.), *Die BASF: Eine Unternehmensgeschichte,* München: C.H. Beck, 2002, p. 355.

[14] Raymond G. Stokes, *Divide and Prosper: The Heirs of I.G. Farben under Allied Authority 1945-1951,* Berkeley: University of California Press, 1988, pp. 164-65.

East.[15] While at the end of World War II capacities for about 30 percent of German chemical production had ended up in the Soviet zone of occupation, by 1995 the territory of the former GDR provided for only 5 percent of chemical production in Germany.[16]

Easy Reconstruction

The undervalued deutsche mark helped exports, and exports helped in modernizing plant. This, in short, was the way Kurt Hansen, a top manager of IGF since the late 1930s and eventually chairman of the Bayer Executive Board in the 1960s and early 1970s, analyzed the German chemical industry's quick recovery in the 1950s.[17] He was not far off the mark.

Reconstruction in a booming economy with exports helped by an undervalued currency was an easy task for management. Almost all products were known, production technology was familiar, and highly qualified labor was abundant. Coal-based "Reppe-Chemie," the technological backbone of the prewar years and the provider of acetylene, the most universal foundation for many plastic materials, continued to dominate and was certainly adequate until well into the 1960s.[18] The transition to petrochemical processes was smooth and did not overtax the adaptive power of research and engineering. Acetylene from coal and lime only gradually gave way to cheaper and more convenient ethylene and propylene from naphtha. With annual growth rates of 20 percent between 1955 and 1965, fed by a booming market for plastics in particular,[19] petrochemicals easily made their way to dominance in the industry and replaced the depreciated plant of 1930s technology. As in most German industry, reconstruction, with its high rates of growth, was the great modernizer that made war losses almost look like a Schumpeterian impulse. Rapid growth was not only forgiving of occasional management errors; it did

---

[15] Rainer Karlsch, "Capacity Losses, Reconstruction, and Unfinished Modernization: The Chemical Industry in the Soviet Zone of Occupation (SBZ)/GDR, 1945–1965," in John E. Lesch (ed.), *The German Chemical Industry in the Twentieth Century*, Dordrecht: Kluwer, 2000, pp. 375–92.

[16] Karlsch, Capacity Losses, p. 367.

[17] Kurt Hansen, "Die chemische Industrie von 1945 bis 2050," in *Nachrichten aus Chemie, Technik und Laboratorium 47* (1999), p. 1039.

[18] Werner Abelshauser, "Die BASF seit der Neugründung von 1952," in Abelshauser (ed.), *Die BASF*, p. 432.

[19] Walter Teltschik, *Geschichte der deutschen Großchemie.Entwicklung und Einfluß in Staat und Gesellschaft*, Weinheim: VCH, 1992, p. 215; Abelshauser, *Die BASF*, p. 438.

not call for much change. What happened, to a large extent, was truly reconstruction.

TECHNOCRATIC CULTURE REAFFIRMED

The blatant success of continuing in the old ways strengthened the technocratic culture in German chemical industry. Ulrich Haberland, executive chairman of Bayer, in his 1955 report to the company's shareholders said: "A number of factors have contributed to the development of a strong demand for chemical products, a demand which can only be satisfied at the cost of considerable efforts. The most prominent of these factors are a favorable, almost uninterrupted development of the economy in the western world together with the evolving of novel products in the chemical realm and the ensuing stimulation of a new need."[20] Still, in 1969, McKinsey at BASF identified what it called a "producers mentality" rooted in the conviction that: "We can sell whatever we can produce."[21] Why question established managerial hierarchies and technological paradigms when the IGF-successors irresistibly were rolling up the field of the largest chemical companies from behind? By 1964, they were back in the top ten and growing faster than any of their immediate competitors.[22]

One side effect of unfettered growth and a notable break with past traditions was the nonrevitalization of the European, and especially German, cartel tradition. This does not mean that there was no more collusive action, there was. Where selling was easy, cartels lost their rationale. Attempts in the early 1950s to regulate the sale of synthetic dyes or to agree on protected fields of research petered out.[23] The unprecedented reconstruction boom of the fifties, in the words of Raymond Stokes, "obviated the need for price-fixing or market-division agreements among competitors: there was simply plenty of business to go around for everyone."[24] And, as the big three IGF-successors did not aggressively compete with each other, but developed quite distinct portfolios from the 1950s

---

[20] Quoted from Teltschik, *Geschichte*, p. 205. ("Die seit Jahren fast ununterbrochen günstige Entwicklung der Wirtschaft der westlichen Welt hat zusammen mit der Entwicklung neuer Produkte auf der Chemie-Seite und der dadurch hervorgerufenen Weckung von neuem Bedarf eine starke Nachfrage nach chemischen Produkten mit sich gebracht, die nur mit Mühe befriedigt werden kann.")

[21] Abelshauser, *Die BASF*, p. 624.

[22] Stokes, *Divide and Prosper*, p. 204.

[23] Abelshauser, *Die BASF*, pp. 459–460.

[24] Stokes, *Divide and Prosper*, p. 208.

on, there was little overlap that gave reason for elaborate collusive action. Bayer and Hoechst, the two pioneers in dyestuffs and chemically related pharmaceuticals, continued to move further into pharmaceuticals and developed a wide array of synthetic fibers, whereas BASF, nicknamed "the feedstock shop" ("Rohstoffladen"), concentrated on its strength as a supplier of raw materials and semi-manufactures to the chemicals processing industry. In 1965, BASF was the world's greatest supplier of fiber intermediates.[25] Through this strategy, BASF eventually ended up with the greatest interconnected chemical plant in the world at its main site in Ludwigshafen.[26]

Markets for plastics boomed in the 1950s and early 1960s, helping all three big companies. The most important domestic market, accounting for almost 30 percent of production, was the building industry, itself one of the great beneficiaries of reconstruction.[27] Demand for both fertilizers and artificial fiber did not pick up at the same rate but was also expected to boom soon because of population increase worldwide. In the 1950s, many economists and industrialists believed that food production was to become the major bottleneck in the decades to come. Artificial fertilizer would be needed to increase agricultural productivity and farmland would have to be used for food rather than fibers in the future.[28] Eventually, only synthetic fibers fulfilled these high expectations for the German chemical industry. Being closer to new consumer markets, as had been foreseen by IGF managers Ambros and ter Meer, Hoechst and Bayer were more successful in their investment policy and grew quicker during the 1950s than BASF, which stayed closer to its prewar tradition of bulk production of intermediate products.[29]

OPTING FOR OIL

The transition from coal and coal gas to oil as the main feedstock of organic chemistry brought about a major and lasting reorientation of industrial culture as well as of market strategies. This conversion was the equivalent of giving up decades of technological leadership during which

---

[25] Abelshauser, *Die BASF,* p. 438.

[26] Ibid., p. 506.

[27] Jochen Streb, *Staatliche Entwicklung und branchenübergreifender Wissenstransfer: Über die Ursachen der internationalen Innovationserfolge der deutschen Kunststoffindustrie im 20. Jahrhundert*, Berlin: Akademie-Verlag 2003, p. 157, fn. 49.

[28] Hansen, "Die chemische Industrie . . . ," p. 1037.

[29] Abelshauser, *Die BASF,* p. 480.

the German chemical industry had mastered the transformation of coal and coal gas to organic products better than any of its competitors. At best, petrochemistry put the works on an equal footing with British and American companies. More often, however, German companies had to import vital technology and know-how. Petrochemistry marked the end of unquestioned German technological leadership. For the first time, the German chemical industry had fallen behind in the very core of its business. It was the end of autarchy in raw materials as much as the end of autarchy in research and technology.

The IGF-successors had an early interest in exploring the potential of petrochemical technology. In view of the insecure raw material supply from coal mines and steelworks where most of German coal and steel capacity was concentrated, which gave rise to the prospects of deconcentration and maybe even nationalization as in the British sector, it seemed wise to look for alternative feedstocks. Petrochemistry in the early 1950s was still largely terra incognita.[30] Most important, the expertise in chemical engineering that was crucial to master huge cracking and refining plants was largely absent in Germany.[31]

When the West German parliament waived the tax on fuel oil in 1953, and abolished the tariff in 1956, it only took the end of the Suez crisis for oil imports to rapidly expand.[32] The enormous expansion of fuel oil sales in Germany from 1957, which came very much to the surprise and eventual decline of the coal mines, had given rise to an oversupply of benzene that was not yet absorbed by gasoline demand for cars. German motorization trailed behind heating houses with fuel oil.[33] This situation both offered an opportunity for, and created a threat to, the chemical industry. Oil refineries had to find an outlet for benzene.[34] They could do this by supplying the chemical industry or by erecting their own plant for processing hydrocarbons. Eventually, oil refineries did both. The IGF-successors knew this situation well from their rocky relationship with coal mines and steelworks before the war. Each of them, therefore, individually began to negotiate patterns of cooperation that would benefit both partners. The profitability of petrochemistry as the main raw material source of chemical production in West Germany, however, was largely

---

[30] Abelshauser, *Die BASF*, p. 441.
[31] Rainer Karlsch and Raymond G. Stokes, *Faktor Öl: Mineralölwirtschaft in Deutschland 1859–1974*, München: C.H. Beck, 2003, p. 294.
[32] Abelshauser, *Ruhrkohlenbergbau*, pp. 89–93. Karlsch/Stokes, *Faktor Öl*, p. 303.
[33] Bäumler, *Ein Jahrhundert Chemie*, p. 161.
[34] Karlsch/Stokes, *Faktor Öl*, p. 317.

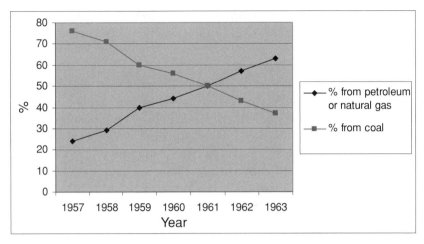

Figure 5.1. Feedstocks used in producing organic chemicals in West Germany, 1957–63. *Source:* Stokes, *Opting for Oil,* p. 235.

dependent on the building of oil pipelines to replace coal supplies by Rhine River navigation Figure 5.1. It is no coincidence, therefore, that the opening of the major oil pipelines connecting the locations of the IGF successors to cheap petroleum in 1958 (Bayer), 1960 (Hoechst), and 1962 (BASF) were parallel to the changeover from coal to oil as dominant feedstock for organic chemistry.

The first of the big three IGF-successors to move was BASF.[35] In 1949, Bernhard Timm, executive chairman of BASF from 1965 to 1974, already had complained about "how all our research has come under the influence of Dr. Reppe" and was looking for alternatives.[36] A year earlier, in 1948, first talks with Standard Oil of New Jersey, the longtime partner of IGF, had taken place. Even more promising seemed an approach by the chairman of the London-based Shell Group, which, in October 1949, led to plans to erect a joint gas-oil cracking unit in Ludwigshafen. BASF, after negotiating with Shell on an equal footing in 1952, entered a joint venture creating the "Rheinische Olefinwerke" that went into operation in 1955. Rheinische Olefin (ROW) mainly produced polyethylene that was marketed under the brand name "Lupolen" by BASF. By

---

[35] Raymond G. Stokes, *Opting for Oil: The Political Economy of Technological Change in the West German Chemical Industry, 1945–1961,* Cambridge: Cambridge University Press, 1994, pp. 133–53.

[36] Quoted from Abelshauser, *Die BASF,* p. 442.

1959, ROW was one of the world's largest polyethylene plants, with an annual capacity of 125,000 tons.[37] Lower costs of feedstock eventually came with the opening of an oil pipeline from Marseilles to Mannheim in 1962.[38]

Bayer had a slower start into petrochemistry but a most favorable environment.[39] Of the big three, it was best situated for the transition to oil as it had a number of refineries in its neighborhood. Still, in 1977, Wirtschaftswoche, a weekly economic journal, saw this locational factor as a major reason for Bayer's leading position among IGF-successors in the 1950s and 1960s.[40] The refineries in Bayer's neighborhood were the coal-based hydrogenation plants of the Nazi era that had to be converted to processing crude oil after the Allies had banned hydrogenation of coal in 1945. In 1954, these former hydrogenation plants accounted for 25 percent of West German refinery capacity and turned out 40 percent of German car fuel.[41] By then, they had entered partnerships with Shell, BP, and Esso (Standard Oil of New Jersey), respectively. With the first major German oil pipeline to come into operation in 1958 and a new deep-water oil harbor at Wilhelmshaven connecting these refineries,[42] they were a formidable backbone to the evolving petrochemistry in the Rhein-Ruhr-area, where Bayer was situated. After an aborted attempt to form a joint venture with Esso in 1952, Bayer in 1954 turned to BP and its German partner to form the "Erdölchemie" (EC) on a wasteland next to a Bayer plant for synthetic fiber in need of acrylonitrile.[43]

Unlike BASF and Bayer, Hoechst tried to go it all alone.[44] The Hoechst board in 1953 decided to build their own cracker that went into operation three years later, to be followed by another more powerful steam-cracker in 1958.[45] Both, however, were only an episode on the way to large-scale cooperation with the evolving refinery industry.[46] Caltex, a latecomer to the German refinery industry that was dominated by foreign companies,

[37] Abelshauser, *Die BASF*, p. 389. Karlsch/Stokes, *Faktor Öl*, p. 295.
[38] Dieter Nagel, *Die ökonomische Bedeutung der Mineralöl-Pipelines*, Hamburg: Deutsche Shell AG, 1968, p. 27.
[39] Stokes, *Opting for Oil*, pp. 154–75.
[40] Wirtschaftswoche No.23 (May 27) 1977, p. 39.
[41] Karlsch/Stokes, *Faktor Öl*, p. 286.
[42] Heiner Holzhausen, 20 Jahre Nord-West Oelleitung, in *Bergbau 27* (1976) 12, pp. 507–11.
[43] Karlsch/Stokes, *Faktor Öl*, p. 316–17. Verg et al., *Meilensteine*, pp. 358–361.
[44] Stokes, *Opting for Oil*, pp. 176–196.
[45] Bäumler, *Ein Jahrhundert Chemie*, pp. 142–155.
[46] Stokes, *Opting for Oil*, p. 193.

opened a refinery in the neighborhood, strategically placed in the triangle between Frankfurt airport (then still the main airbase of American troops in Europe), the city of Frankfurt, and the main Hoechst production site. The Caltex refinery started operations in 1961 after the Rotterdam-Rhine-Pipeline (RPR) had come into operation, and two years later, Hoechst was supplied with ethylene by a short pipeline.[47] It was synergy at its best. The Knapsack-works of Hoechst, situated in the Rhineland not far from the Bayer plant and a vital supplier of carbide to the Hoechst ethylene-based processes,[48] contracted hydrocarbons from a local lignite-producer at Wesseling,[49] one of the former hydrogenation plants, which itself was to become a major refinery in the 1950s.

The Buna-manufacturer Hüls, jointly owned by the IGF-successors and the state-owned coal mine Hibernia, changed from coal to oil as Hibernia itself erected an oil refinery.[50] By diversifying into oil, a number of coal companies continued to be suppliers of hydrocarbons to the chemical industry. But they never again played the dominating role of raw material suppliers as before the war.

Oil derivates now were abundant, and the earlier view that processes for oil-based and coal-based hydrocarbons would run side by side for the foreseeable future was quickly abandoned. After 1956, there was no doubt that the feedstock for organic chemistry would be oil. The slow transition to petrochemistry that had begun in the early 1950s was now hastened. While in 1957, more than three-quarters of all chemicals had been produced from coal, in 1963 it was a mere 37 percent.[51] On the eve of the oil price crisis, in 1973, 90 percent of the industry's feedstocks were oil-based.[52] The old foundations of autarchy had been irrevocably erased.

Although export drives and growing dependence on export revenues were not new to the German chemical industry, dependence on foreign supplies of raw material as in the case of oil marked a new departure.[53]

---

[47] Bäumler, *Ein Jahrhundert Chemie*, p. 162.
[48] Ibid., p. 122.
[49] Ibid., p. 162.
[50] Wolf, "Unternehensstrategien . . . ," p. 1044.
[51] Karlsch/Stokes, *Faktor Öl*, p. 317.
[52] Ashish Arora and Alfonso Gambardella, "The Dynamics of Industry Structure: The Chemical Industry in the U.S., Western Europe, and Japan in the 1980s," in Ashish Arora, Ralph Landau, and Nathan Rosenberg (eds.), *Chemicals and Long-Run Economic Growth*, New York: Wiley, 1998, p. 421.
[53] Stokes, *Divide and Prosper*, p. 203.

For the first time in its history, the German chemical industry allowed itself to become fully dependent on world markets. This did not change only the fabric of the industry; it came closer to becoming a cultural revolution among management. The industry had left the "Wagenburg" for good.

LEAVING THE TABOR

The 1960s saw the first major change in the fabric and strategies of the German chemical industry since its phenomenal rise before World War I. The wholehearted shift from coal to oil was immediately followed by a shift to internationalization of production instead of forced exports from the well-protected home base. For the first time in the history of the German chemical industry, production followed markets for reasons other than patent-law requirements and tariff barriers. The internationalization of the 1960s was very different from the internationalization of the 1920s.[54] Expansion was driven by rising market demand in light of eroding trade barriers, rather than by finding export valves. By the early 1970s, every third new plant of the "big three," BASF, Bayer, and Hoechst, was erected abroad.[55]

Some of the expansion abroad was through restitution of German property, notably in the United States.[56] Most of it, however, was bona fide new foreign direct investment in the most promising markets. The export boom of the 1950s and early 1960s continued as a foreign investment boom in the late 1960s and early 1970s. Management's focus now turned from collusive action at home to market opportunities worldwide. Direct investment mainly went in three directions: the EEC, North America, and South America.

The EEC countries were the most favored field of expansion. The emerging Common Market substantially lowered the risk of direct investment across borders. With integration of the European market progressing, it seems debatable whether expansion within the EEC really should be regarded as "foreign" investment. The "foreignness" of European economies to each other was gradually eroding during the second half of the twentieth century until, in 2002, the Association of the German Chemical Industry (VCI) could bluntly state: "The German chemicals

---

[54] Schröter, "Auslandsinvestitionen . . . ," p. 190.
[55] Wirtschaftswoche, No. 24 (6 June) 1975, p. 12.
[56] Schröter, "Auslandsinvestitionen . . . ," p. 192.

industry regards the EU as its domestic market."[57] France was the most important sales market in the EEC and attracted the most direct investment in the 1950s and early 1960s. By 1964, it accounted for more than half the capital invested in EEC countries. Italy, the other big sales market, followed with about a third.[58]

The Netherlands and Belgium with their direct access via the Rhine River, where all the major companies, Bayer, Hoechst, BASF, Hüls, Degussa, Henkel, and so on were situated, were to become the low-cost coastal site for export oriented production but also were to serve as suppliers to processing plants in Europe. Their importance grew through the late 1960s and the 1970s. German ports were much less attractive since their waterways led into the GDR. Only Bayer went there in 1973 to erect a plant near Hamburg.[59] With the EEC becoming a success story, the domestic sites offered no more specific advantages. With growth rates of West German chemical production close to 10 percent through the 1960s and early 1970s, more than either the United States, the United Kingdom, or France,[60] heavy investment in favorably placed production sites within the common European market was the most obvious strategy for further expansion.

Not all companies were quick to implement this expansion strategy into the EEC, however. Already in the 1960s, Henkel, the major producer of detergents, soaps, and related consumer products had considered erecting standard production plants elsewhere in Europe. It took until the 1990s before the company eventually moved to build what they called "Euro-plant" (Euro-Fabriken).[61]

By 1965, about half the FDI of the German chemical industry had gone to America; only one third of this, however, had gone to the United States.[62] Early hopes rested on South America. What seemed to be a promising market in the early 1960s had lured the IGF-successors to revitalize old contacts from the prewar years in Brazil and Argentina.[63] It turned out to be the least fortunate expansion. The South American

[57] VCI, Chemiewirtschaft in Zahlen 2003, Frankfurt/Main: VCI 2003, p. 11.

[58] Schröter, "Auslandsinvestitionen...," p. 195.

[59] Teltschik, *Geschichte*, p. 241.

[60] Ibid., p. 251.

[61] Susanne Hilger, "American Consultants in the German Chemical Industry: The Stanford Research Institute at Henkel in the 1960s and 1970s," in Enterprises et histoire 25(2000), p. 64.

[62] Schröter, "Auslandsinvestitionen...," p. 195.

[63] Teltschik, *Geschichte*, p. 241.

market never grew to the extent anticipated in the early 1960s. Sales in this region in the 1980s and early 1990s only occasionally exceeded 6 percent of total sales of the major companies.[64]

The U.S. market, by contrast, in the long run was to become the main focus of investment and expansion. In an almost unanimous strategic decision, the major German chemical companies set off to "Americanize" their operations. After early acquisitions in the 1950s and 1960s, the main move to the United States took off in the 1970s. By 1977, German direct investment in the American chemical industry equaled American investment in the German chemical industry.[65] In 1979, Bayer, BASF, and Hoechst had climbed to ranks 10, 13, and 17, respectively, in the United States. The American subsidiaries were the "favorite children of German big chemical industry."[66]

Looking back on three decades of direct investment in the United States, in 1984 the executive chairman of American Hoechst, Dieter zur Loye, saw three distinct phases. In phase one, until the mid-1970s, it was important just to be present in the American markets for pharmaceuticals and chemicals. In phase two, from the mid-1970s until the mid-1980s, the aim was to make operations reasonably profitable. In phase three, the decade from the mid-1980s to the mid-1990s, the goal was to achieve profits like American chemical companies.[67] Only in phase three was Americanization achieved. The German chemical industry's operating margins and return on assets did not yet approach American standards in the United States or in Europe.[68] The German chemical industry had mastered all new production processes coming from America but still had to learn to make the same profits.

Expansion abroad through direct investment was not limited to the big three, nor was their strategy unique. Degussa, to name one example, also heavily invested in production facilities for its staple products, hydrogen peroxide, sodium perborate, and aerosil in Belgium (1968) and in Texas (1973). The United States eventually became the most important field of investment for Degussa. The company's strategy, as stipulated by its executive director, was "not further diversification but

[64] Harald Bathelt, "Global Competition, International Trade, and Regional Concentration: The Case of the German Chemical Industry during the 1980s," in Environment and Planning C: *Government and Policy 13* (1995), p. 411–12.
[65] Wirtschaftswoche No. 52 (Dec 16) 1977, p. 29.
[66] Wirtschaftswoche No. 49 (Dec 5) 1980, p. 70.
[67] Wirtschaftswoche No. 20 (May 10) 1985, p. 182.
[68] Teltschik, *Geschichte*, p. 252.

geographical expansion in its most competitive products."[69] In a similar vein, Konrad Henkel of Henkel, the detergent manufacturer, in the early 1970s decided: "Europe is not big enough."[70] The company, Henkel management believed, had to grow to stay in the market. It was a striking deflection from the earlier generation of Jost Henkel, who, in 1958, still had feared that "the shop is getting too big."[71]

Perceived comparative advantages on the level of the company as incorporated in know-how and product-specific experience were the driving motifs to go abroad in a world where foreign investment had lost the smack of adventure. At the same time, this was the strategy adopted to eventually catch up with American levels of profitability. The way toward this goal, however, went through the detour of intensified vertical integration to reach out for enticing markets downstream. Before the German chemical industry concentrated on doing what it did best, it did what it could to grow.

INTEGRATING FORWARD: THE END OF THE PARADIGM OF NONINTERFERENCE WITH CUSTOMERS' MARKETS

Foreign direct investment provoked another major change in the structure and strategies of the industry. It had been a long-standing policy among the major companies to avoid interfering with customers' markets. One of the great strengths of the major German chemical producers was their well-developed applications technologies ("Anwendungstechnik"), by which they researched the potential applications of their products to help industrial customers, often SME in Germany and abroad, to process intermediate products into the finished product. This arrangement had its roots in synthetic dyestuffs that could not simply be shipped to dye works but had to be accompanied by detailed technical information not available to traditional dyers. Application technologies were part of the package and often made using German products more profitable, even if they came at a higher price. Application technologies turned the research potential of the major companies to the advantage of smaller companies close to the market. It was a shrewd division of labor that, from the perspective of the big companies, protected and fostered their outlets, which in turn would pay for the extra service through higher

---

[69] Wolf, "Unternehmensstrategien . . . ," p. 1042.
[70] Susanne Hilger, "Der Zwang zur Größe . . . ," p. 227.
[71] Susanne Hilger, "Unternehmen im Wettbrewerb . . . ," p. 298.

prices. In the political arena, being flanked by a range of small companies, whose interests were always well represented, although not very powerful, in the national association of the chemical industry (VCI) provided stability and calculability for the "big three."[72] This arrangement would only work, however, as long as companies processing intermediates from one of the big German chemical companies wouldn't have to fear competition from their suppliers. In a kind of collusive understanding, the big players in German had refrained from entering end markets.[73]

With market opportunities developing downstream, in places and to a dimension the traditional clients, many of them SME, could not cover, the IGF-successors began to unfold a full-range portfolio. From relying on applications technologies to expanding sales via SME-processors of their basic and intermediate products, the big three began to integrate forward. It was not the end of application technologies playing a major role in developing markets, but their role was limited to products that would not constitute mass markets likely to offer vast economies of scale.[74] Nor was it the end of the "associative order" in the VCI, but from the 1980s, the big companies relied less on the industry's association and began to develop independent political strategies and capabilities.[75] Integrating forward had given rise to forms of individualization among major German chemical companies in ways unknown in the past.

NEW MANAGEMENT STRUCTURES

Through this phase of rapid expansion the big three shed their established hierarchical management structure and, in what looked like a concerted action, adopted the M-form in 1970/71.[76] Not only newly created divisions at home, but also regional profit centers especially in North America, won more independence in their operations. Structure had to follow strategy. And successful this strategy was. The big three, Hoechst, BASF, and Bayer, once again dominated export markets and climbed to rank 1, 2, and 3 among the world's chemical producers

---

[72] Wyn Grant, Alberto Martinelli, and William Paterson, "Large Firms as Political Actors: A Comparative Analysis of the Chemical Industry in Britain, Italy and West Germany," in *West European Studies 12/2* (1989), pp. 75-6.

[73] In Degussa: Wolf, "Unternehmensstrategien . . . ," p. 1041. In BASF Abelshauser, *Die BASF,* p. 429.

[74] Wolf, "Unternehmensstrategien . . . ," p. 1041.

[75] Grant/Martinelli/Paterson, "Large Firms . . . ," pp. 77-78.

[76] Teltschik, *Geschichte,* p. 250. Abelshauser, *Die BASF,* pp. 571-75.

by the early 1970s.[77] Thereafter, the major companies saw a reform of their management structure about once every decade.[78] Change was institutionalized.

WEATHER THROUGH THE OIL PRICE CRISIS

The oil price shock in the 1970s led to a short decline in growth rates but did not break the overall pattern of continued expansion.[79] Disastrous collapse of business was restricted to a few branches of the chemical industry. The "fiber crisis," the decline of the once thriving synthetic fiber industry, received the most attention, not least because it gave rise to intervention from the Brussels bureaucracy to stabilize the European synthetic fiber industry. The fiber cartel of 1978, however, failed to break the downward trend, and, by 1982, the German trade press wryly asserted that at all events only members of executive boards still had safe jobs.[80] Nevertheless, Günter Metz, president of Association of the German Chemical Fiber Industry (IVC) and chief executive of synthetic fiber production at Hoechst, maintained that the profits his industry had made before 1975 still superseded the losses thereafter.[81] Another big loser of the 1970s was fertilizers, which, in 1978 didn't sell more than in 1969.[82]

In the meantime, the plastics sector that had fared relatively well during the first oil price shock of 1975 also got into deep trouble.[83] The second rise in oil prices in the early 1980s left the industry in disarray. BASF, the major producer, had to scale down the capacity of its steamcrackers for ethylene at ROW by 40 percent and capacity of its LDPE plant by almost 50 percent. Hoechst did not fare much better in its HDPE production.[84] At the height of the petrochemical crisis in the summer of 1982, Matthias Seefelder, executive chairman of BASF, complained: "The calamity resides in overcapacities in fibers, standard plastics, refineries,

---

[77] Teltschik, *Geschichte*, p. 251.

[78] Abelshauser, *Die BASF,* p. 581–83.

[79] Georg Müller-Fürstenberger, *Kuppelproduktion: Eine theoretische und empirische Analyse am Beispiel der chemischen Industrie,* Heidelberg: Physica, 1995, pp. 51ff. (Kop. Bathelt, p. 112).

[80] Wirtschaftswoche, No. 3 (Jan 19) 1982, p. 85.

[81] Ibid.

[82] Fonds der chemischen Industrie (ed.), *Die chemische Industrie - der forschungsintensivste Industriezweig in der Bundesrepublik Deutschland* (= *Schriftenreihe, Heft 15*), Frankfurt am Main: Fonds der chemischen Industrie, 1979, p. 9.

[83] Wolf Rüdiger Streck, *Chemische Industrie. Strukturwandlungen und Entwicklungsperspektiven,* Berlin: Duncker & Humblot, 1984, pp. 195-96.

[84] Teltschik, *Geschichte,* p. 263.

and crackers. In plastics we have conditions like in the coal and steel industry."[85] To find outlets for underutilized plant, the oil companies intensified their production of basic petrochemicals and thereby further aggravated the situation. Massive layoffs and concentration were the inevitable outcome. H. Willersinn, president of the Association of Plastic Manufacturers (Verband kunststofferzeugende Industrie), in April 1983 claimed that between 1980 and 1983 the number of plastic producers in Western Europe had declined from eighty-five to fifteen.[86] Bayer was among those who had given up early on. In 1979, the company sold its share in Hüls, a producer of standard plastics, to the state-owned VEBA, while BASF stuck to its heavy investment in feedstocks and tried to balance what was felt to be just temporary losses in this sector by integrating forward into highly refined products.[87] At the end of the oil price crisis, BASF and Bayer were even more different than before.

The rising costs of feedstocks during the 1970s further contributed to the internationalization of the German chemical industry in surprising ways. Rather than concentrating on traditional, location-specific strengths like a highly diversified domestic market and a supreme workforce on the shop floor, as much as in research, the industry turned to raw material and energy intensive processes. Economies of scale rather than diversity and novelty of product was the leading strategy. The rationale of this reorientation lay in the revaluation of the deutsche mark vis-à-vis the U.S. dollar. Price increases for oil-based hydrocarbons in relation to other factor costs were less dramatic in Germany than in the United States. Although the price of naphtha in U.S. dollars had increased twelve times from 1970 to 1980, in deutsche mark the increase was only sixfold.[88] Because the cost of labor had continued to increase through the 1970s, cost advantages swung from labor intensive, including research, to raw material and energy intensive. Whereas the terms of trade between West Germany and the United States for simple plastics like PVC and PE had dramatically swung against Germany in the five years from 1968 to 1973, in the mid-1970s, much of this effect was wiped out again.[89]

In this situation, the German chemical industry whose self-image had always been to compensate for high costs of energy and raw materials by

[85] Quoted from Teltschik, *Geschichte*, p. 263. "Die Misere liegt in den Überkapazitäten bei Fasern, Standardkunststoffen, Raffinerien und Crackern. Bei den Kuststoffen herrschen Verhältnisse wie bei der Montanindustrie."
[86] Handelsblatt 25.4.1983, quoted from Streck, *Chemische Industrie*, p. 197.
[87] Teltschik, *Geschichte*, p. 264.
[88] Streck, *Strukturwandlungen*, p. 154.
[89] Streb, *Technologiepolitik*, p. 166.

continuously developing a highly innovative range of research and skill intensive products, turned to cheap mass production and economies of scale. Plant size grew and concentration was accelerated. The export position in pharmaceuticals and pesticides, both research-intensive products, declined, while the export position of hydrocarbons, alcohols, synthetic rubber, and synthetic fibers improved.[90] Total output shows a similar picture. The share of organic bulk chemicals grew through the 1970s, while the share of the still most important group, special products for processing (plastics, synthetic rubber, synthetic fiber), declined. Another group to expand its share was special products for consumption, including pharmaceuticals.[91] The sale of pharmaceuticals, however, only fared well on the home market; its export shares were well below the average of the industry.[92]

The years of high oil prices heralded the new strategies dominating from the mid-1980s until the present, when low-cost production benefiting from available economies of scale at optimal sites took precedence over full-range presence in most markets. This was not the end of specialties production, to be sure, but from then on, diversification and expansion to get a foothold into attractive foreign markets had served their purpose and gave way to more intensive profit seeking. Step by step, scale began to prevail over scope. But it took another major and even more painful crisis for the chemical industry to finally abandon the technological model in R&D strategies and to fully bring about the transition to what was perceived to be an American strategy.

PATH DEPENDENCE IN CHEMICAL SYNTHESIS

In view of the stunning success of reconstruction, internationalization, and product differentiation, it is striking that the German chemical industry failed to recognize and grasp the great potential of biotechnology. This is ever more surprising since Germany had been the most successful country in fermentation technology at the beginning of the twentieth century.[93] Early leadership was allowed to slip and the lessons taught by penicillin at the end of World War II seemed forgotten in the 1960s

---

[90] Harald Legler, *Internationale Wettbewerbsfähigkeit der westdeutschen chemischen Industrie* (=Beiträge zur angewandten Wirtschaftsforschung, vol. 10), Berlin: Duncker & Humblot, 1982, pp. 131-164.

[91] Streck, *Strukturwandlungen*, p. 171.

[92] Ibid., pp. 222-23.

[93] Luitgard Marschall, *Im Schatten der chemischen Synthese: Industrielle Biotechnologie in Deutschland (1900-1970)*, Frankfurt am Main: Campus, 2000, pp. 25-86.

and 1970s. By that time, biotechnology was clearly underdeveloped in Germany.[94]

In fact, some sort of active unlearning of biotechnology had occurred in the meantime. Such was the confidence in the industry's competence in chemical synthesis that even companies successful in biotechnological production made every effort to convert their production line to organic chemical synthesis and its "German" core of catalytic high-pressure technology. It was only after World War II that the German chemical industry became, or rather made itself, fully path-dependent on chemical synthesis at the expense of earlier excellence in biotechnology.[95] Catalytic organic chemical synthesis had consolidated into a technological trajectory through the interwar years. Biotechnology was restricted to the fewer and fewer fields where this trajectory failed. And, more important, biotechnology in the eyes of the dominating elite among German chemists was stigmatized as low-tech and doomed to decline.[96] Early criticism of the industry's leaders notwithstanding, the old spirit of "Reppe-Chemie" continued to haunt R&D strategies as it had merged so easily into petrochemistry. Confronted with this general mind-set, companies like Merck, Röhm & Haas, and Boehringer Ingelheim, which had been quite successful in their biotechnological processes, wherever possible converted product lines and R&D strategies to organic chemical synthesis well into the 1980s. Ironically, it happened at the same time when Hoechst, Bayer, and BASF were about to enter biotechnology belatedly and at high costs.[97]

Earlier than industry, private research foundations and the federal government were aware of the great potential of biotechnology. Volkswagenstiftung (Volkswagen Foundation), in 1965 helped to create a research center for molecular biology, and already in the early 1970s, the Federal Ministry for Research started programs to help develop expertise in biotechnology in Germany.[98] Still, in 1984 a German pioneer of biotechnology stated: "in recent years, all impulses on the sector of

[94] Klaus Buchholz, "Die gezielte Förderung und Entwicklung der Biotechnologie," in: Wolfgang van den Daele, Wolfgang Krohn, and Peter Weingart (eds.), *Geplante Forschung*, Frankfurt: Suhrkamp, 1979, p. 71.

[95] For a full account, see the pathbreaking study by Luitgard Marschall, *Im Schatten der chemischen Synthese: Industrielle Biotechnologie in Deutschland (1900-1970)*, Frankfurt am Main: Campus, 2000.

[96] Buchholz, *Die gezielte Förderung*, p. 69.

[97] See the case studies in Marschall, *Im Schatten*, pp. 203-349, esp. p. 349.

[98] Jörg Munzel, *Ingenieure des Lebendigen und des Abstrakten, Die Entwicklung der Biotechnologie und der Informatik an der Technischen Universität Carolo-Wilhelmina zu Braunschweig*, Hildesheim: Georg Olms, 1998, pp. 60-79.

biotechnology emanated from the Federal Ministry of Research." In the same article, titled "the sluggishness of the leaders," the author complained that: "the history of modern biotechnology was . . . no glorious chapter for [Hoechst, Bayer, and BASF], rather a fiasco of missed opportunities because academia and industry got along with each other too well by professional inbreeding."[99] The federal ministry for research continued its efforts to bring modern biotechnology, especially molecular biology and genetic engineering to Germany in setting up a 1.5 billion deutsche mark program in 1990.[100]

It is still difficult to decide whether this was just shortsightedness on the side of decision makers in the industry, or whether it was an expression of a well-founded and century-old conviction that R&D would eventually find a way to synthesize substances more cheaply. Opting for oil in the 1950s and 1960s had come easier than opting for biotechnology in the 1970s and 1980s. It took for the merger of biotechnology with genetic engineering to overcome the reservations of science-oriented R&D in the major companies. By then, in a déjà vu from the early years of penicillin, German shortfalls in molecular biology had created a situation where know-how had to be brought in from America.[101] "Learning from the USA," the leitmotif of German reconstruction, was back again. In 2001, the Ministry of Research was still complaining about persistent deficiencies because of the late start into "new biotechnology."[102] But, although the concerns of the federal government are limited to the national territory, industry has more geographical options and can choose where to grow and where to buy expertise. In an unparalleled shopping tour, led by Hoechst and Bayer, German companies bought the knowledge base they had failed to generate in the first place.

Its late start in biotechnology notwithstanding, the pharmaceutical branch of the chemical industry was a success story in the 1980s. In 1988, the German pharmaceutical industry was leading world exports with a share of 15.1 percent,[103] certainly not a sign of crisis. Between

---

[99] Innovationen in Deutschland III, "Die Trägheit der Großen," in Wirtschaftswoche, No. 18 (April 27) 1984, pp. 77, 82.

[100] Teltschik, *Geschichte*, p. 322.

[101] For Hoechst, see Barbara J. Culliton, "The Hoechst Department at Mass General," in *Science 216* (1982), pp. 1200–1203.

[102] Bundesministerium für Bildung und Forschung, *Zur technologischen Leistungsfähigkeit Deutschlands, Zusammenfassender Endbericht 2000, Gutachten im Auftrag des Bundesministeriums für Bildung und Forschung*, Bonn: BMBF, 2001, pp. 77–87.

[103] Robert Ballance, János Pogány, and Helmut Forstner, *The World's Pharmaceutical Industries*, Edward Elgar, Aldershot, 1992, quoted after Achilladelis and Antonakis (2001), p. 569.

1982 and 1992, its share of all chemical production grew from 15 percent
to 20 percent, equal to organic and inorganic basic chemicals combined,
whose share had dropped from 29 percent to 20 percent during the same
years.[104] Of the big three, Bayer and Hoechst were particularly successful
in pharmaceuticals and managed to expand their already strong position
with shares of 18 percent of their total production. BASF, being a late
entrant via the acquisition of Knoll, never got to a breakthrough.[105] Not
surprisingly, BASF eventually gave in and in 2000 sold out its pharmaceutical business to Abbot Laboratories, USA.[106]

<h2 style="text-align:center">NEW WAVE OF CONCENTRATION IN 1990S</h2>

The struggle to catch up with biotechnological development abroad coin-
cided with the severest recession for the chemical industry since the end
of World War II. From 1990 to 1993, the chemical industry, top ranking
among German sectors during most of its history, went into a deep cyclical downturn culminating in negative growth of 3.6 percent in 1993.
Recovery in 1994 was short-lived, as 1995 again saw zero-growth.[107] But,
by this time the German market no longer held the largest share. At the
end of the 1980s, BASF, Bayer, and Hoechst each sold more in the United
States than in Germany. Important acquisitions like Celanese by Hoechst
(1987) and Miles by Bayer (1978) had very much helped this accelerated penetration of the American market. The combined sales of the big
three in the United States stood at $17 billion in 1989, more than DuPont's
chemicals sales of in the same year.[108] Following the so-called triad rule
(the assumption that only those companies that are firmly embedded in
all three major markets, Europe, the United States, and Japan, will survive), all three had also heavily invested in Japan during the 1980s, where
they effectuated combined sales of DM 6.5 billion in the same year.[109]
Heavy foreign direct investment notwithstanding, the German chemical
industry continued to dominate world exports and was still leading in
international patent statistics.[110]

[104] Harald Bathelt, *Chemiestandort Deutschland: Technologischer Wandel, Arbeitsteilung
und geographische Strukturen in der chemischen Industrie*, Berlin: edition sigma, 1997,
pp. 100–101.
[105] Abelshauser, *Die BASF*, p. 620. Teltschik, Geschichte, p. 318.
[106] Abelshauser, *Die BASF*, p. 622.
[107] Bathelt, *Chemiestandort*, p. 112. VCI, The German Chemical Industry (July 2002), p. 2.
[108] Teltschik, *Geschichte*, p. 331.
[109] Ibid., p. 333.
[110] Ibid., p. 334.

FROM DIVERSIFICATION TO CONCENTRATION
OF PORTFOLIOS

The 1990s definitely witnessed the reorientation of strategies among the major companies toward high levels of profitability rather than just scale and scope. The highly diversified structures of the 1960s and 1970s were resolutely dismantled and replaced by best-practice profit centers in least-cost locations. Kurt Hansen, honorary chairman of the supervisory board (Aufsichtsrat) of Bayer, in 1999 stated: "In order to adapt to a globalizing world, many companies concentrate on their core business. Moreover, one tries to expand one's market share in these core fields either by acquisition of firms or by mergers."[111] For Hoechst and Bayer, this stood for a concentration on pharmaceuticals and specialties, whereas BASF continued the massive integration of processes at its site in Ludwigshafen.

Although the big three German chemical companies were impressive in size, their profitability was much less so. "Phase three" of Dieter zur Loye, American Hoechst chairman in the 1980s, had not yet been achieved in the United States or at home. It took another American Hoechst chairman, Jürgen Dormann, to involve the whole company in trying to make it competitively profitable. Dorman had been very successful in integrating Celanese, at $3 billion the greatest FDI-endeavor of a German company in the United States when it was bought in 1987.[112] After rumors circulated that Hoechst might be taken over by its competitors because of its low profitability, Dormann set out to dramatically change its company culture and cut it into pieces that would prosper. According to Dorman, who in 1994 was the first nonchemist to become executive chairman of Hoechst, "Hoechst was deeply rooted in Germany, very research intensive but little market oriented, very introvert and very academic. This we have to change."[113] Following the example of Novartis (merger of Ciby Geigy and Sandoz), in December 1998, the executive chairmen of Hoechst and of Rhône-Poulenc announced the imminent merger of their companies to form "Aventis," a "life-sciences" company, based in Strasbourg. Because it had to be a merger of equals, Hoechst had to be downsized. By selling a great number of plants and spinning off

---

[111] Hansen, "Die chemische Industrie . . . ," p. 1039. ("Um sich den Anforderungen einer globalisierenden Welt anzupassen, konzentrieren sich viele Unternehmen auf die Kerngebiete ihrer Aktivitäten. Ferner versucht man, sich durch Zukauf von Firmen oder Fusionen auf diesen Kerngebieten zu vergrößeren und den Marktanteil zu steigern.")

[112] Heribert Klein, *Operation Amerika: Hoechst in den USA*, München: Piper, 1996.

[113] Quoted after A. Berthoin Antal, *The Transformation of Hoechst to Aventis: Case Study*, *Wissenschaftszentrum Berlin*, Berlin: WZB, 2001, p. A1.

Celanese as an independent company, Hoechst between 1994 and 1999, when the merger was completed, reduced its worldwide workforce from 172,000 to 97,000. Although criticized by some analysts as "the merger of a blind to a lame,"[114] Aventis, unlike many mergers of the 1990s, was successful in increasing its stock value in the years to follow, ranking among the top twenty European companies in 2002. Dormann, German "Manager of the Year" in 1995, left Aventis in 2002 when the new company was among the most profitable in pharmaceuticals markets, with expected annual growth of 11-12 percent.[115]

In late 2003, Bayer embarked on a very similar strategy, concentrating most of its chemical production in a new company, NewCo, which was to be burdened with debts and spun off, while Bayer concentrated on pharmaceuticals and more profitable specialties. According to Executive Chairman Werner Wenning, Bayer anticipated becoming "a medium-sized European pharmaceuticals enterprise."[116] Only five years earlier, Bayer, then the world's greatest producer of basic polyurethanes, was still looking forward to a great future in plastics.[117] The stock market approved of this move to concentrate business and sent Bayer stocks up 7.5 percent within hours after the announcement. What still looked irresponsible, iconoclastic, and almost suicidal in the case of Hoechst five years earlier, has become a pattern. The pursuit of profitability eventually won out over the pursuit of size.

Today only BASF is left intact in its original size. BASF proceeded on a completely different path after World War II by developing what is called "Verbund," the interconnection of all processes to bring down unit costs. The core business of BASF is process management utilizing synergies among a great variety of product lines. BASF plant worldwide is connected vertically and/or horizontally by at least one product to other company plants. The culmination of this principle is the home plant in Ludwigshafen, the most integrated chemical plant in the world. This pattern is being exported to other BASF locations worldwide. Appropriately,

[114] A. Ruess, "Eine Ebene zuviel: Der Umbau der Organisation sorgt für Unruhe: Deutsche und Franzosen kämpfen um die besten Posten - gegeneinander," in Wirtschaftswoche No. 7 (Feb 7) 2002, pp. 52-56.
[115] Christian Keun and Jürgen Dormann, "Der Umbaumeister," in Manager-Magazin (March 6, 2002) http://www.manager-magazin.de/koepfe/mdj/0,2828,179004-3,00.html (site accessed Nov. 7, 2003).
[116] Financial Times Deutschland (Nov. 7, 2003), http://www.ftd.de/ub/in/1067671010807, html (site accessed Nov. 7, 2003).
[117] Kristin Mädefessel-Herrmann, "Meilensteine der Kunststoffchemie," in Nachrichten aus Chemie, Technik und Laboratorium 47 (1999), p. 1068.

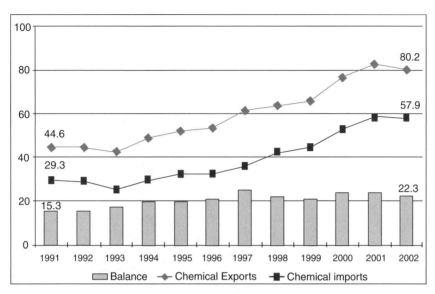

Figure 5.2. Development of German foreign trade in chemical products (bn €).
Source: VCI, The German Chemical Industry, August 2003, p. 7.

they have been labeled "decal pictures of Ludwigshafen" in management journals.[118] Only late in the 1990s, BASF moved out of the pharmaceuticals sector, concentrating on chemicals while Hoechst and Bayer did everything to focus on pharmaceutical production and get out of chemicals. Looking back to more than a century of large-scale chemical production in Germany, BASF is still closest to its roots – with great success at the time of writing. Jürgen Strube, chief executive of BASF, was elected German "Manager of the Year" in 2002 for successful management strategies "against all fashionable trends."[119]

During these years of reorganization and concentration, a number of domestic markets were abandoned to better placed competitors abroad, whereas company-specific strengths were exploited fully at home and – even more so – abroad. The German market itself was fully integrated into the international division of labor (see Figure 5.2). The effect of this policy becomes visible when Germany's share in global chemical sales is compared with the share of German chemical companies in global chemicals

---

[118] Heide Neukirchen, Mühsamer Prozess. BASF: Ludwigshafen wird umgebaut, in *Manager Magazin,* vol. 32, No. 10 (October 1), 2002, pp. 46–49.

[119] Heide Neukirchen and Thomas Werres, Jürgen Strube: Chef des Chemiekonzerns BASF: Mann des Jahres, in *Manager Magazin* (1.12.2002).

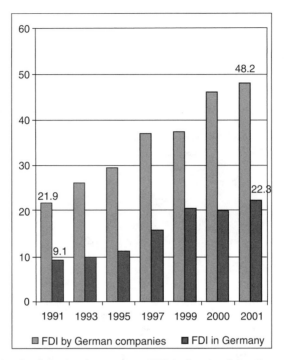

Figure 5.3. Stock of foreign investment (FDI in bn €) of the German chemical industry. *Source:* VCI, The German Chemical Industry, August 2003, p. 8.

sales, including sales of subsidiaries of German companies abroad. While Germany's share in global sales has declined from 10.5 percent in 1991 to just 7.3 percent in 2001, the share of German chemical companies was almost stable with 12.5 percent in 1991 and 11.5 percent in 2001. The German chemical industry keeps moving abroad. Its stock of foreign direct investment more than doubled during the same decade from €21.9 billion in 1991 to €48.2 billion in 2001[120] (see Figure 5.3).

The German chemical industry, like many other European industries today, is more global than national.[121] The striking case of Hoechst, splitting itself up in a French company, "Aventis," and an "American" company, Celanese, with its seat in Frankfurt, both of which have their biggest market in the United States,[122] suggests the final abandonment of the

---

[120] VCI, The German Chemical Industry, August 2003, pp. 7–8.

[121] Achilladelis and Antonakis (2001), p. 572, come to the conclusion that the pharmaceutical industry that has been the most dynamic branch of the German chemical industry.

[122] See respective annual reports at the companies' Web sites.

concept of a national industry. Managers today and over the last two decades were picking the most appropriate "national container" for individual branches of production. National resources, national regulations, and national infrastructure play an important part in these decisions, which, however, no longer relate to anything like a "national character" of industry. At the same time, the very diverse paths taken by the major German chemical companies remind us that there is no "one best way" to be successful in an industry. As BASF shows, even continuity can be a very good strategic choice in changing environments.

6

The American Chemical Industry
Since the Petrochemical Revolution

JOHN KENLY SMITH, JR.

... I think [petroleum intermediates] poses one of the great problems before our company today. Should we get into it at all; if so, how deeply? What are our advantages or disadvantages opposite the many others who are already interested in that field among both chemical manufacturers and petroleum manufacturers.

Walter S. Carpenter, Jr., DuPont President, 1946[1]

It is now timely to consider the problem of "what went wrong" in the petrochemical industry at some point along its development path.

Peter Spitz, 1988[2]

Some visionaries predict that most industries will be absorbed into the chemical industry.

Fortune, 1950[3]

Or *vice versa*?

---

[1] W. S. Carpenter to C. H. Greenewalt, August 16, 1946, Records of the E. I. du Pont de Nemours and Co., Series II, Part 2, Box 829, Hagley Museum and Library, Wilmington NC.

[2] Peter H. Spitz, *Petrochemicals: The Rise of an Industry* (New York: Wiley, 1988), p. 537.

[3] "The Chemical Century," *Fortune* (March 1950), p. 70.

INTRODUCTION

World War II launched the American chemical industry into several decades of growth at twice the rate of GNP growth,[4] and with profits that averaged 25 percent higher than those in manufacturing generally.[5] That growth was largely based around polymers, pesticides, and pharmaceuticals made mostly from petrochemicals. Government-led initiatives in all these areas during the war set the stage for their rapid proliferation afterward. After the war, R&D laboratories added numerous new products and processes to the industry's overall portfolio. Eventually, competition in research became as intense as competition for markets. Unlike in the prewar era, it was increasingly difficult for any one firm to dominate a particular field because of strict government antitrust policy, widespread technology licensing, easily avoidable patents, and extensive product substitutability. By 1960, a wide variety of polymeric materials had invaded the material realm across a broad spectrum of applications ranging from paint, to films of all kinds, toys and appliances, and textile fibers. During that decade, new plastics were beginning to compete with each other, rather than with older materials. Not surprisingly, the introduction of new polymer products declined substantially. Industry growth rates were sustained by an increase in petrochemical process innovations that lowered the cost of polymers and end products. The chemical industry was also buoyed by the spectacular growth of the pesticide – especially herbicide – market in the 1960s. Pharmaceuticals also were a rapidly growing and highly profitable business, but generally in the United States, drugs had become an industry that was distinct from chemicals.[6] As most chemical markets became saturated in the 1970s, growth rates dropped toward that of GNP, profitability declined but generally was still above manufacturing, and product and process innovation declined dramatically. During the 1980s, sluggish growth and declining profitability led chemical company executives to attempt to shift product mixes away from commodities and toward specialties.[7] This effort

---

[4] Ashish Arora and Nathan Rosenberg, "Chemicals: A U.S. Success Story," in Ashish Arora, Ralph Landau, and Nathan Rosenberg, eds., *Chemicals and Long-Term Economic Growth: Insights from the Chemical Industry* (New York: John Wiley & Sons, 1998), p. 90.

[5] U.S. Department of Commerce, *1988 Industrial Outlook* (Washington, DC, 1988), p. A-3-6.

[6] John Kenly Smith, Jr., "The End of the Chemical Century? Organizational Capabilities and Industry Evolution," *Business and Economic History* 23, no. 1 (Fall 1994): 152–61.

[7] Kikor Bozdogan, "The Transformation of the US Chemicals Industry," *MIT Commission on Industrial Productivity: Working Paper*, Vol. 1 (Cambridge, MA, 1990).

failed to alter the overall trajectory of the industry but did initiate the mass movement of assets between established companies and to new ones. This trend has continued as the industry generally has moved away from diversification and toward concentration on fewer key products and markets. Current strategy appears aimed at obtaining significant global market share. The American chemical industry finds itself today in peculiar circumstances. It is a large and important sector of the economy; it generates a significant trade surplus; it is technologically sophisticated; it supports a creative R&D establishment; and because its returns are not much above the cost of capital, it has fallen out of favor with investors. How the industry fell into this unusual and perhaps unprecedented set of circumstances is the result of historical as well as market forces.

Historical Structure of the Chemical Industry

The chemical industry was so successful for so long because it developed the ability to apply chemistry and chemical engineering to the manufacture of new compounds and materials. This was not a case of simply exploiting science for technological purposes. In fact most important technological innovations are initially not very well understood scientifically. This is almost a necessary condition for a radical innovation. If it were a straightforward application of existing science, then the innovation would be obvious to many investigators.

The relationship between the structure and properties of chemical compounds has not fully become a science, even today. Chemists became very clever at making new molecules, but the process of determining efficacious uses of those molecules still required an unusual degree of insight. The history of chemical research is rich with tales of serendipity because of this fact. The stories are somewhat misleading because they emphasize the creation of a new chemical compound instead of the *recognition that the new compound has a combination of unusual properties that might find commercial applications*. Knowing what those potential applications are requires familiarity with far-flung networks of chemical users across a broad spectrum of industries. Historically, this has been an important function of firms, gathering information from the broader network and using it to guide research and development.

After a laboratory breakthrough had been made and a potential use identified, developing a process and scaling it up required other skills. Controlling chemical reactions was one major challenge. Of course, entrepreneurs do not have to know what they are doing to be success-ful, but they do have to be able to produce repeatable results. Even if a chemist could make a reaction go the way he wanted in the labo-ratory, the scaling up to a large process involved numerous technical problems that had to be solved empirically. These technical obstacles created formidable barriers to entry for many materials processors who might have considered integrating backward into chemical production. A few companies, such as DuPont, were forced to make this transition when new chemical-based products undermined their existing business. The company's original product, black powder, which was made by mix-ing three ingredients, was replaced by dynamite and smokeless powder, which were made by chemical reactions.

The chemical industry became an entirely new industry that found a niche for itself between suppliers and processors of natural or raw materials and downstream producers of consumer goods.[8] By the late nineteenth century, there emerged a clearly articulated vision for the "chemicalization" of industry generally. Arthur D. Little, an expert on the new chemical process for making paper, was one of the major visionaries. Chemicalization consisted of three principles. First, that the application of basic chemical principles, such as the law of conservation of matter, and standard chemical techniques, such as analysis, would lead to sig-nificant improvements in virtually all processing industries. Second, that the natural chemicals that were used in processing industries could be made synthetically or replaced entirely with new synthetic ones. Third, that actual *materials* could also be replaced by new or synthetic ones.[9]

The strategy of chemicalization led to fraternal relationships between chemical companies. This was true for both markets and production. Opportunities for growth appeared to be so immense that there was little reason to compete for particular bits of turf. There was a whole world of industry to be chemicalized. Another reason for cooperation was a tech-nical one. Individual companies could not economically manufacture the large number of compounds used to make its products. Therefore, it did not make sense for a company to invest in the same chemicals that others made, especially in an era when chemicals were made from a wide vari-ety of raw materials and by many different types of processes. Instead,

[8] John Kenly Smith, Jr., "The Evolution of the Chemical Industry: A Technological Perspec-tive," in Seymour H. Mauskopf, ed., *Chemical Sciences in the Modern World* (Philadelphia: University of Pennsylvania Press, 1993), pp. 137-57.
[9] Smith, "The End of the Chemical Century," pp. 154-55.

each company tended to specialize around certain technologies, which represented the accumulated chemical and engineering knowledge that existed in each firm. Thus, chemical companies, who depended on each other for certain products, were each other's best customers. This interdependency promoted cooperative rather than competitive behavior.[10]

Because the prewar chemical industry consisted of a diverse range of products made using a wide variety of techniques, most companies organized themselves into either product- or process-based divisions. DuPont had pioneered in this form of organization in 1921 because diversification had made the older functional approach with one production and one sales department obsolete. The new form, with semiautonomous operating divisions, also made acquisitions easier, because the acquired company could be attached as a new division. Most of the major chemical companies grew this way during the 1920s and 1930s. At the same time that the industry diversified its products and processes, researchers were beginning to explore the fundamentals principles underlying the industry.

In the years after World War I, the chemical network in the United States began to expand beyond the industry itself to include academia. These changes would have profound influences in both areas. Progressive elements within industry and academia were aware that chemical technology had run far ahead of science and that the gap was getting wider. It was not that academic chemists were not working on important problems in chemistry; it was just that the problems they selected were not relevant to the work of industry. Some companies began to do more academic-style research on subjects of industrial significance. There also was a growing realization in academe that the resources of the chemical industry were enormous and could be tapped to support industrially relevant research.

When an industrial technology becomes a big business, academic researchers are attracted to it because the new technology usually poses interesting scientific questions and offers the potential for financial support. Rather than solve particular problems, university researchers preferred investigating more generalized phenomena (science) underlying their technologies. What corporations did not fully realize is that scientific principles help organize disparate information, thus making it more easily transmitted and learned by others. It reduces the amount of knowledge that potential competitors have to assimilate before entering

[10] Ibid.

a particular business. Ironically, the progressive industrial patronage of academic research would eventually help to undermine the unique capabilities of the leading edge firms.

One important rapprochement between the chemical industry and academia occurred in the chemical engineering department of MIT. Led by Warren K. Lewis, MIT chemical engineers created a research agenda to establish a more generalized, systematic (scientific) underpinning to their art. The MIT creed soon spread to other colleges and universities through Lewis's pioneering textbook, *The Principles of Chemical Engineering*. The long-term importance of this agenda was to make the generic component of chemical engineering grow at the expense of specific practice. In other words, when designing a chemical plant, chemical engineers now had a set of generalized principles that applied to all chemical processing. Thus, the amount of specific knowledge needed to build a particular plant decreased. As the modern academic profession of chemical engineering grew and prospered, the university emerged as a center of generalized knowledge and matriculated chemical engineers spread the gospel to a host of processing industries. Chemical engineering capabilities were becoming relatively more common and inexpensive to acquire.

In chemical research, a few large companies, DuPont and I. G. Farben especially, also began to initiate programs to improve the scientific understanding of their own products and processes. The most dramatic example was polymers or long-chain molecules. One of the most important materials used in the chemical industry was cellulose derived from wood. After processing, it was sold as rayon fibers, cellophane film, celluloid plastics, movie film, and fast-drying lacquers. DuPont during the 1920s had become primarily a cellulose processing company selling all the above products. In spite of its importance, cellulose remained a mysterious molecule. This was also the case for the increasingly numerous synthetic resins, such as commercially successful Bakelite.

To remedy this situation in polymers and other important areas of chemical technology, DuPont central laboratory research director, Charles M. A. Stine, in 1926, proposed that the company hire prominent academic chemists and put them to work on industrially relevant topics. When it proved impossible to attract established academics into industry, Stine settled for younger PhDs. In two areas, polymers and chemical engineering, Stine succeeded in hiring outstanding young researchers. It was in the just emerging discipline of polymer chemistry that DuPont struck gold – or more precisely nylon. Stine hired

Wallace H. Carothers, an assistant professor of organic chemistry at Harvard, to work on polymers. In his research at DuPont, Carothers elegantly and convincingly demonstrated that polymers were not mysterious entities but just longer versions of ordinary organic molecules. To prove this assertion, he developed techniques for making polymers out of ordinary organic molecules. While working out this general scientific research program, Carothers' team began to discover polymers, such as neoprene synthetic rubber (1930) and nylon (1934), which had major commercial potential. From the broader technological perspective, these techniques could be used to make innumerable new polymers that might have useful properties. DuPont had hoped to capitalize on this new capability. However, the uniqueness of this capability began to erode at the same time that it was being established. One aspect of the scientific research program at DuPont was publishing, which kept chemists happy and created good will in the larger chemical community, especially in academia. But Carothers's published papers taught other chemists how to make polymers and directly led to the discovery of a different type of nylon and polyester by other researchers. Although losing some of its research lead, DuPont also pioneered in the development of its two sophisticated polymer products.

Once its chemists had shown that neoprene and nylon had some potentially useful properties, the chemical engineers had to figure out how to manufacture these materials on a commercial scale. The intermediate chemicals used to make the polymers were laboratory curiosities, and methods for controlled large-scale polymerization did not exist. To develop these products successfully, DuPont had to make significant innovations in the entire vertical chain of production from benzene and acetylene raw materials to actually making end products to show potential customers how to process the new materials. The neoprene and nylon experiences gave DuPont a tremendous new organizational capability that it hoped it would be able to exploit for decades.[11]

WORLD WAR II

The diffusion of polymer science and technology was greatly abetted by World War II. Despite the discovery of many polymers with potentially useful properties during the 1930s, the commercialization of polymers

---

[11] David A. Hounshell and John Kenly Smith, Jr., *Science and Corporate Strategy: DuPont R&D, 1902-80* (New York: Cambridge University Press, 1988), Chapters 12-13.

remained very risky in terms of both technology and markets. *Plastics Horizons* (1944) noted that "few manufacturers in the prewar era, even on the basis of careful market analyses conducted, had the industrial courage shown by DuPont when it spent tens of millions of dollars to produce nylon... scarcely five years after its initial discovery."[12] This situation changed during the war when the United States government vigorously promoted the use of polymer products to replace scarce or unavailable materials. Production of vinyl resins went from 5 to 220 million pounds during the war. Acrylic polymers found widespread use in aircraft, raising their output by a factor of ten. Two new exotic polymers – polyethylene and Teflon – played essential roles in radar and in the Manhattan project. The most dramatic example, however, was synthetic rubber. When the Japanese attacked Pearl Harbor, the United States depended on natural rubber from Southeast Asia. During the war, an industry-government-university effort established a giant synthetic rubber industry, which produced two million tons of rubber for the war effort. In addition to creating a new industry, the synthetic rubber project and other similar ones acted to accelerate the development of polymer science and the diffusion of polymer engineering. After the war, many American chemical, oil, and rubber companies had organizational capabilities to manufacture, fabricate, and market polymer products.[13]

World War II unleashed three other technological trajectories that would carry the industry for decades: pesticides, pharmaceuticals, and petrochemicals. The second major growth area – pesticides – offered chemical companies the opportunity to develop significant organizational capabilities, primarily in research. DDT, the first modern pesticide, was discovered by Paul Mueller in 1939. Its insect-killing capability gave it wonder-chemical status during the war.[14] At about the same time, the first organic herbicide, 2,4-D, an organic compound developed by DuPont to stimulate the growth of plants, turned out to actually kill them. After the war, chemical companies developed elaborate sets of screens to determine the physiological activity of thousands of organic compounds and began to discover new insecticides, herbicides, and fungicides.[15]

---

[12] B. H. Weil and Victor J. Anhorn, *Plastics Horizons* (Lancaster, PA: Jaques Cattell, 1944), p. 74.

[13] John Kenly Smith, Jr., "World War II and the Transformation of the American Chemical Industry," in Everett Mendelsohn, Merritt Roe Smith, and Peter Weingart, eds., *Science, Technology and the Military* (Boston: Kluwer, 1988), pp. 307–22.

[14] Smith, "The End of the Chemical Century," p.158.

[15] Hounshell and Smith, *Science and Corporate Strategy*, Chapter 20.

The third wartime development, the mass production of penicillin, sparked the transformation of the pharmaceutical industry from a primarily chemical-based industry to one centered more on biological science and biochemical engineering. Most of the companies that made this transition were already in the pharmaceutical and/or fine chemical business. For chemical companies, the entry into this business was by acquisition, which was a diversification into a new type of business that required different organizational capabilities from traditional chemical companies.[16]

The demands and dislocations of World War II established the basic parameters around which petrochemicals would expand in the postwar decades. The war greatly accelerated the deployment of petrochemical processes, the development of products made from petrochemicals, and the geographical relocation of the chemical industry to oil and gas producing regions, notably the Gulf Coast region of Texas.[17] (This latter factor would give an important advantage to smaller companies such as Dow, who could build integrated petrochemical complexes, to produce a variety of products. Larger companies, such as DuPont, had already made major investments in plants dispersed around the country, many inherited from earlier enterprises.) Before the war, petrochemical production had been limited by a lack of end products that could be made from them. One success was Union Carbide's use of natural gas to make ethylene glycol, a new radiator coolant antifreeze.[18] In the late 1930s, Standard Oil developed a reforming process that could convert specific crude oil fractions into aromatic compounds. This process was used during the war to make toluene for TNT high explosives. The most important petrochemical project was the manufacture of butadiene and styrene to be polymerized into synthetic rubber. Most of the butadiene plants were built and operated by oil companies using petroleum derived feedstocks, while chemical companies, notably Dow and Monsanto, made styrene from ethylene and benzene, the latter still recovered from coking of coal.[19] Overall, World War II established the patterns of innovation and growth that the chemical industry would follow for decades.

[16] Smith, "Then End of the Chemical Century," p. 158.
[17] Spitz, *Petrochemicals*, Chapter 3.
[18] Ibid., Chapter 2.
[19] John Kenly Smith, Jr., "Patents, Public Policy, and Petrochemical Processes in the Post-World War II Era," *Business and Economic History* 27, no. 2 (Winter 1998): 413-19.

THE CHEMICAL CENTURY

From the mid-1930s to the mid-1960s, the chemical industry outperformed virtually every other sector of American industry. In 1950, *Fortune* proclaimed the twentieth century the "Chemical Century," noting that "Chemicals must now be considered the premier industry of the U.S.... the chemical industry cannot be matched by any other in dynamics, growth, earnings, and potential for the future."[20] Entering the postwar era, the chemical and allied industries manufactured a wide range of polymers that replaced other materials in a vast array of uses from toys to packaging and structural uses. Polymers would be the major growth area of the chemical industry throughout the 1940s, 1950s, and 1960s. For example, at Dow, sales of plastics products increased from 2 percent in 1940 to 32 percent in 1957.[21] In the 1960s, 85 percent of DuPont's earnings came from polymers and intermediates.[22] Polymer research became intensively competitive. Ziegler-Natta type catalysts, which led to the development of linear polyethylene and polypropylene, were discovered in a number of laboratories at approximately the same time. DuPont employed a massive R&D effort to commercialize its Delrin polyacetal resin, which replaced metals in many applications, only to lose half the market to Celanese, which exploited a loophole in DuPont's patent to produced a similar product.[23] Nearly every type of plastic used today was developed before 1960.[24] Afterward, competition for market share became intense.

Although the development of new polymeric materials became commonplace, companies did maintain significant firm specific organizational capabilities through relationships with downstream fabricators. DuPont's knowledge of textile fibers and close relationships with textile companies gave it the dominant position in the development of the synthetic fiber industry generally. Union Carbide drew upon its prewar expertise in ethylene and aggressive wartime polyethylene production to become the major producer of the new plastic. Dow combined its long experience with chlorine and its new ethylene capability to take a leading position in PVC.[25] However, as early as the 1950s, these particular

[20] "The Chemical Century," p. 69.
[21] Don Whitehead, *The Dow Story* (New York: McGraw-Hill, 1968), p. 236.
[22] Hounshell and Smith, *Science and Corporate Strategy*, p. 579.
[23] Ibid., Chapter 21.
[24] Walter S. Fedor, "Thermoplastics: Progress Amid Problems," *Chemical and Engineering News* 39 (May 29, 1961): 80–92.
[25] Spitz, *Petrochemicals*, Chapter 6.

advantages began to erode when new competitors invaded polymer markets. By the 1960s, the ongoing development of polymer science and technology had led to a flood of new products, which increasingly competed with other polymers for market niches.[26] Everyone jumped on the polymer bandwagon, and soon it began to sag from the weight of all the riders. By the late 1960s, polymers alone accounted for one-third of the industry's $20 billion sales. If the value of the intermediates used to make them is included, polymers probably accounted for about half the industry's sales.[27]

During the 1950s and 1960s, chemical companies discovered dozens of chemicals that exhibited useful insecticidal, herbicidal, and fungicidal properties. Just when the polymer-petrochemical complex showed signs of maturity, agricultural chemicals gave the industry a boost in the 1960s and 1970s.[28] In recent years, however, innovation and sales growth have declined.[29] In addition, the high cost and unpredictability of research in this area, the problems of regulation and liability, the saturation of markets, and the lack of growth in farming have combined to take the luster off agricultural chemicals.

In the decades after the end of World War II, the American chemical industry became increasingly dependent on raw materials derived from oil and natural gas. By 1950, half of the American output of organic chemicals was made from petrochemicals, and by the end of the decade, that fraction had increased to five-sixths. Cheap petrochemicals intermediates enabled the expansion of the chemical industry.[30]

THE PETROCHEMICAL REVOLUTION

The shift of the basis of the organic chemical industry from cellulose and coal can be called a petrochemical revolution, which had some unique characteristics. First, the demand for petrochemicals is highly derived; they must be transformed into something people actually use. A remarkable historical coincidence placed the polymer revolution just

[26] Hounshell and Smith, *Science and Corporate Strategy*, Chapters 19, 21, 22.

[27] Jules Backman, *The Economics of the Chemical Industry* (Washington, DC: Manufacturing Chemists Association 1970), p. 33.

[28] Basil Achilladelis, Albert Schwartzkopf, and Martin Cines, "A Study of Innovation in the Pesticide Industy: Analysis of the Innovation Record of an Industrial Sector," *Research Policy* 16 (August 1987): 175–212.

[29] "Agchem Producers Sow Plans for a Rich Harvest," *Chemical Week* (August 18, 1993): 33–35.

[30] Arora and Rosenberg, "Chemicals," p. 94.

slightly before the petrochemical one. Needless to say, the proliferation of polymers, the driving force of postwar chemical industry growth, was made possible by the rapid development of processes for making low-cost petrochemical intermediates. For example, by 1965 the production of aromatic compounds from petroleum was ten times what would have been available from coal processing.[31] Second, petrochemicals were always a secondary concern for both chemical and petroleum companies. The chemical companies were pursuing profits in polymers, pesticides, and pharmaceuticals, whereas the oil companies focused on the discovery of new sources of supply. Both of these industries were large and powerful, and both were unsure about the terrain that lay between them. The third peculiar feature of petrochemicals was the important role that specialized engineering firms came to play in the creation of an international market for petrochemical plants. They played a key role in making petrochemical technology available at reasonable prices, thus increasing competition between existing producers and lowering barriers to entry for new ones.

Historically, as oil companies concentrated on finding and producing oil, there were opportunities for independents, such as J. Ogden Armour, Jesse and Carbon Petroleum Dubbs, and Eugene Houdry, to improve refining processes. They developed processes for "cracking" large molecules into smaller ones to increase the overall yield of gasoline. Armour's Universal Oil Products Company, before being bought by a consortium of oil companies in 1931 (to avoid paying royalties) began to do research on catalytic petrochemical processes. During World War II, UOP made important contributions to the development of fluidized bed catalytic cracking of crude oil for gasoline production – the technology still used today. After the war, the oil companies spun off UOP to avoid antitrust problems. The new independent company soon stunned the oil industry with its new catalytic reforming process, Platforming, which dramatically improved the octane rating of gasoline by converting cyclic saturated hydrocarbons into aromatic ones such as benzene, toluene, and xylene. In addition to being burned in gasoline, these important building-block chemicals, formerly extracted from coal, now could be produced from oil. UOP had developed expertise in the key capabilities in catalysis and process development ahead of most of the oil companies. To build plants, they teamed up with oil companies or construction companies.[32]

---

[31] Spitz, *Petrochemicals*, pp. 184–91.
[32] Ibid., pp. 165–82.

At the same time that UOP was developing Platforming, chemical companies were attempting to determine how much of a stake they would take in petrochemicals. Dow was particularly aggressive setting up its own oil and gas subsidiary in 1946. The company had moved to the Gulf Coast in 1940 to extract magnesium from seawater. The plant was fueled by natural gas supplied by pipeline. In 1943, the company built an ethylene pipeline to supply its new styrene facility.[33] DuPont also moved to the Gulf Coast to build plants to supply the intermediates for its synthetic fibers. On whole, however, DuPont's enthusiasm for petrochemicals was mixed. For example, in 1946, the president of the DuPont Company, Walter S. Carpenter, Jr., interestingly characterized petrochemicals as a "problem" instead of as an opportunity. At this point, the future evolution of the industry was unclear for a number of reasons. The chemical industry had prospered using sophisticated processing to transform cheap widely available raw materials into much more valuable products. Synthetics were the latest and most rapidly growing segment of the industry.[34] The industry had limited interest in backward integration because its suppliers were much less profitable, especially in terms of investment turnover and return on investment.[35] Also, the prospect of the oligopolistic oil industry controlling the raw materials for chemical production was unsettling.

For the petroleum industry, chemical production would complicate an already complex business. Because most oil companies were fully integrated, they engaged in a wide variety of activities. First and foremost was the discovery and development of oil fields. Once found and recovered, oil had to be transported, processed, distributed and sold.[36] Historically, the industry tended to share refining technology, especially after the patent battles between Burton and the Dubbs over thermal cracking of crude oil in the early 1920s. Also, with relatively weak in-house research capability the oil industry relied on outside sources for new technology.[37] With limited experience in process innovation the oil companies could not easily move into petrochemicals, which would move them into new markets involving new customers anyway.

[33] E. N. Brandt, *Growth Company: Dow Chemical's First Century* (East Lansing: Michigan State University Press, 1997), pp. 184-91.

[34] Ibid., Chapter 6.

[35] Development Department to the Executive Committee, "The Petroleum Industry," February 24, 1966, Accession 1850, Hagley Museum and Library, Wilmington, Delaware.

[36] Ibid.

[37] Spitz, *Petrochemicals*, Chapter 6.

While the oil and chemical companies eyed one another some-what suspiciously, a few entrepreneurial firms, such as Scientific Design founded by chemical engineers Ralph Landau, Harry Rehnberg, and Bob Egbert, saw an opportunity to develop processes and construct plants for producing chemicals from petroleum. With limited capital and no intention of building large plants, the engineering firms depended on working closely with their clients to develop new processes. Its first process, ethylene oxide, was developed from pilot to full-scale facil-ity by Petrochemicals Ltd., who had received an exclusive license for Great Britain. Scientific Design had licensed the fully developed pro-cess over one hundred times by the 1980s. The development of a suc-cessful business strategy for selling technology to an industry with mas-sive R&D establishments posed a real challenge for the new engineering firms.[38]

A real danger for the engineering firms was threat of the loss of control over their process technologies. How could they package their technol-ogy so that customers would know what they were getting yet still have to pay for it to get it? Part of this was made possible by the maturation of the discipline of chemical engineering, which developed a generalized vocabulary that permitted discussion of processes without divulging pro-prietary aspects of them. Another key aspect for both seller and buyer was patent protection. Strong patents were essential to making chemical processes a salable commodity.[39] Yet, strong process patents were diffi-cult to construct, especially when compared to product patents. Some of the chemical engineering entrepreneurs developed processes by work-ing around claims in others' patents.[40] Probably the most patentable part of a process is the catalyst used in the chemical reaction. Yet, because there is no general theory of catalytic action, it is difficult to extend claims beyond the specific invention. This, of course, opens the door for competitors to find similar yet legally differentiable substitutes.[41]

The irony in the petrochemical industry is that process patents proved to be adequate for developing a well-defined market for buying and selling

---

[38] Ibid., pp. 318–30. See also Ralph Landau and Nathan Rosenberg, "Successful Commer-cialization in the Chemical Process Industries," in Nathan Rosenberg, Ralph Landau, and David C. Mowery, eds., *Technology and the Wealth of Nations* (Stanford, CA: Stanford University Press, 1992).

[39] Ashish Arora, "Patents, Licensing, and Market Structure in the Chemical Industry," *Research Policy* 26 (1997): 391–403.

[40] Spitz, *Petrochemicals*, p. 315.

[41] Ibid., pp. 331–38.

processes. The obvious question was why somebody should pay for something that could be obtained for free. Theft of processes is difficult to document because companies can hide plant technology from competitors. It appears that the incentive for the buyer to go along with the deal is the prospect of getting the technology in place cheaper than if he did it himself. This is perhaps reflected in the fact that most licensed processes are relatively bargain priced.[42] Higher prices would encourage companies to develop their own processes.

As they gained experience, the engineering firms developed at least two critical advantages over their clients. First, by selling the same process many times, the engineering firms benefited from constant feedback concerning process improvement. Also, in the postwar period, the scale of individual plants increased dramatically. For example, from 1952 to 1970, the size of a typical new vinyl chloride monomer plant increased from thirty million pounds per year to one billion pounds per year, a more than thirtyfold increase.[43] Because of all the problems caused by scaling-up, the engineering firms could leverage their experience from one generation of plants to the next one.

Even if a handful of engineering firms were able to successfully sell petrochemical processes, this might not have made a major impact on the industry as long as the firms in the industry decided to keep their own processes proprietary. However, this turned out generally not to be the case. A few important exceptions were Dow and DuPont, who did not license everything.[44] The oil and chemical companies entered into the market for petrochemical processes selling them at the same bargain prices as did the engineering firms. In the period from 1951 to 1971, chemical and oil companies generated one-third and one-quarter of new process developments. Specialized engineering firms accounted for 18 percent; foreign firms and a few from other industries supplied the remainder.[45] Although the incentive for marketing processes may have come from the engineering firms, obviously the chemical and oil companies decided to participate in this activity.

There are several explanations for this behavior. One is the obvious incentive to make a high profit by selling the same good over and over again, sacrificing whatever in-house advantages the process

---

[42] Ibid., pp. 540–43.

[43] Ibid., p. 395.

[44] Ibid., p. 547.

[45] Edwin Mansfield et al., *The Production and Application of New Industrial Technology* (New York: Norton, 1977), Chapter 3.

contained. For example, if useful life of a given process was very short, then firms would attempt to maximize the value of a process by wide spread licensing. One analysis showed that it took competitors nearly six years to respond to an innovative process.[46] It is uncertain whether this time frame is short enough to encourage immediate licensing of processes.

Finally, the strict antitrust climate of the postwar era strongly encouraged companies to license their technology instead of keeping it proprietary. After World War II, the U.S. Justice Department decided to attack the gentleman clublike atmosphere of the prewar industry both domestically and internationally.[47] The revelation of the chemical industry's participation in international cartels, especially its entanglements with I. G. Farben, were a source of ongoing embarrassment. One source of collusive behavior was the tangled web of patents that surrounded the burst of innovation of the 1930s. The bartering of patent rights had been the legal basis of the cartelization of the industry before the war.[48] Compulsory licensing of patents, it was argued, could end this type of control over innovation by a few large companies. On the domestic front, less-than-competitive attitudes reflected the fact that chemical companies were each others' best customers.[49] By allowing each company to make its own chemical raw materials and intermediates, petrochemicals promised to decouple companies from each other, allowing a more sincere form of competition to occur.

The rapid growth of the chemical industry in the postwar decades reinforced the licensing of chemical technology. One of the most unusual cases is that of DuPont actively recruiting a competitor, Chemstrand – a joint venture of Monsanto and American Viscose – to set up in the nylon business.[50] Avoiding future antitrust problems was one reason for doing this, but business concerns were also involved. DuPont did not want to have to invest most of its capital into the rapidly expanding nylon business when it had other new products that might turn out to be new nylons.

---

[46] Robert Stobaugh, *Innovation and Competition: The Global Management of Petroleum Products* (Boston: Harvard Business School Press, 1988).

[47] Arora, "Patents," p. 397; Hounshell and Smith, *Science and Corporate Strategy*, pp. 346-47.

[48] John Kenly Smith, Jr., "National Goals, Industry Structure, and Corporate Strategies: Chemical Cartels between the Wars," in Akira Kudo and Terushi Hara, eds., *International Cartels in Business History* (Tokyo: University of Tokyo Press, 1992), pp. 139-58.

[49] Backman, *Economics*, pp. 91-93.

[50] Hounshell and Smith, *Science and Corporate Strategy*, p. 347.

In spite of its importance to both the oil and chemicals industries, both regarded petrochemicals as secondary to their major strategies, producing high value-added chemicals and gasoline, respectively. In the mid-1960s, DuPont was still trying to decide whether it should integrate backward into petrochemicals or continue to buy feedstocks from others. A company report stated that petroleum and natural gas provided the base for over 80 percent of the organic chemicals produced in the United States, whereas chemicals represented only 4 percent of the materials handled by the petroleum industry. Even though chemical sales of petroleum companies doubled between 1960 and 1964, they still accounted for less then 10 percent of all chemical sales. DuPont maintained that backward integration into oil would put the company into a highly competitive business that produced commodity products. Rather than investing in less profitable oil companies, DuPont decided to continue its traditional strategy of introducing high value-added new products.[51] The fact that DuPont was seriously thinking of investing in an oil company indicates that the faith in the nylon paradigm may have been weakening, however. In the 1960s, industry leaders began to be nostalgic for the good old days of plentiful proprietary and profitable new products. In a dramatic move, DuPont launched an intensive and expensive effort to produce new nylons. When this strategy failed to distance DuPont from its competitors, it reluctantly began to look for other opportunities such as buying an oil company.[52] But petrochemicals had become commodities in markets much more competitive than those for chemical products. By making process technology widely available, the industry had lowered the cost of all chemical products, thus encouraging rapid – and sometimes dizzying rates of growth.

In the 1970s, when product markets began to falter, petrochemicals got some of the blame for ruining the industry. Industry participants and analysts such as Peter Spitz looked back on the chaotic heyday of petrochemical process development and licensing with some regrets. The prewar regime of closely held technology appeared to be a more orderly and sustainable industrial model.[53] The unusual place of petrochemicals, between two large industries, entrepreneurial engineering firms, and the antitrust policy of the United States government, had combined to create a highly competitive chemical industry.

[51] Development Department, "The Petroleum Industry."
[52] Hounshell and Smith, *Science and Corporate Strategy*, Chapter 22.
[53] Spitz, *Petrochemicals*, Chapter 13.

CHEMICALS: THE BALL IS OVER

In 1961, *Fortune* magazine writer Perrin Stryker published a piece entitled "Chemicals: The Ball Is Over," warning that the industry was rapidly maturing.[54] The main problem at this point was intense competition, even in new products such as polypropylene, which was leading to excess capacity, falling prices, and slim profit margins. During the 1950s, the number of companies making polyethylene had expanded from two to thirteen; vinyl polymers from six to nineteen; and polystyrene from four to sixteen.[55] The newcomers represented diversification by chemical companies and a few new firms from outside the industry. The 1960s turned out to be a pivotal decade for the chemical industry. Overall growth continued to be strong, supported by a number of important process innovations. However, the number of important new products introduced declined significantly from previous decades, in spite of DuPont's massive R&D program aimed at new products.[56] In the 1970s, continuing maturity was exacerbated by slowing growth rates, erratic petrochemical costs, and environmental regulation. In response to so much uncertainty, the chemical industry conservatively sought incremental learning curve-type improvements in both products and processes.[57] Cutbacks in R&D both reflected and caused declining numbers of new process and products. Patent statistics compiled by Achilladelis show a significant drop-off in patenting in the late 1960s by American firms in virtually every product group, including plastics, pesticides, and synthetic fibers.[58]

In the 1980s, the industry sought to rejuvenate itself through increased R&D spending, diversification into pharmaceuticals and biotechnology, and shifting its focus from commodity to speciality chemicals. American companies also sought growth by increasing globalization, a trend which had been gathering momentum since the 1950s. In spite of these initiatives, growth rates and profitability continued to decline, making the turbulent 1970s look good by comparison. Some realignment of the industry began in the 1980s, as firms began to cast off commodity products. Also, corporate raiders began to attack chemical companies such as Union Carbide, which had been weakened by the disaster at Bhopal,

---

[54] Perrin Stryker, "Chemicals: The Ball Is Over," *Fortune* (October 1961): 125–27, 207–18.

[55] "How Do You Measure Up?" *Chemical Week* (November 12, 1960): 88.

[56] Michael J. Bennett, Andrew A. Boccone, and Charles H. Kline, "The New Chemicals Enterprises," *Chemtech* (March 1988): 162–64.

[57] Hounshell and Smith, *Science and Corporate Strategy,* Chapter 25.

[58] Basil Achilladelis, Albert Schwarzkopf, and Martin Cines, "The Dynamics of Technological Innovation: The Case of the Chemical Industry," *Research Policy* 19 (1990): 14.

India, in 1984.[59] This trend accelerated in the 1990s, as buying and selling chemical assets became the prime activity of executives in an industry that had definitely matured into mediocrity or worse. The major new product push of genetically modified seeds, often to be used with specific pesticides, faltered after highly publicized opposition in Europe caused farmers to hedge their bets. In the late 1990s, the annual value of asset exchanges represented about one-quarter of the chemical industry's assets worldwide.[60] The impact of all this churning on the American chemical industry was to reduce the number of major chemicals producing firms; *Chemical and Engineering News* has reduced its Top 100 to Top 75 in 1998 because "so many large and medium-sized companies were devoured."[61] The other major effect has been that companies are attempting to increase focus by increasing market share and actual production in their core businesses. Statistics put together by Arora and Gambardella on several basic petrochemicals show that although the number of producers declined between 1973 and 1990, so have concentration ratios, indicating that the major producers have shut down some capacity.[62]

In the late 1990s, however, a new round of consolidations in petrochemicals has significantly raised concentration ratios. What was a "14 player lineup of leading companies is down to a six-member team in less than two years and the shoot out is far from over," according to *Chemical Week*. In North America, the top five ethylene producers accounted for 69 percent of regional capacity in 2000, compared with 43 percent in 1993.[63] Dow has been particularly aggressive in ethylene and polyolefins; its recent merger with Union Carbide will make Dow the world leader in each area. It is highly likely that this consolidation scenario will be followed in other sectors of the industry. Although the industry is not likely to grow much faster than economies generally and generate profits much greater than the cost of capital, the chemical industry is still a large and critical part of the world's technological and economic infrastructure.

Chemical industry R&D continues to develop extremely sophisticated technologies. The recent development of metallocene catalysts

---

[59] Bozdogan, "Transformation."

[60] "Feeding Frenzy," *Chemical Week* (March 8, 1999): 27–29.

[61] "Change at the Top 75," *Chemical and Engineering News* (May 1, 2000): 21–25.

[62] Ashish Arora and Alfonso Gambardella, "Evolution of Industry Structure in the Chemical Industry," in Arora, et al., *Chemicals and Long-Term Economic Growth*, p. 401.

[63] "Petrochemical High Noon: The Supermajors Ride In," *Chemical Week* (March 29, 2000): 31.

demonstrates some of the pitfalls of innovation in a mature industry. Like many chemical breakthroughs, metallocenes had a long incubation period. These unusual compounds, which consist of a sandwich of cyclic organic compounds with a metal ion in the center, were synthesized by academic chemists in America and Germany in the early 1950s, leading to a Nobel Prize in 1973. Three years later, other academic researchers accidentally discovered that metallocenes are highly active catalysts for making polyolefins.[64] In the past decade, a number of companies have developed metallocene catalysts and patent lawyers have attempted to stake out some proprietary claims. These catalysts can create polymers with remarkable degrees of uniformity in molecular weight and orientation. For example, Dow has been developing stereoregular forms of polystyrene, which transform the normally brittle polymer into a much tougher one that can become an engineering plastic used in structural applications. Metallocene R&D has consumed several billion dollars so far, but it is still uncertain whether customers will pay premium prices for plastics with improved properties.[65] As DuPont discovered in the 1960s, radically new materials, even those with remarkable properties, find applications slowly, something that can be financially devastating if large sums have already been spent on R&D and commercialization. When it became apparent to DuPont's management that its new product efforts were absorbing huge amounts of money with very little return, company analysts began to apply discounted cash-flow models to their product portfolio. What they discovered was that the company's remarkable high-technology materials, such as Kevlar, would probably never break even because of heavy front-end investment in R&D and commercialization, coupled with slow market growth.[66]

With the pipeline of innovation virtually shut off, the industry had to rely on its existing portfolios, which also were experiencing intense competitive pressures. Although the chemical industry continues to be an integral part of the global economy, its capitalist foundation is at best shaky when financial instruments such as internal rate of return or discounted cash flow analysis are used to evaluate specific products or the industry generally. Returns to the mature chemical industry have been cycling around the cost of capital with an increasing amplitude from

---

[64] W. Kaminsky, "Synthesis of Polyolefins with Homogeneous Ziegler-Natta Catalysts of High Activity" in Raymond B. Seymour and Tai Cheng, eds., *History of Polyolefins: The World's Most Widely Used Polymers* (Dordrecht: Kluwer, 1986), pp. 257–70.

[65] *Chemical Week* (September 30, 1998): 17.

[66] Hounshell and Smith, *Science and Corporate Strategy*, pp. 532–35.

peak to trough. Profit cycles peak when economic activity is brisk and current capacity is fully utilized. At that point, marginal costs skyrocket and producers commit to additional capacity in the form of large plants intended to capture additional economies of scale and market share. When these big new lumps of capacity come on-stream several years later while economic activity has slowed, supply exceeds demand and prices plummet.[67]

For over a century, the chemical industry found a lucrative niche between the producers of basic commodities and end-product manufacturers. Chemicals revolutionized many industries, such as textile fibers, tires, and packaging. With the coming of petrochemicals after World War II, chemical companies generally integrated backward into more basic intermediates. The petrochemical trajectory along with polymers, pesticides, and pharmaceutical ones focused the American chemical industry on a common set of strategies and technologies. Increasingly, companies began to look like one another. The reasons for the rapid convergence and maturing of the chemical industry have been discussed earlier. Within the constraints of the postwar American political economy, could the industry have followed different pathways? The other obvious strategic move was forward integration into specific end-product manufacture.

If any chemical firm could have integrated forward, DuPont was well positioned to do so, having had considerable experience in developing and fabricating products. It also had no peer in inventing novel materials, Teflon and Kevlar being only two of the better known ones. Yet, in the critical period of the 1950s and 1960s, the company held tight to its traditional strategy. By doing this, it lost several opportunities to integrate forward. In the late 1950s, DuPont allowed chemist W. L. Gore to develop a Teflon fabrication technology and business that has become a highly successful several billion dollar per year business. With regard to fabricators, DuPont had a policy of not-competing-with-its-customers; however, with some new materials, such as Teflon, there was an opportunity to integrate forward. Later, DuPont did establish its Corian cast acrylic business, which included product fabrication and more recently took over Ford's automotive painting business. Very recently, DuPont bought into the seed business to create an outlet for its genetically modified seeds. These moves were too small and too late to change the overall trajectory

---

[67] Albert D. Richards, "Connecting Performance and Competitiveness with Finance: A Study of the Chemical Industry," in Arora et al., *Chemicals and Long-Term Growth*, Chapter 14.

of the company. When one looks at what DuPont made, the potential for semiconductor manufacture stands out. Could DuPont have been Intel? The former was the pioneer in the production of pure silicon, photopolymer masks, and fluorine chemicals, all of which are necessary to make semiconductors. In the early 1960s, DuPont management had considered investing in fields as diverse as theme parks and aircraft manufacture, so semiconductors should not have been beyond the company's horizon. For a variety of reasons, including hidebound upper management who were mired in past glories, DuPont did not take any bold new steps. Sticking to what it did best, DuPont, like most of the chemical industry, has embarked on a long downhill slide from prominence toward becoming an undistinguished producer of commodities.[68]

This is perhaps the inevitable fate of any industry. Nevertheless, the chemical industry rushed toward that end with reckless abandon. The trajectories established by World War II, the shift to petrochemicals, entrepreneurial engineering firms, a stringent antitrust environment, and weak intellectual property protection all contributed to making the American chemical industry highly competitive. The headlong rush into chemicals demonstrated the power of markets to reduce prices but also contributed to wasteful overinvestment, and undermined valuable organizational capabilities.

---

[68] John Kenly Smith, Jr., "The Nylon Syndrome: Will the DuPont Company Survive Its Greatest Achievement," unpublished manuscript.

Competitors

7

The Export-Dependence of the Swiss Chemical Industry and the Internationalization of Swiss Chemical Firms (1950–2000)

MARGRIT MÜLLER

INTRODUCTION

The expansion and internationalization of the Swiss chemical industry in the postwar period is described by focusing on the level of the industry, and of the firm.[1] In 1999, about 90 percent of total production of chemicals in Switzerland was exported,[2] and the dependence on imports must have been even higher.[3] In the following section I shall, therefore, describe the evolution of the industry by focusing on chemical exports and imports. Subsequently, I shall concentrate on the growth pattern of two chemical firms that merged into one in 1970, and that can be considered typical of the development of the chemical industry in Basle, still the most important sector of the Swiss chemical industry. The microlevel

[1] I would like to thank the participants of the International Colloquium, "The Global Chemical Industry since the Petrochemical Revolution," organized by ASSI and the Economic History Institute of Bocconi University in Milan, October 2000, especially Harm G. Schröter, for valuable comments. I am also grateful to Pedro Abreu, Manuel Hiestand, Cécile Steiner, and Friedrich von Gusovius, who have contributed to this study with seminar papers and statistical work.
[2] *The Swiss Chemical-Pharmaceutical Industry,* Swiss Society of Chemical Industries, Zurich, 2000, 40–41.
[3] According to the Annual Report of the SSCI of 1980, 20, over 95 percent of basic material (raw material and intermediary products) were imported.

approach is important because to a large extent the growth of the Swiss chemical firms took place in foreign countries. With this form of internationalization, they conformed to the general pattern of growth in the main Swiss export industries. The firms focused on technically advanced high-quality products for which domestic demand was too small almost from the start. How did the Swiss chemical firms manage to compensate for the "location specific disadvantages" of being situated in a country with a small home market and largely without adequate raw material and still internationalize successfully?[4] In each section, a short review of the historical background will explain the rather unusual beginnings of this industry and the development of the firms up to World War II. The growth of the firms depended on their being internationally competitive and able to cope with barriers for international trade. These conditions fostered specialization and diversification with regard to both products and markets. What was the impact of the liberalization of world trade and, therefore, major changes in location-specific advantages and disadvantages on chemical exports and other forms of internationalization? How did the firms cope with the rapid pace of scientific and technical progress and manage to remain internationally competitive?

In the mid-1970s, after two decades of continuous expansion, overall economic growth slowed and the Swiss economy suffered a deep recession, followed by a few years marked by moderate growth rates and a rather weak downswing in 1982/83.[5] The subsequent period of expansion was interrupted by a rather strong and persistent economic downturn in the 1990s. Periods of growth and major recessions marked the economies of other countries too, but not everywhere at the same time and to the same extent. The overall trend toward reducing international barriers to trade and integrating the world economy slowed down somewhat in the 1980s, but persisted and accelerated in the 1990s. What was the impact of economic fluctuations and changing conditions for

---

[4] *The Swiss Chemical-Pharmaceutical Industry,* 2000, 41: "The World Trade Organization's statistics assign Switzerland 8th place among the major exporting countries of chemical and pharmaceutical products, with a share of more than 4 percent in world chemical exports." For the concepts of "location specific advantages," see John H. Dunning, "Trade, Location of Economic Activity and the MNE: A Search for an Eclectic Approach," in Bertil Ohlin, Per-Ove Hesselborn, and Per Magnus Wijkman, eds., *The International Allocation of Economic Activity* (London: Macmillan, 1977), 395-418.

[5] An overview of the economic history of Switzerland with bibliography is given in Hansjörg Siegenthaler, "Die Schweiz (Switzerland) 1914-1984," in Wolfram Fischer et al., eds., *Handbuch der europäischen Wirtschafts- und Sozialgeschichte,* Vol. 5 (Stuttgart: 1985), 443-73.

international competitiveness on the industry level and on the performance and strategies of the firm?

THE SWISS CHEMICAL INDUSTRY (1950–2000)

Unlike the chemical industries in other countries, the early development of the Swiss chemical industry was not resource based.[6] The incentive to produce chemicals came from the demand side, namely, from the domestic textile industry. When the first firms in basic chemicals were founded in the early nineteenth century, the high transport costs and the irregularity of trade flows sheltered these firms from import competition. In the second half of the nineteenth century, only a few enterprises producing basic chemicals were able to survive and compete with cheaper imports.[7] The lack of an adequate resource base clearly limited the development of an industry in mass chemicals and favored those products that were more knowledge-based, namely, dyes and pharmaceuticals.

The main firms manufacturing dyes were all located in and around the town of Basle.[8] Plausible explanations for this regional concentration are the favorable political environment, relatively low transport costs, and the abundant water supplies. Links with universities were important, especially with regard to the development of chemists, but these links were mainly with German universities and the "Eidgenössische Technische Hochschule (ETH)" in Zurich, and not with the University in Basle.[9] Regional concentration also was typical for other industries in Switzerland and for other chemical products. The scent and perfume industry, for example, was concentrated in and around Geneva. The production of pharmaceuticals was more decentralized initially, but the most successful firms on an international scale all would be ultimately

---

[6] The only resource-based sector of the Swiss chemical industry was electrochemicals.

[7] One of these firms was the "Chemische Fabrik Uetikon" located at Lake Zurich, today "Chemie+Papier Holding" (CPH). Ulrich Geilinger-Schnof, *175 Jahre Chemie Uetikon. Die Geschichte der Chemische Fabrik Uetikon von 1818 bis 1993 (175 Years Chemicals Uetikon. The History of the Chemical Works Uetikon from 1818 to 1993)*, Uetikon, 1993. See also Margrit Müller, "Good Luck or Good Management? Multigenerational Family Control in Two Swiss Enterprises since the 19th Century," *Entreprises et Histoire*, 12 (1996): 19–47.

[8] Rudolf Baumgartner, *Die wirtschaftliche Bedeutung der chemischen Industrie in Basel (The Economic Importance of the Chemical Industry in Basle)* (Dissertation, Basle 1947), 47. In 1939, there were fifteen enterprises manufacturing dyes in Switzerland, of which six enterprises with a share of about 85 percent of total employees were located in Basle.

[9] Christoph Tamm, "Universität und Industrie (University and Industry)," in Thomas Busset et al., eds., *Chemie in der Schweiz. Geschichte der Forschung und der Industrie (Chemicals in Switzerland. History of Research and of the Industry)* (Basle, 1997), 59–75.

located in Basle. Hoffmann-La Roche specialized in pharmaceuticals from the start; and the main dye manufacturing firms – Ciba, Geigy, and Sandoz – took up the production of pharmaceuticals in the late nineteenth century and in the interwar period. Production of dyes and pharmaceuticals was science-based, which may explain the early scientific orientation of the Swiss chemical firms and the traditionally high importance attached to firm-centered R&D activities.[10]

Up until World War I, the chemical industry's share in total industrial production and exports was very small. Its relative importance increased during the war and interwar periods, especially in the 1930s. The share of chemical products in the total value of industrial exports increased from less then 3 percent in 1892, to almost 5 percent in 1913, 9.4 percent in 1920, 15.4 percent in 1935, and 19.7 percent in 1939.[11] In 1939, the share of tar-dyes was 59 percent and the share of pharmaceuticals (including scent and perfume) 28 percent of the total export of chemicals. The chemical industry in Basle contributed about 55 percent to the total production of chemicals and about 70 percent to total chemical exports.[12]

The chemical industry was less affected by the depression of the 1930s than were the other main export industries, namely, the machine, watch, and textile industries. In the case of the chemical industry in Basle, this was partly because the impact of the depression was dampened by national and international cartels. With the formation of the "Interessen-Gemeinschaft" (IG) in 1918, competition among the three dye manufacturing firms, Ciba, Geigy, and Sandoz, was eliminated. The firms joined forces in securing supplies and facilitating sales, and together established some chemical works and sales organizations abroad. In 1929, the IG joined the international dyestuff cartel, which regulated exports as well as the expansion of production and the establishment of sales

---

[10] German and Swiss firms were rather successful in introducing science into industry. See, for example, John J. Beer, "Coal Tar Dye Manufacture and the Origins of the Modern Industrial Research Laboratory," in David E. H. Edgerton, ed., *Industrial Research and Innovation in Business* (Cheltenham, UK/Brookfield, VT: Edward Elgar Pub, 1996), 55–63. First published in *Isis, 49, Part 2* (June 1958): 123–31.

[11] Baumgartner, *Bedeutung der chemischen Industrie (Importance of the Chemical Industry), 64–66*. In 1939, the export shares of the other main industries were: machines 17.5 percent; watches 15 percent; textiles 17.8 percent. With reference to the *Statistisches Jahrbuch der Schweiz (Statistical Yearbook of Switzerland) 1939*, 268–69.

[12] *Ibid.*, 67–69. In 1939, the chemical industry had 22,428 employees, only about 1.7 percent of total employees in Switzerland (2.2 percent of workers). With regard to Basle, this share was almost 10 percent (14 percent workers). *Ibid.*, Table 7, 170–71.

organizations abroad.[13] Increasing barriers to trade in the 1930s and restrictions imposed by the international cartel were a strong incentive to diversify into new product lines, shift production to foreign countries, and expand sales to new foreign markets.[14]

The immediate postwar years were marked by the shortage of supplies, especially of sulphur; export opportunities were still limited by import restrictions in other countries; and international trade continued to be regulated by bilateral trade and payment agreements.[15] Nonetheless, Swiss chemical exports rose continuously, especially from 1953 onward, when supplies of raw material were normalized. The number of European countries able to trade free from bureaucratic controls also increased, and within the OECD the objective of abolishing restrictions for international trade was institutionalized. Efforts to liberalize trade through international organizations and agreements shifted from eliminating quantity restrictions to reducing duties. But liberalization of international trade remained an aim in the long term, and barriers to trade persisted in various forms. Furthermore, new barriers for international trade were established with the division of Europe into the EEU and the EFTA.[16] The main concern of the SSCI was to avoid any competitive disadvantage for the Swiss chemical industry in comparison with firms

[13] Harm G. Schröter, "Kartelle als Form industrieller Konzentration. Das Beispiel des internationalen Farbstoffkartells von 1927-1939 (Cartels as a Form of Industrial Concentration. The Example of the International Dyestuff Cartel 1927-1939)," *Vierteljahrschrift für Sozial- und Wirtschaftsgeschichte,* Vol. 74 (Stuttgart 1987): 479-513.

[14] With regard to Ciba, see Margrit Müller, "Coping with Barriers to Trade: Internationalisation Strategies of Swiss Firms in the Interwar Period," in Hubert Bonin *et al.*, eds., *Transnational Companies, 19th-20th Centuries* (Paris: P.L.A.G.E., 2002). The development of Geigy in this period is analyzed in Andrea Rosenbusch, *Organisation und Innovation. Die Entwicklung der J.R. Geigy A.G. (Organisation and Innovation. The Development of J.R. Geigy AG) 1923-1939,* unpublished thesis (Lizentiatsarbeit der Philosophischen Fakultät der Universität Zürich), 1995.

[15] The development of the industry is quantified with figures on chemical exports and imports, their share in total industrial exports and imports, and their geographical distribution. This section draws mainly on the annual reports of the Swiss Society of Chemical Industries (SSCI), Annual Reports, 1947/48-2000. The society was founded in 1882 in Zurich.

[16] Switzerland joined the OECE in 1948, the European Payment Union in 1950, the GATT in 1958, and the EFTA in 1960. Joining the EEU was completely out of question and not a political option in Switzerland at that time. The position of the SSCI was clear already in 1957. One objection against joining the EEU was the common external tariff, which would have meant a considerable increase of duties on imported raw material and intermediary products. The total liberalization of trade was not perceived as a major threat to the domestic economy, provided that it was carried through at a moderate pace. SSCI, Annual Report 1956/57, 46-55; SSCI, Annual Report 1957/58, 64-68.

of the EEU countries. This was achieved through the foundation of the EFTA, as well as by signing bilateral agreements with the EEU to reduce duties to the same low level effective for producers within the EEU.[17]

The rapid expansion of the Swiss chemical industry up until the middle of the 1970s was not seriously hampered by such constraints. Limitations on growth were, rather, felt on the supply side because the shortage of labor became very acute in the 1960s. Another problem was rising costs and, even more disturbing, new demands with regard to environmental protection. Still, in nominal values, exports of chemical products rose from 590 million CHF in 1950, to 1,600 million CHF in 1960, 4,750 million CHF in 1970, and 9,500 million CHF in 1980. Growth rates were positive, except in 1975, when chemical exports decreased by over 10 percent. In the 1960s and up to 1974, the yearly growth rates of exports fluctuated between a minimum of 7.3 percent and a maximum of 24.8 percent, with an average of 11 percent; in the second half of the 1970s, the average yearly growth rate was only about 2.5 percent. Employment increased from 35,500 in 1950, to 50,900 in 1960, and 67,100 in 1970. By 1980, the number of employees in the chemical industry had declined to 64,800.[18]

In the early 1980s, the European chemical industry felt the impact of the economic recessions in Western Europe and North America. The Swiss chemical industry performed rather well, and was only slightly affected by the fall in demand in its main export markets. Chemical exports rose from about 9,500 million CHF in 1980, to 17,800 million CHF in 1989. Yearly growth rates were positive throughout the decade, but varied considerably between a minimum of 1.2 percent in 1986 and a maximum of 12.3 percent in 1989, mainly because of currency fluctuations. Since the beginning of the 1980s, further steps toward abolishing restrictions for international goods and capital flows were initiated. The main objective shifted from lowering tariffs to eliminating technical and administrative trade barriers, which implied the internationalization of rules and procedures, especially of property rights (mainly with regard to patents, trademarks and direct investments). Notwithstanding

---

[17] SSCI, Annual Report 1958/59, 59–71, 77–79. About one-third of total exports of the Swiss chemical industry were delivered to the EEU countries. Furthermore, the main competitors of the Swiss chemical firms were located in these countries. SSCI, Annual Report 1957/58, 61. The bilateral agreement, concluded in July 1972 and accepted by the Swiss voters in December, largely eliminated trade barriers for Swiss chemicals to this important market (about a third of total exports). SSCI, Annual Report 1972, 28–31.

[18] Heiner Ritzmann-Blickensdorfer, ed., *Historical Statistics of Switzerland* (Zurich: Chronos, 1996), 397.

all these activities, the overall pace of market liberalization began to lag and progress within the GATT and later the WTO was very slow.[19] Persistent current account deficits destabilized exchange rates and were countered with protectionist policies. Furthermore, multilateral negotiations were increasingly bypassed with regional and bilateral agreements.

The main concern of the SSCI was the impact of the process of European integration on the competitiveness of the Swiss chemical industry in this important market. The SSCI pursued a policy of adapting domestic rules and procedures in the production and distribution of chemicals and pharmaceuticals to conform with those applied within the European Union. In the domestic sphere, complaints about rising costs and the shortage of qualified labor continued. The SSCI was in favor of Switzerland joining the EEA (European Economic Area) in the early 1990s, as a first step toward joining the European Union. One main advantage perceived was the higher flexibility in the labor market. After the rejection of the EEA, the same request was made on the level of the bilateral agreements. Once these agreements were concluded, the question of European integration disappeared from the SSCI's agenda.

Hostility to the chemical firms, already a topic in the 1970s with regard to pollution, suddenly increased after the fire in the warehouses of Sandoz in Schweizerhalle near Basle, and the subsequent pollution of the Rhine in 1986. The extraordinary frequency of grave accidents (Seveso in Italy, and Bhophal in India) demonstrated to the whole world the high risks incurred with the production and storage of chemicals. Environmental questions, an area perceived as being largely under control with the environmental law implemented the year before, came to the fore again.[20] An increasing number of international organizations were concerned with environmental issues and active in developing programs to reduce pollution and the risks of contamination. The SSCI, had difficulty keeping up with all these activities and concentrated instead on participating actively within the CEFIC (Conseil Européen des Fédérations de l'Industrie Chimique). The fear of becoming isolated within the European institutional setting and of losing its influence on the topics dealt with on the level of international organizations clearly increased.

---

[19] The policy of the SSCI was to influence this process on two levels: as a member of the CEFIC (Conseil Européen des Fédérations de l'Industrie Chimique) on the European level, and as a member of the "Schweizerischer Handels- und Industrie-Verein" (Swiss Association for Trade and Industry) on the national level.

[20] SSCI, Annual Report, 1986, 5, 8, 15–17.

The composition of exports changed considerably in the last decades of the twentieth century. Although the share of dyes, the main product in the period before World War II and up until the 1960s, decreased to only about 13 percent in 1980 and 7 percent in 1997, the share of pharmaceuticals doubled in the 1970s and increased to over 50 percent in the 1990s. Complaints about the increasing interference of the state with regard to prices of pharmaceuticals were frequent since the 1970s, but it was precisely this field of activity that expanded most rapidly. Given the rising importance of pharmaceuticals, the repeated political initiatives against methods and objectives of R&D - especially in biotechnology, the most promising new field of research - were a new and very disturbing phenomenon. Toward the end of the 1980s and in the early 1990s, the SSCI intensified its influence on national political issues, supporting campaigns in favor of the freedom of research. The main argument was that the consequence of such restrictions would be the transfer of research and related activities to other countries. The Swiss voters repeatedly rejected these initiatives; nonetheless, the traditional consensus within the Swiss society on the priority of R&D was shaken and prospects for the future remained uncertain. Switzerland remained the most important location for R&D activities of the chemical industry, but in the 1990s the foreign R&D expenditures were higher than the domestic ones, and in the last years of the decade the domestic expenditures decreased while the expenditures abroad increased.[21] Location-specific advantages and disadvantages for the chemical industry in Basle became an important topic since the late 1980s. The aim of the SSCI was to make R&D as well as production in Basle more competitive with regard to foreign locations by reducing costs and avoiding rules that lowered the attractiveness of Basle as a research center.[22] But there were other reasons for expanding R&D activities abroad.

A good indicator of the performance of the chemical industry as compared with the other main Swiss export industries is its share in total industrial exports. This share was about 16 percent in the 1950s and 20

---

[21] *Forschung und Entwicklung in der schweizerischen Privatwirtschaft (Research and Development in the Swiss private Economy)*, ed. Schweizerischer Handels- und Industrie-Verein (Vorort) (Zurich, 1976, 1980, 1983, 1986, 1992, 1996, 2000). The R&D expenditures were estimated about every three years, on the basis of inquires. The samples of the different years are only partially comparable.

[22] Competitiveness understood as "Europafähigkeit" is a regular topic of the Vorwort des Präsidenten (Foreword of the President) since the late 1980s. SSCI, Annual Reports, 1988–2000.

percent in the 1960s; it reached a maximum of 23 percent in 1974. In that year, and in 1972, chemical exports were even slightly higher than machine industry exports. In the subsequent years, this share fell to a minimum of about 19 percent in 1980. So, although growth of chemical exports was somewhat more rapid up until the mid-1970s, it was largely in line with the average growth rates of the main Swiss export industries.[23] During the 1980s, the export share of the chemical industry stagnated at about 21 percent, but in the 1990s it rose continuously to over 28 percent in 1999. Obviously, the chemical industry performed exceptionally well during that decade, notwithstanding the deep recession in Switzerland and other European economies. The yearly growth rates of exports were positive throughout the decade and varied between 2.7 percent and 14.2 percent. With about 34,000 million CHF in 1999, the nominal export value almost doubled, and the average growth rate (6.5 percent) was only slightly lower than in the 1980s (7 percent).

The geographical distribution of exports remained quite stable.[24] But the yearly growth rates of exports to these regions differed considerably each year. Geographical diversification to some extent dampened the impact of regional economic fluctuations. The geographical distribution of imports was very stable too, with about 85 percent imports from European countries.[25] Because of the increasing share of specialities produced, the composition of imports changed considerably, and dependence on foreign suppliers increased, also because stocks of chemicals were drastically reduced after the fire in the warehouses of Schweizerhall mentioned above.[26] But the import side was never a prominent topic throughout the period and can be largely neglected in this study. Quite generally, supplies were easily available, although sometimes at rising costs.

[23] In the list of the most important export industries, the chemical industry became second in 1959, after the machine industry and before the watch industry. SSCI, Annual Report, 1959/60, 36, and *Historical Statistics of Switzerland*, 683, 687.

[24] This result is in line with the results of a study on total industrial exports. Stephan Mumenthaler, *Die geographische Struktur des Schweizer Aussenhandels. Historischer Ueberblick, theoretische und empirische Analyse sowie Szenarien für die Zukunft (The Geographical Structure of Swiss Foreign Trade. Historical Overview, Theoretical and Empirical Analysis as well as Scenarios for the Future)* (Zürich: Chur, 1999), 15–30.

[25] The share of imports from Europe increased continuously from about 55 percent in 1951 to 85 percent at the end of the 1960s, whereas the share of "America" declined from about 35 percent to 12 percent. In the following decades, 84 percent to 88 percent were imported from Europe and about 7 percent to 10 percent from "America."

[26] According to the SSCI, Annual Report of 1981, 18, and 1986, 14, the number of different primary products was estimated over 10,000 in 1981 and about 20,000 in 1986.

The figures describing the development of export and imports of the Swiss chemical industry show a largely continuous and stable pattern of growth, interrupted but not diverted by a decrease in the mid-1970s, and hardly affected by the economic recessions of the early 1980s and in the 1990s. What we cannot perceive from these figures is how the Swiss chemical firms reacted to the liberalization of world trade in the postwar decades. Was the pattern of internationalization, marked by the period of protectionism of the 1930s discontinued in the 1950s and 1960s? How did the chemical firms react to the fluctuations of overall economic growth and the new economic and political instabilities since the 1970s? In 1999, the chemical industry employed sixty-eight thousand persons, the same number as in 1983; about two hundred thousand persons were employed in foreign countries, compared with 140,000 in 1991.[27] Do these figures indicate an important change in the role of the home country for the multinational chemical firms in Basle, and what are the implications of such changes for the prospects of the Swiss chemical industry in the future? In order to find answers to these questions, the inquiry now shifts to the microlevel of the firm.

CASE STUDIES

This section concentrates on two of the four main chemical firms in Basle, Ciba Aktiengesellschaft Basel (referred to as Ciba), and J. R. Geigy AG (referred to as Geigy). Both firms were diversified and produced dyes, pharmaceuticals, fine and speciality chemicals, and agrochemicals. At the beginning of the 1950s, Geigy was about half the size of Ciba with regard to total sales; by the end of the 1960s, the sales of Geigy were slightly larger than those of Ciba, and in autumn 1970, Ciba and Geigy merged to become "Ciba-Geigy AG." In the following decades, Ciba-Geigy (Ciba since 1992) figured regularly among the largest industrial firms in Switzerland (with regard to total turnover), together with Sandoz and Hoffmann-La Roche, the other two large chemical firms in Basle that continued to prefer internal growth, joint ventures, and takeovers to merger. In 1996, Ciba and Sandoz merged and established Novartis, one of the largest pharmaceutical firms worldwide.

---

[27] The figures for 1983 and 1991 are indicated in the annual reports of the SSCI for the years 1983, 4, and 1991, 5; the figures for 1999 in *The Swiss Chemical-Pharmaceutical Industry,* 2000, 21.

Both firms had common roots in the tar-dyes manufacture beginning in the 1870s and 1880s.[28] By the turn of the century, they had already become multinational firms.[29] The deterioration of international trade in the interwar period was aggravated by the worldwide economic depression of the 1930s. Depression and increasing barriers to trade promoted innovations, diversification into new product lines, and all forms of internationalization, including the expansion of exports to new and more distant markets, joint ventures with foreign firms, and FDI.[30] In both firms, expenditures for R&D increased, and more weight was given to systematic scientific research. The different growth path of Ciba and Geigy in the interwar decades was mainly because of the "IG" agreement concluded in 1918 among the three dye producers, Ciba, Geigy, and Sandoz, which regulated the production of dyes. Given that growth in these product lines was restricted, at first by this agreement and later also by an international cartel that the "IG" joined in 1929, all firms intensified their R&D in nonregulated sectors. Ciba had taken up the production of pharmaceuticals by the late 1880s, but on a rather small scale. In the 1920s, R&D and production capacities in pharmaceuticals were expanded. In 1928, the firm took up the production of auxiliary and refining products for textiles, and in 1933, it ventured into the production of synthetics. Geigy gave up the production of pharmaceuticals when it joined the IG. In 1925, somewhat earlier than Ciba, the firm took up the production of auxiliary and refining products for textiles, and in the 1930s, it was the first of the chemical firms in Basle to start R&D for insecticides. Geigy decided to diversify into the production of pharmaceuticals in January 1939,

---

[28] Geigy was established as a trading firm for drugs in 1758. The origin of Ciba was a silk dye-house that began to produce "Fuchsin" in 1859. In 1873, the firm was sold to Bindschedler & Busch and in 1884, it was transformed into the "Gesellschaft für Chemische Industrie in Basel" (CIBA). An overview on the development of the main chemical firms in Basel up to the 1950s is given in *Herkunft und Gestalt der Industriellen Chemie in Basel (Origin and Form of Industrial Chemistry in Basle)*, Herausgegeben von der CIBA aus Anlass ihres 75jährigen Bestehens als Aktiengesellschaft (ed. by CIBA at the occasion of 75 years as a corporate company) (Olten and Lausanne, 1959), 100–101.

[29] See Harm G. Schröter, "Unternehmensleitung und Auslandsproduktion: Entscheidungsprozesse, Probleme und Konsequenzen in der schweizerischen Chemieindustrie vor 1914 (Governance of the Firm and Foreign Production: Decision-Making Processes, Problems and Consequences in the Swiss Chemical Industry before 1914)," in *Schweizerische Zeitschrift für Geschichte (Swiss Journal of History)* 44 (1994): 14-53.

[30] For Ciba, see Margrit Müller, "Coping with Barriers to Trade;" for Geigy, see Andrea Rosenbusch, "Das Ende des 'frisch-fröhlichen Erfindens.'" Die Entwicklung einer neuen Organisationsstruktur in der J. R. Geigy A.G. 1923-1939 ("The End of 'Cheerful Inventing': The Development of a New Organizational Structure at J. R. Geigy A.G. 1923-1939)," in *Chemie in der Schweiz (Chemicals in Switzerland)*.

deliberately risking a conflict with its partner firms in the IG. But the days of the agreement concluded in 1918 were numbered anyway. With the beginning of the war, the international cartel broke down, and during the war, the firms – especially Ciba – followed a policy of protecting their foreign subsidiaries by excluding them from the IG, especially in the United States. In 1950, the IG was definitely dissolved, and the two firms were free to embark on their own growth paths.

Because of the partition of product lines within the IG, the firm-specific capabilities with regard to technology and markets differed considerably at the beginning of the period.[31] But in the postwar decades, capabilities and experience became more similar, because the firms diversified into the same main product lines. Before World War II, Geigy already had decided to take up the production of pharmaceuticals, and Ciba began to produce agrochemicals in the 1950s. For both firms, the postwar years until 1952 were marked by several short downturns, but from then on, turnover and employment increased continuously. The importance of the "Konzern" increased, but the "Stammhaus" still maintained a dominant position, especially with regard to production and to R&D. In the next section, the figures of the "Stammhaus" are indicated in brackets.

In the case of CIBA, total sales were 532 million CHF (218 million or 41 percent) in 1950; 1,130 million CHF (446 million or 39.5 percent) in 1960; and 3,092 million CHF (1,162 million or 37.5 percent) in 1969. Employment increased from about 17,400 (6,000 or 34.5 percent) in 1951 to almost 21,900 (7,900 or 36 percent) in 1960, and about 39,000 (9,900 or 25 percent) in 1969.[32] Clearly, the share of the Stammhaus

---

[31] This section draws mainly on the annual reports of Ciba and Geigy, which contain quantitative data on different dimensions of the firms' development, both for the "Konzern" as well as for the "Stammhaus" in Switzerland (these figures are mentioned in brackets). The reports also give a good overview of the main topics and problems dealt with. Ciba's reports contain more detailed figures; those of Geigy are more explicit about important aspects of the common overall economic context, such as the conditions for international trade, domestic and foreign economic policies, the main problems and opportunities perceived, and so on. Of course, it should be remembered that that these comments are made from the perspective of the firm and with the intention of being communicated. The 75th annual report of Ciba in 1958 gives an overview of the quantitative development of the firm since its beginnings. The "Stammhaus" comprises Ciba's chemical works in Basle and in Monthey/VS. The "Stammhaus" of Geigy comprises the works in Basle and Schweizerhalle (J. R. Geigy AG, Basel, Geigy-Werke, Schweizerhalle AG). Usually, the works in Grenzach (Germany) nearby were also included. The works in Grenzach were, by tradition, considered as belonging to the "Stammhaus."

[32] Employment in the foreign affiliates was distributed as follows: at the end of 1951 (1969): Europe 12,791 (27,876); North and South America 3,708 (4,135 North America, 3,159 South America); other regions 890 (3,775).

in total sales was almost maintained, while in the 1960s employment increased much more rapidly abroad. This was partly because of the extreme shortage of labor in Switzerland but also to the fact that more and more functions were shifted from the Stammhaus to the foreign affiliates. The general view, shared also by Geigy, was that the main functions of the Stammhaus were the production of specialities and R&D, those of the affiliates the production of basic and intermediate products and distribution (sales and marketing).[33] But, because of the limited capacities in Switzerland, the established chemical works abroad began to produce specialities too, and soon the management found that by establishing R&D centers abroad it was possible to take advantage of the scientific capacities and the advanced scientific research in foreign countries. The growth of R&D expenditures is impressive (figures for the Stammhaus in brackets). In 1955, expenditures for R&D were 38 (25.3) million CHF, in 1960, 86.2 (52.2) million CHF and in 1969, CHF 251 (143) million. In the early 1950s, R&D was about 4 percent to 5 percent of total sales; by 1960, this share had risen to 7.6 percent, and the average rate in the 1960s was 8.5 percent. The contribution of the foreign firms to R&D increased from about a third to almost half of total R&D expenditures. In addition to Basle, there were R&D centers in the US and England.[34]

The share of exports in the total sales of the Stammhaus remained stable at about 90 percent, and the share of sales of the Stammhaus to the Konzern at about 60 percent. Building investments were high, especially in the 1950s, when a renewal of practically all the old works of the Stammhaus was carried through. In percent of total sales, building investments fluctuated between 9.5 percent and 12.8 percent in the 1950s, and between 8.3 percent and 12.8 percent in the 1960s. Until 1955, investments were about equally distributed between Switzerland and abroad, but from then on, foreign investments were higher with domestic investments sinking to an average of about 40 percent of total investments in the 1960s. The Konzern comprised about fifty foreign affiliates in 1958 and, ten years later, sixty-five foreign affiliates (thirty-nine in Europe, fifteen in North and South America, five in Asia, four in Africa, and two in Australia).[35] Among these firms there were also

---

[33] In 1957, the functional distribution of employment in the foreign affiliates and in the Stammhaus (in brackets) was as follows: production 56 percent (67 percent), distribution 23 percent (9 percent), administration 13 percent (7 percent), and R&D 8 percent (17 percent).

[34] Especially R&D in pharmaceuticals was increasingly shared with the affiliates in the United States.

[35] Ciba, annual reports 1958 and 1968.

the three joint works (Gemeinschaftswerke) of Ciba, Geigy, and Sandoz, established within the IG during the interwar period.[36]

In the 1950s, the main product lines remained the same as in the interwar period, namely, pharmaceuticals (including cosmetics), dyes and other "technical applications," and "technical synthetics," mainly specialities based on synthetic resin. But, of course, there were continuous product and process innovations within these product lines, and their share in total sales changed. In 1950, dyes and other applications contributed 52 percent; pharmaceuticals (including cosmetics) 40 percent; technical synthetics 8 percent. By 1954, the shares of pharmaceuticals and dyes equalized, and, in 1960, total production was divided among pharmaceuticals – 46 percent, dyes and other applications – 34 percent, and technical synthetics – 20 percent. In the 1960s, there were regular reports on the still rather small but rapidly growing agrochemicals department (mainly insecticides and some herbicides), and on photochemicals. R&D in agrochemicals had been begun in 1944, but sales started only in the 1960s and increased somewhat toward the end of the decade because of new application methods.[37] By 1969, the main product lines were regrouped into pharmaceuticals – 39 percent, dyes and other applications – 31 percent, and other products (including synthetics, agrochemicals, photochemicals, and rare metals) – 30 percent.

In the case of Geigy, total sales (figures of the Stammhaus in brackets) were about 250 million CHF (80 million or 32 percent) in 1950, 365 million CHF (314 million or 36.3 percent) in 1960, and 3,161 million CHF (830 million or 26.35 percent) in 1969.[38] Total employment within the whole group was 12,779 (4,064 or 31.8 percent) in 1960, and about 25,675 (7,569 or 29.5 percent) in 1969. In contrast with Ciba, the share of employment in Switzerland was almost maintained during the 1960s,

[36] Cincinnati Chemical Works, Inc., Cincinnati (USA); Clayton Aniline Company Ltd., Clayton, Manchester (GB); Società Bergamasca per l'Industria Chimica, Seriate (I).

[37] Ciba, Annual Report 1969, 18: The marketing strategy was to offer complete "agroprojects," including not only the delivery of the insecticides or herbicides but also their application with small aeroplanes of the CIBA-Pilatus AG. It is mentioned that "die behandelte Gesamtfläche umfasst rund 850,000 Hektaren, nahezu das Dreienhalbfache des offenen Ackerlandes der Schweiz. Die auf diese Weise erzielte beträchtliche Verbesserung der Ernteerträge darf gleichzeitig als wirksamer Beitrag an die privatwirtschaftliche Entwicklungshilfe betrachtet werden (the total area comprises about 850,000, almost three and a half times the open agricultural land of Switzerland. The considerable improvement of crop yields achieved by this treatment can also be considered an effective contribution to private economic development aid)."

[38] Sales figures were not published in the annual reports until 1957. These approximate figures are taken from a diagram in the Annual Report 1957.

whereas sales increased more rapidly abroad. The main reason probably was that R&D was still more concentrated on the Stammhaus than was the case with Ciba. The foreign works also increasingly produced specialities, and some R&D centers were established in the United States and in Great Britain, although apparently these centers were limited to specific product lines. In the 1960s, building investments were on the average 13 percent of total sales per year, and about two thirds of the new investments were made in foreign countries.[39] As was the case with Ciba, the share of exports in total sales from the Stammhaus remained stable at about 90 percent, but the share of total exports delivered to the affiliates was somewhat smaller – about 50 percent. Unfortunately, no figures are given in Geigy's annual reports with regard to R&D expenditures, and there is only scattered information about foreign affiliates. However, from the number of countries mentioned, this network must have been quite extensive and was continuously expanded.[40]

The main difference between the two firms was that Geigy was growing more rapidly than Ciba, especially in the 1960s.[41] This was probably because in the case of Geigy the new pharmaceutical department as well as agrochemicals expanded very quickly. The broader diversification of Ciba into agrochemicals and other new product lines (photochemicals, rare metals) implied a wider array of different technologies and markets, and this diversification of know-how may have lowered the capacity for growth. But, except for the more rapid pace, the pattern of growth of Geigy was just as *continuous* and *additive*. The main product lines remained the same as in the late 1930s: dyestuff, auxiliary and refining products for textiles, synthetic tanning agents, pharmaceuticals, and insecticides. At the end of the 1960s, these divisions were regrouped into dyes and textile chemicals, industrial chemicals, pharmaceutical specialities, and agrochemicals.[42] The *geographical diversification* followed about the same pattern as Ciba's, with the difference that Geigy concentrated more on the U.S. market than Ciba.[43]

[39] Geigy, Annual Report 1968, 44. No figures were published for the earlier period.

[40] See, for example, Geigy's annual reports for the years 1955, 9 ff., 1968, 17 ff.

[41] Throughout the interwar period Ciba was considered, within the IG, about twice the size of Geigy. When the two firms merged in 1970, the relation of their inner worth was assumed as equal.

[42] No figures are given in the annual reports of Geigy with regard to the shares of the different product lines in total sales.

[43] According to the first annual report of Ciba & Geigy AG for 1970, the geographical distribution of sales in 1969 was as follows: In the case of Ciba, Europe 49 percent (EEU 26 percent, EFTA 17 percent), North America 24 percent, Latin America 10 percent, Asia

The main topics raised and the main problems perceived were also largely similar. Immediately after the war and in the early 1950s, the major concerns were political and monetary instability, the varying conditions for international trade and increasing competition. There were no domestic barriers to trade or capital and payment flows; but, given that about 90 percent of total sales were exported (of which 50 percent to 60 percent to the affiliates), dependence on trade restrictions in foreign countries was very high. In the course of the 1950s, some barriers to trade were abolished, but with the division of Europe into the EEU and EFTA, new ones were established. Furthermore, the U.S. tariffs for chemicals remained prohibitive.[44] Because of the overall context of growth and the rapidly rising demand, export possibilities were not seriously hampered, especially for the specialities. Barriers for trade were overcome by expanding and establishing new production facilities abroad, especially in the United States. Already in 1959, Ciba had decided to expand its chemical works in the countries belonging to the newly formed EEU. These investments were largely self-financed so that both firms were quite independent from the capital market, both at home and abroad.

In both firms and both decades, expansion was larger abroad than in the Stammhaus, and this certainly put a strain on the organizational structure. Both firms went through a period of reorganization before the merger. In countries with several affiliates, holding companies were established in order to have more control over activities in these countries. The motives for the strong drive toward internationalization in the postwar decades were twofold. On the one hand, barriers to trade persisted in various forms, notwithstanding the movement for trade liberalization within the OECD, the GATT, and the various Kennedy-Rounds. In the 1960s, the tendency to expatriate production because of barriers to international trade was reinforced by the division of Europe into two separate trading blocks, and other new restrictions such as the changing attitude of domestic governments toward international trade and multinational firms in Asian and South American countries. By contrast, and in addition to these "pull factors," the boundaries for growth in Switzerland became a major "push factor" for shifting production abroad. In the

---

11 percent, Africa, Australia and Oceania 6 percent; in the case of Geigy, Europe 35 percent (EEU 18 percent, EFTA 12 percent), North America 50 percent, Latin America 8 percent, Asia 3 percent, Africa, Australia and Oceania 4 percent.

[44] This was caused by the American Selling Price System introduced after the war.

1960s, the perceived problems clearly shifted from the demand to the supply side. The main bottleneck for growth was the extreme shortage of labor within Switzerland, especially qualified personnel, notwithstanding the relatively large number of "Grenzgänger" (employees working in Basle and living just over the border).[45] Toward the end of the decade, the shortage of labor became even more pronounced because the political pressure to restrict the number of foreign employees in Switzerland increased. Large countries with a high standard of living and strong positions in chemical research, such as the United States and Great Britain, were the most attractive locations for direct investments, especially in the case of Geigy.

In the 1950s and 1960s, the two firms expanded in an evolutionary, path-dependent way, diversifying into activities that were to some extent different but required similar capabilities.[46] Even the merger concluded in 1970 fits into this pattern of continuous and additive growth. The main objective of the merger was to maintain the potential for further growth by reducing costs, raising productivity, and counteracting the increasing shortage of resources, especially of qualified labor and scientific personnel.[47]

Although the activities of the two firms were largely similar and complementary, consolidating their activities in Switzerland and around the world took several years. It was accomplished step by step, avoiding sharp discontinuities. Basically, the organizational structure remained decentralized, leaving ample room for maneuver for the foreign affiliates. Some of Ciba's holdings in the United States had to be sold in the early 1970s because of antitrust laws, but total sales continued to increase up until 1974. The integration of Ciba and Geigy accompanied the expansion of the network of foreign subsidiaries and production facilities. Foreign sales agencies were transformed into subsidiaries; production facilities

---

[45] The shortage of labor is mentioned for the first time in the annual report of Ciba, 1959. To the limits of growth within Switzerland, see, for example, the Annual Report of Geigy, 1965, 6 ff. According to the Geigy management, the domestic economic policies, especially with regard to the labor market, were causing additional shifts of production to foreign countries. See also the Annual Report of Geigy for 1964, 5 ff.

[46] G.B. Richardson, "The Organization of Industry," *The Economic Journal* (Sept. 1972): 883–96.

[47] The story of the merger is told in Paul Erni, *Die Basler Heirat, Geschichte der Fusion Ciba-Geigy (The Marriage of Basle. History of the Merger Ciba-Geigy)* (Zurich: Buchverlag der Neuen Zürcher Zeitung, 1979.) For our purpose, it is not of much value, because it contains little information about the background, the motives, and the consequences of the merger. The chronology of the merger and a selection of letters and newspaper articles are useful.

and research centers were established and expanded, both in Switzerland and abroad – mainly in the United States, Great Britain, and India.

Notwithstanding this continuous growth of the firm, the view of the overall economic context began to change. Already in the annual report of 1971, the sociopolitical environment was perceived as increasingly restricting the firm's room for maneuver. In addition to the usual complaints about labor shortage and high inflation rates, the outstanding event of the breakdown of the Bretton Woods system presented a new problem. Because of what was at first perceived as a temporary overvaluation of the Swiss franc, the competitive advantage of the Stammhaus in the production of new specialities was questioned, and some investments planned for the expansion of production facilities in Switzerland were postponed. In the following years, the share of the Stammhaus in total investments continued to decline.

There was a wide array of other problems that came to the fore rather suddenly, although they had been around for several years.[48] One was the increasing gap between the rising costs of R&D in pharmaceuticals, and the pressure of the public health services on reducing patent protection and prices, which was experienced as progressive degeneration of patent laws. Another disturbing development was linked with a strange disease in Japan that had been noticed for some years but that was now linked, rather convincingly, to pharmaceuticals sold by Ciba.[49] The most disturbing new phenomenon was the society's increasing concern for environmental protection as a precondition for preserving the health and well-being of the population. It seems that the firms, and the chemical industry in general, had to a considerable extent neglected the ecological and toxicological dimension of their products and production processes during the decades of rapid growth. Geigy's annual reports mention the protection of the environment only from the middle of the 1960s onward, and mainly with regard to water and air pollution.

From 1971 onward, the environmental dimension of chemical products and processes begin to invade all areas of activities. Gradually, the way chemical products were perceived shifted from stressing their benefits to enhancing their external costs and unintended effects. The new rules and social norms for the protection of the environment and the well-being of the population implied high investments in purification of

---

[48] All these problems are mentioned in the Annual Report of 1971.
[49] This story is told by Olle Hansson, *Inside Ciba-Geigy*, Universitets Forlages, Oslo. German translation: *Ciba-Geigy Intern* (Zurich, 1987).

waste water and combustion plants. That part of environmental demand was accepted with much good will, as the firm was convinced that it could cope by relying on its technical competencies and large financial means. It was a costly demand, but it did not increase uncertainty and was expected, in the long term, to reduce risks and create new business opportunities. The more disturbing effects were, on the one hand, that the new aspirations made R&D activities more complex and, on the other hand, that it fundamentally changed the internal and external selection criteria for innovative activities.[50] The first herbicides and insecticides had been largely developed by chemists and were based on highly specialized but also very limited scientific knowledge. The experience with these products made it necessary to include other disciplines in R&D activities with a very different basis of knowledge, and they called forth the interventions of governments and other organizations. The increasing complexity of R&D and regulating efforts can be perceived in other product lines too, especially in pharmaceuticals. Some of these developments can be traced back to the 1960s, but it seems that only in the 1970s did the far-reaching consequences of all these tendencies begin to have an impact on the prospects of the firm and on the management's vision for the future.

Up until 1974, the forces supporting growth were still strong. Cash flows were high notwithstanding lower profits, and the positive attitude clearly prevailed. But, from 1975 onward the downward pressures became dominant. Between 1975 and 1980, total sales fluctuated between 9,000 and 10,000 million CHF (Stammhaus: between 3,500 and 4,000). Only at the end of the 1970s was there again a marked increase of total sales. Investments dropped to about 60 percent of the maximum invested in 1975 (both in Switzerland and abroad), and employment stagnated at around seventy-five thousand persons (twenty-one thousand in Switzerland) until 1978. Declared profits regained the level of 1973 only in 1981.[51]

The obvious explanation for the economic downturn is the impact of the oil-price shock of 1974. But, from the perspective of the firm this

[50] For the concepts of internal and external selection criteria, see Maureen McKelvey, "R&D as Pre-market Selection: Of Uncertainty and Its Management," Paper written for the 6th International Joseph A. Schumpeter Society Conference, "Entry, Competition and Economic Growth – The firm, the Innovator, the Entrepreneur and Market Competition." June 3-5, 1996, Stockholm, Sweden. With reference to Maureen McKelvey, *Evolutionary Innovations: The Business of Biotechnlogy* (New York: Oxford University Press, 2000).

[51] Ciba-Geigy, Annual Report, 1980 and 1981.

was a minor and temporary disturbance, largely confined to that year. The overvalued Swiss franc, and high inflation rates in many countries were perceived as the main causes for the fall of demand and low profits. In order to get a clearer picture of the real performance of the firm, accounting systems were introduced that eliminated the impact of inflation and exchange-rate fluctuations.[52] These new accounting methods not only revealed that losses were, to a large extent, mainly due to exchange rate fluctuations, they also revealed the difference between nominal and real profits in periods with high inflation.[53] But the dominant growth strategy of the firm did not change: diversification of products and markets as well as the intensification of R&D, now on a broader basis, continued to be perceived as the best means of coping with dependence on foreign markets and foreign policies. Expenditures for R&D increased from 6.4 percent at the beginning of the decade, to an average of over 8 percent in the late 1970s.[54] There were only minor modifications in the composition of the main product lines, except that the share of other products increased and the dyestuff division decreased considerably.[55]

From 1978 onward, the annual reports contain a list of the main subsidiaries and partnerships as well as their functions. Compared with similar lists published in the 1960s, the process of "globalization" in the sense of geographical expansion of the Konzern continued during the 1970s. In 1971, eighty-eight companies are mentioned in forty-four countries, of which fourteen were in Switzerland.[56] In 1980, the Konzern consisted of 114 companies in 56 countries, of which 17 were in Switzerland.[57] With 60 percent of R&D expenditures and 25 percent of total employees (29 percent in 1971), the role of the Stammhaus within the Konzern remained quite important.[58] The share of Switzerland in total investments

---

[52] Since the first drastic upgrading of the Swiss franc in 1971 the annual reports usually indicated growth rates of annual sales with and without a change of exchange rates.

[53] Except for Switzerland, where inflation rates dropped in the sharp recession of 1975, inflation rates in other countries remained at a high level.

[54] Ciba-Geigy, Annual Report 1980, 62.

[55] Ciba-Geigy, Annual Reports 1971, 24, and 1980, 11.

[56] Ciba-Geigy, Annual Report 1971, 63–65.

[57] Ciba-Geigy, Annual Report 1980, 63–65: seventy-nine companies were owned to 100 percent, twenty to more than 50 percent, fifteen to less than 50 percent; twenty companies were active in R&D, production and distribution (sales and marketing), forty-three in production and distribution, three in R&D and production, one only in R&D, nineteen only in distribution, twenty only in production, and eight in special functions (financing, insurance).

[58] Ciba-Geigy, Annual Report 1980, 15. According to the Annual Report of Ciba-Geigy for the year 1980, 9, the Stammhaus consisted of Ciba-Geigy AG, Basle, the works in

(building and equipment) was about 40 percent during the 1970s (Ciba in the 1960s – 40 percent; Geigy 33 percent), and the share of sales from Basle even increased slightly to 35 percent of total sales (32 percent in 1971).[59] In 1980, the geographical distribution of sales reveals no major changes compared with the figures of 1971, except that sales to North America declined because of the already mentioned divestments after the merger. The share of sales within the Swiss market was merely 2 percent. Ciba-Geigy was, in its own perspective, a truly multinational firm and practically without a home market.[60]

The performance of the firm in the early 1980s was perceived as quite satisfactory when compared with the rather depressed state of the European chemical industry.[61] From 1984 until 1989, a series of very good years were registered. Ciba-Geigy raised its total sales considerably, from about 12,000 million CHF in 1980 to a maximum of over 20,000 million CHF in 1989. Yearly growth rates of total sales became rather unstable mainly because of currency fluctuations and varied between a minimum of −12 percent in 1986 and +19 percent in 1984. The total number of employees varied between seventy-nine thousand and eighty-one thousand until 1985, then rose to a maximum of about ninety-four thousand in 1990. New investments in building and equipment were 6 percent of total sales in the early 1980s and increased to 10 percent at the end of the decade. Also, total expenditures in R&D increased from about 8 percent of total sales in the first half of the 1980s to over 10 percent at the end of the decade.

The most remarkable feature of the 1980s is that the activities of the Stammhaus expanded almost at the same pace as those of the Konzern.

Switzerland and in Grenzach (Germany); Ciba-Geigy Werke Schweizerhalle AG, Ciba-Geigy Werke Kaisten AG; Ciba-Geigy Münchwilen AG; Schweizerische Handels- und Immobiliengesellschaften.

[59] Ciba-Geigy, Annual Report 1980, 8–9, 14.

[60] Ciba-Geigy, Annual Report 1980, 14, and 1981, 19: "Die Ciba-Geigy verfügt praktisch über keinen Heimmarkt (. . .) und ist als echt multinationales Unternehmen voll auf die von den Konzerngesellschaften erzielten Ergebnisse angewiesen (Ciba-Geigy is practically without a home market (. . .) and depends, as truly multinational enterprise, completely on results achieved by the Konzern)."

[61] The following figures are recapitulated in the Annual Report of Ciba-Geigy for the year 1989, 64–65, and 1990, 58–59. The development of Ciba-Geigy and Sandoz since the 1970s and the formation of Novartis in the middle of the 1990s is described extensively in Christian Zeller, *Globalisierungsstrategien – Der Weg von Novartis (Globalisation Strategies – The Path of Novartis)* (Berlin; New York: Springer, 2001), Chapter 5. The author focuses mainly on the path of development and the internationalization strategies adopted for the pharmaceuticals divisions.

Sales from Basle increased from under 4,000 million CHF in 1980 to almost 8,000 million CHF in 1990, and in both years the share in total sales was about 34 percent. At 38 percent, Switzerland's share in new investments in the 1980s was about the same as in the 1970s (40 percent).[62] The Swiss share of employment in the total employment of the Konzern remained stable at about 25 percent. Although employment in the other European countries fell by 5.1 percent, employment in Switzerland even rose by 7.1 percent).[63] The geographical distribution of sales shifted from Europe and Latin America to North America and Asia.

The first signs of a somewhat modified strategy of growth already can be perceived in the early 1980s. The pattern of growth was rather stable with regard to the main product lines. The share of pharmaceuticals increased moderately, while the share of "dyes and chemicals" continued to decline. Of course, the composition of products within each division changed, especially with regard to "other products."[64] Within pharmaceuticals, the categories "generics" and "OTC-products" became more important. Each division was evaluated with regard to its future prospects of growth, and a few marginal products were discarded, but the thrust remained on discovering promising new areas for further expansion and diversification. The main guidelines – priority to innovation and R&D, diversification into related activities and new markets – were not questioned throughout the 1980s, but were applied in a more flexible and more decentralized way.

Internationalization accelerated because diversification into new activities requiring new capabilities was often achieved by buying foreign firms already active in the field, especially in the United States. The new policy of takeovers widened the scope for further R&D activities, whereas development and production of new products were increasingly

---

[62] Ciba-Geigy, Annual Report 1989, 7, and 1990, 7, 19.

[63] Ciba-Geigy, Annual Report 1989, 22. This result contrasts with the results of Christian Zeller, *Globalisierungsstrategien (Gobalisation Strategies)*, 228, in which the figures for Switzerland and for the rest of Europe must have been exchanged by mistake. Consequently, employment for the rest of Europe remains quite stable and employment in Switzerland declines, which gives a distorted view on the role of the home country in this period.

[64] In 1970, the other products were synthetics and additives, consumer articles and photochemicals; in 1980 consumer chemicals had been regrouped within Airwick and photochemicals within Ilford. A separate division was established for electronic instruments, a product line that Ciba had taken up in the 1950s; by 1990 the industrial chemicals had been regrouped into additives, synthetics, pigments, Mettler-Toledo, Gretag/Spectra, Physics (sold in summer 1990) and Ciba Vision (taken up in the early 1980s). Airwick was sold in 1988.

managed by the foreign firms specialized in this field. Employment share in North America increased by about 50 percent, while the share of expenditures in R&D in Switzerland fell to 50 percent (1980 – 60 percent) of world wide R&D.[65] Total expenditures increased continuously from 1,076 million CHF in 1981 to over 2,000 million CHF at the end of the decade.[66] The process of "globalization" in the sense of geographical expansion of the Konzern continued during the 1980s. In 1990 (figures for 1980 in brackets), the Konzern consisted of 130 (114) companies in fifty-nine (fifty-six) countries, of which twenty-one (seventeen) were in Switzerland.[67]

In the early 1990s, expansion slowed and major structural changes were initiated. In the case of Ciba-Geigy, new guidelines already had been introduced in 1990 as a means to assure the future of the firm. The economic and political environment was perceived as increasingly difficult because of worldwide competition and new kinds of instability, especially the interference of governments in the health sector and exchange rate fluctuations.[68] Major objectives of the new "Vision 2000" were, first, focusing on the main product lines and on those activities where synergies could be exploited, and, second, accelerating the development of new products by expanding and establishing new research facilities in the United States, England, France, Germany, and Japan.[69] Location-specific advantages abroad in the form of innovative capacities were an increasingly important motive for FDI and alliances with foreign firms and universities. However, it was certainly not a new motive. Expanding R&D activities abroad was partly explained using the same arguments as in the 1970s and 1980s: the short supply of qualified personnel in Switzerland and the possibility of participating in the innovative

---

[65] Ciba-Geigy, Annual Report 1990, 12.

[66] Ciba-Geigy, Annual Report 1990, 59.

[67] Ciba-Geigy, Annual Report 1990, 79–82. In 1990 (figures for 1980 in brackets) ninety-two (seventy-four) companies were owned to 100 percent, nineteen (twenty) to more than 50 percent, and nineteen (fifteen) to less than 50 percent; twenty-three (twenty) companies were active in R&D, production and distribution (sales and marketing); forty-eight (forty-three) in production and distribution; two (three) in R&D and production; three (one) only in R&D; one (one) in R&D and distribution; nineteen (nineteen) only in distribution; twenty (twenty) only in production; twelve (eight) in special functions (financing, insurance, services) and two (–) in distribution and services/financing.

[68] Ciba-Geigy, Annual Report 1990, 3. Compared with 1989, total sales had decreased by about 5 percent and total profit by about a third. But with 1,032 million CHF, total profit was still quite high.

[69] The different product groups were reorganized into more homogeneous divisions in order to obtain a clearer picture of their performance.

capacities located in foreign countries. New arguments included the opposition of some political groups to methods and objectives of research (as already mentioned) and, especially to Ciba-Geigy's efforts to establish a research center for biotechnology in Basle. More generally, it was the uncertainty about future legislation in this field in Switzerland, because the traditional, widely shared consensus on the importance of R&D for the international competitiveness of the Swiss industry had been questioned with regard to biotechnology. Tired of waiting for the final decision of the local authorities, Ciba-Geigy decided to establish the new center of research just over the border, in France.[70] The trend toward expanding R&D activities abroad can also be perceived on the level of the chemical industry.

Another disadvantage of the home country, which may have promoted internationalization of R&D, was the rejection of the adherence to the European Economic Area, mainly because of its effect on the flexibility of the labor market, especially the supply of qualified personnel. With regard to production, it was of minor importance because the large chemical firms were all well placed in the European countries. Furthermore, the corner location of Basle between three countries facilitated shifting activities over the border and locating activities in an EU country. The management of Ciba-Geigy justified the expansion of R&D and other activities abroad with exchange rate instabilities and the rising gap between the share of costs in Switzerland (25 percent) and the share of sales invoiced in Swiss francs (8 percent) or sold in the Swiss market (2 percent).[71] This was hardly a new phenomenon, but it was now perceived as a disadvantage because of its impact on profits. Given the new financial policy described below, declaring high profits was more important now than in earlier decades.[72]

A third main objective was to take advantage of the deregulation of global capital markets, mainly by participating in the large U.S. capital

---

[70] Neue Zürcher Zeitung, 24.4.1991, Nr. 94, 33; Tages-Anzeiger, "Ciba-Geigy baut Biotechnikum ennet der Grenze (Ciba-Geigy is building a 'Biotechnikum' over the border)," 17.12.1991.

[71] For example, in Neue Zürcher Zeitung, "Ein Plädoyer für ertragsstarke Stammhäuser (A plea for profitable 'Stammhäuser')," Nr. 132, 11.6.1982.

[72] Finanz und Wirtschaft, 24.4.1991, Nr. 31, 15; Neue Zürcher Zeitung, 24.4.1991, Nr. 94, "Ciba-Geigy und der Standort Schweiz. Zunehmende Internationalisierung – Rolle der Heimbasis (Ciba-Geigy and location in Switzerland. Increased internationalisation – Role of the home basis)."

market.[73] This decision implied a completely new financial policy with considerable consequences for the ownership structure of the firm. Throughout the 1980s, nominal share capital and long-term debt had remained stable at about 5,400 and 3,000 million CHF, and investments had been largely self-financed.[74] The firm was quite independent from the capital market, and national ownership and control was assured by registration rules that allowed the exclusion of foreign shareholders. Shareholders' interests were hardly mentioned in the annual reports, and certainly not in an outstanding way.[75] Already in 1991, new accounting standards were introduced for the whole Konzern, and ROE increased from 6.4 percent in 1991 to 10.9 percent in 1993.[76] In 1992, the total number of registered shares suddenly increased from about 5.5 to 29.4 million, mainly because each share was split into five shares, a clear sign that the management wanted to expand demand for Ciba shares by lowering their price.[77]

Restructuring soon expanded to the whole Konzern. A new organizational structure was implemented, the number of divisions increased from eight to fourteen, and the number of employees in Basle was reduced by sixteen hundred. Even the name of the Konzern was changed to Ciba in 1992, and, in the same year, the financial results were presented to the press in London and not as usual in Basle.[78] The market rate of registered shares rose by almost 47 percent, mainly because of the new registration rules, and in 1993, foreign ownership reached one-third of total share capital. With "proactive communication" allowing all stakeholders to judge the activities of the firm on the basis of clear facts, the firm aimed at correcting prejudices and wrong perceptions, especially of investors.[79] In effect, it was not just a problem of communicating

[73] Ciba-Geigy, Annual Report 1990, 10–11.
[74] The cash flow fluctuated between 8.7 percent and 13 percent of total turnover. Ciba-Geigy, Annual Report, 1989, 64–65, and 1990, 58–60.
[75] The new financial policy was also explained as a means to raise the price of shares and prevent unfriendly takeovers. But because voting power for every Swiss and foreign shareholder was limited to 2 percent of total registered share capital, this argument is not very convincing. Tages-Anzeiger, "Auch bei Ciba-Geigy fallen Mauern (Also at Ciba-Geigy walls are crumbling)," 22.2.1990.
[76] Ciba-Geigy, Annual Report 1991, 3, and Ciba, Annual Report 1993, 41–42. Ciba introduced the International Accounting Standards (IAS) in 1993.
[77] Ciba, Annual Report 1992, 64–66.
[78] Ciba-Geigy, Annual Report 1991, 4–5. Finanz und Wirtschaft, Alex Krauer: "We are looking for shareholders," 28.3.1992, Nr. 25, 17.
[79] Ciba, Annual Report 1992, 4.

the new identity of the firm to its environment, but also of creating such a new identity within the firm.

Rising sales and profits seem to confirm the successful transformation of "Vision 2000" into adequate strategies, structures, and procedures.[80] Declared total earnings rose continuously from 1,033 million CHF in 1990, to 2,156 million CHF in 1995. But in the year before the merger with Sandoz in 1996, total sales dropped by about 6 percent. Total employment decreased by about 7 percent since 1992 and new investments fell from over 10 percent in the late 1980s to 6.2 percent in 1995. R&D expenditures almost remained at the high level reached at the end of the 1980s. Unfortunately, the role of the Stammhaus cannot be compared with the earlier periods, because figures on its share within the Konzern are omitted since 1993. The new global perspective is also reflected in a broader division of the geographical sales areas: Europe, Western Hemisphere, Eastern Hemisphere.[81] Each year, the report is presented in a new form, some information is added and some omitted, which makes comparisons with previous years rather difficult. Divisions are regrouped into sectors, product lines are expanded in one year and given up in the next. The frequent changes, especially with regard to the industrial divisions, are a clear signal that the three sectors – health, agriculture, and industrial – were getting out of balance. "Vision 2000" became increasingly blurred, and broad product diversification was no longer perceived as a means for spreading risks and stabilizing performance.[82] The process of globalization in the sense of geographical expansion of the Konzern continued. In 1995 (figures for 1990 in brackets), the Konzern included 156 (130) companies in sixty-four (fifty-nine) countries, twenty-one (twenty-one) in Switzerland.[83] The six new countries were Poland, Czech Republic, Romania, Russia, Slovenia, and Hungary.

---

[80] The following figures are recapitulated in Ciba Finanzübersicht (Ciba Financial Review) 1995, 41. R&D expenditures were 10.6 percent of total sales in 1992 and 9.6 percent in 1995.

[81] Ibid., 19, we are informed that "Western Hemisphere" stands for North America and Latin America, "Eastern Hemisphere" for Africa, Asia, Australia, and Oceania.

[82] The reports for the years 1993, 1994, and 1995 contain explicit statements in favor of the industrial divisions, but the figures reveal their relatively weak performance.

[83] Annual report Ciba-Geigy 1995, 36–40 (figures for 1990 in brackets): 112 (92) companies are owned to 100 percent, 25 (19) to more than 50 percent, and 19 (19) to less than 50 percent; 28 (23) companies were active in R&D, production and distribution (sales and marketing), 59 (48) in production and distribution, 2 (2) in R&D and production, 2 (3) only in R&D, 4 (1) in R&D and distribution, 28 (19) only in distribution, 16 (20) only in production, 16 (12) in special functions (financing, services), and 1 (2) in distribution and services/financing.

In the middle of the 1990s, the core of the chemical-pharmaceutical industry in Basle was thoroughly restructured according to a completely different view on global competitiveness. Ciba and Sandoz merged into Novartis and, concomitantly, two new firms, Ciba Specialty Chemicals and Clariant, were established for those product lines that no longer fit into the production program of the new firm.[84] The Konzern consisted of 275 subsidiaries in 142 countries. Novartis started with those three main product lines, which were perceived as having the best prospects for growth and high profitability: health, agribusiness, and nutrition.[85] The number of employees was about 116,000 and total sales were 36,233 million CHF (Europe 39 percent, North and Latin America 44 percent, Asia, Australia, and Africa 17 percent).[86] Already in 1999, the decision was taken to divest the agribusiness (without animal health). The new operating sectors in the year 2000 were pharmaceuticals, generics, consumer health, Ciba Vision, and animal health. The number of total employees decreased to about sixty-eight thousand, and the share of "The Americas" in total sales increased to 50 percent.[87]

What were the reasons for the major changes of strategy and structure of the main chemical firms in Basle? On the basis of the sources used in this study, the answer is that these changes were mainly a result of the orientation toward the U.S. capital market in the late 1980s. The impact on product diversification and on the new external and internal communication policies was a consequence of the new financial policies, especially the decision to internationalize ownership structures. The traditional additive growth pattern and broad product diversification had to be gradually adapted to the preferences of investors in the U.S. market. In the course of this learning process, the expected costs and benefits of internalizing activities based on very different technical knowledge

---

[84] Swiss Chemicals in Perspective, Pictet & Cie publications, Geneva (April 1999), 24: Clariant (annual sales of CHF 10 billion in about 60 countries and some 31,000 employees in 1999) was a spin-off from Sandoz in summer 1995. It integrated Hoechst's speciality chemicals businesses two years later. Ciba Speciality Chemicals (CHF 9 billion in 117 countries and some 21,000 employees) was spun off in spring 1997 as part of the merger between Ciba-Geigy and Sandoz.

[85] Novartis, Annual Report 1996, 10–11. Health (Pharma 75 percent, Consumer Health 10 percent, Generics 8 percent, Ciba Vision 7 percent); the Agribusiness (Crop Protection 73 percent, Seeds 16 percent, Animal Health 11 percent); and Nutrition (Infant and Baby Nutrition 37 percent, Health Nutrition 35 percent, Medical Nutrition 21 percent, Others 7 percent).

[86] Novartis, Annual Report 1996, 5.

[87] Novartis, Annual Report 2000, 1–2.

and sold in different markets were reevaluated and replaced by a new strategy combining a global market orientation with the concentration on specific product segments.[88]

CONCLUSIONS

The chemical industry has been a particularly dynamic part of the Swiss economy. Between 1975 and 1995, production increased at an average annual rate of 5.7 percent, while the average growth rate of all industries was only 3.6 percent. At the end of the 1990s, the share of chemicals in total industrial exports reached 28 percent.[89] Distribution of world exports with Western Europe as the main market remained extremely stable, but the composition within the main product lines changed considerably. At the beginning of the twenty-first century, about 90 percent of the products of the Swiss chemical industry are specialities.[90]

The 1950s and 1960s were years of rapid and continuous growth at the industry level as well as at the two firms, Ciba and Geigy. Exports and the network of subsidiaries established in foreign countries were continuously expanded, which implies that what FDI decided in a context of protectionism before the war was perceived as rational also in a context marked by increasing but rather slow liberalization. The main discontinuity compared with the interwar period was the dissolution of cartel agreements, already initiated in the late 1930s. In both decades, the firms expanded more rapidly abroad than at home. In the 1970s, economic performance was more mixed and influenced by major recessions, the growing instability of the world economy and new restrictions imposed by the sociopolitical environment. Not only the economic context, but also the basis of scientific research and the demand of society with regard to chemical products became more complex. New categories of uncertainty and risks had to be taken into account. The merger of Ciba and Geigy, which had been concluded as a means of shifting the boundaries

---

[88] See, for example, the comment of Dr. Alex Krauer (President of the Board) and Heini Lippuner (CEO) to the Annual Report 1992, or the interview with Dr. Alex Krauer in Neue Zürcher Zeitung, Nr. 94, 23.4.1992.

[89] *The Swiss Chemical-Pharmaceutical Industry,* 2000, 42, 49.

[90] The worldwide sales of the "top ten" Swiss chemical companies are differentiated, according to the areas of application, into the following main product groups: pharmaceuticals (55 percent), vitamins and fine chemicals (10 percent), plant protection agents and animal medicine (9 percent), and speciality chemicals, including dyestuff (23 percent). Ibid., 2000, 8, 12.

for growth, became, in effect, an advantage for coping with a context less favorable to expansion. The process of globalization continued, but the role of the Stammhaus remained quite important with regard to R&D and the production of specialities.

In the 1980s and 1990s, there was a remarkable growth of nominal exports on the level of the industry. The influence of economic fluctuations both in the domestic economy and in the main export markets was even less pronounced than in the 1970s. For Ciba-Geigy, the satisfactory growth rates achieved in the second half of the 1980s seemed, at first, to confirm once more the viability of the traditional strategy of broad diversification with regard to both products and markets. The share of the Stammhaus in the growth of the Konzern declined but less than in the other Western European countries. In these decades, expansion clearly shifted to North America and Asia. The forces behind this new growth pattern were rather complex: a thorough reevaluation of the costs and benefits of internalization with regard to different product lines, reinforced by a mixture of location specific disadvantages at home and advantages abroad.

Location, specific advantages and disadvantages in Switzerland became a major topic only since the beginning of the 1990s, when the process of emancipation from the national context shifted to new areas of decision making, namely, financial policies.[91] In the middle of the 1990s, broad product diversification, one of the main growth strategies that had been adopted in the interwar period and that had dominated the postwar decades, was replaced by a strategy of focusing on core sectors. Geographical diversification, which had proved so successful in absorbing risks, was maintained and even intensified. In this study, this fundamental change in the strategy and structure of the Swiss chemical industry in Basle is traced back to the deregulation of global capital markets and the decision of the firm to internationalize ownership. But the impact of this decision was caused by the integration of new standards, norms and aspirations of foreign shareholders and potential investors, which reshaped the main objectives, and, consequently, the strategies and policies of the firm. The internationalization of ownership can be

---

[91] For the concepts of "location specific, ownership specific and internalizing advantages," see John H. Dunning, "Trade, Location of Economic Activity and the MNE;" John H. Dunning, "Multinational Enterprises and the Globalization of Innovatory Capacity," *Research Policy*, 23 (1994): 67–88; John H. Dunning, *Alliance Capitalism and Global Business* (New York: Routledge, 1997).

observed in other firms, too, but the impact of this process on the strategy and structure of the firms may differ considerably.[92]

The role of the home base for the multinational firms in Basle during these decades is difficult to assess. At the end of the twentieth century, only 2 percent of worldwide sales of the top ten Swiss chemical-pharmaceutical companies were made within the domestic market; only about 14 percent of their employees were working in Switzerland; and more than three quarters of their exports were supplies to subsidiaries.[93] The share of the chemical-pharmaceutical industry in total Swiss FDI is estimated at about 40 percent.[94] Still, this industry employs 32 percent of total research personnel in Swiss industry and contributes about 41 percent to total R&D expenditures in Switzerland[95] but also about 52 percent to total R&D expenditures in foreign countries.[96] In the past, the Swiss chemical firms' "ownership specific advantages" were clearly based on R&D, which was perceived as vital for the firms' international competitiveness. This advantage was firmly grounded on a broad consensus within the Swiss society, shared by all groups of stakeholders (the management, the employees and their unions, as well as investors and shareholders). The new forms of internationalization (decentralization of R&D and internationalization of ownership) have weakened this special type of "national competitive advantage." Furthermore, the new strategies of the large multinational enterprises show that expected costs and benefits of product diversification; that is, the "internalizing advantages" have changed. New core competencies and new firm boundaries will inevitably have a major impact on location-specific advantages at home and in foreign countries.

What do these major changes of the last decade of the twentieth century imply for the future prospects of the Swiss chemical industry? At the beginning of the twenty-first century, there are again four large

---

[92] Nestlé was one of the first movers and a model for other Swiss firms.

[93] *The Swiss Chemical-Pharmaceutical Industry,* 2000, 10, 20, 43.

[94] Ibid., 18. The value of foreign direct investments of the Swiss chemical industry was estimated at about 42,000 million CHF in 1998, distributed on more than one hundred countries.

[95] The share of the sector "Machinery, Metal," in total research personnel is 23 percent and in total R&D expenditures 17.6 percent; the share of the sector "Electrical Engineering" in total research personnel is 19 percent, and in R&D expenditures 17.8 percent. Ibid., 34, 36.

[96] *Forschung und Entwicklung in der schweizerischen Privatwirtschaft (Research and Development in the Swiss Private Economy)* 2000, ed. Economiesuisse (Zurich, 2000), 16.

multinational chemical firms in Basle, each with a somewhat different profile with regard to size and product lines, but all with the objective of expanding rapidly and on a global scale, using all forms of internationalization. The tendency to grow outside of the country will hardly decline. But, besides and partly as spin-offs of the large multinationals, about fifty new biotechnology firms and quite a number of new IT firms attracting know-how and expertise from various countries have been established in Basle.[97] Furthermore, another kind of location-specific advantage, already important at the beginning of the twentieth century, may have been reaffirmed: the advantage of Basle with regard to other European locations, namely its position in the "Dreiländereck," at the border of Switzerland with Germany and France. This corner position allows the exploitation of different market conditions, especially with regard to the labor market and production costs quite generally, as well as different national institutions and attitudes toward new fields of research. So, Basle is about to confirm its reputation as regional center of the chemical-pharmaceutical industry within Switzerland. Of course, prosperity and growth of new firms in the new technologies as well as location-specific advantages have become highly unstable. But, if Swiss firms will participate in developing the new capabilities required in the future chemical-pharmaceutical industry, there is good reason to expect that they will again be located in Basle.

---

[97] NZZ, "Klumpenrisiko oder Chancenvielfalt? Zur Dynamik des Wirtschaftsstandorts Basel (Lump Risk or Manifold Chances? The Dynamics of Basle as Economic Location)," 12.9.2000. The Swiss chemical-pharmaceutical industry is still composed of a large number of small and medium-sized companies. Only 9 out of about 340 companies have a workforce of more than 1,000 and there are almost 200 companies with 20–99 employees. *The Swiss Chemical-Pharmaceutical Industry,* 2000, 52–53.

8

Development Patterns in the Petrochemical Industry in the Nordic Countries, 1960–2000

GUNNAR NERHEIM

The rapid growth in oil consumption in the 1950s and 1960s was accompanied by an even greater increase in demand for a range of synthetic materials. These appeared in all manner of guises, as rainbow-hued plastics, sheer nylons, wrinkle-proof fabrics, lather-free soaps, phonograph records, toys, floor tiles, carpets, electrical fittings, insulation, packaging, car tires, rainwater pipes, guttering, and countless others.

The growth rate of synthetic materials after World War II was extremely high because synthetic materials brought outstanding technical and cost advantages to a wide range of applications. Plastics were light, easy to fabricate and install, frequently had good insulation, and had excellent resistance to corrosion and pests. Some of the newer plastics materials could be used at very high temperatures and were extremely strong but still relatively inexpensive. By contrast, their virtues also led to problems of waste disposal and pollution. Plastics are today one of the world's main groups of industrial materials. World plastics consumption by weight exceeded that of nonferrous metals by 1970 and is far greater in volume. Synthetic fibers accounted for nearly half of total fiber consumption in 1990. From the 1940s to the 1970s, substitution for older materials played a large part in the very high rates of production and consumption growth. When synthetic materials had reached a sufficiently high level of consumption, their rate of growth slowed down. This happened in the 1970s.

During the two decades from 1950 to 1970, the petrochemical industry grew to maturity at an astonishing pace, easily outstripping the general rate of economic growth. Great increases in demand stimulated the construction of plants of ever greater size and the search for new products and processes. The resulting economies of scale, combined with a high rate of technological innovation, brought down costs and increased the range of synthetic materials, sold at prices that more and more people could afford. This stimulated further increases in demand and led to further rounds of expansion.

Many oil companies entered the petrochemicals business in the 1950s. They might have lacked technical and marketing know-how in chemicals but possessed other important attributes. One of the most important was size. The economies of scale available from increases in petrochemical plant size could only be realized by massive investments, which, together with the high costs of research, could be afforded only by a few very large companies. The industry, therefore, came to be dominated by large firms, which became increasingly multinational as they sought access to foreign markets. Their operations were generally integrated vertically in great chemical complexes, often clustered around oil refineries, the source of their main raw material.

In the 1960s, more than three billion dollars had been invested in petrochemical plants in Europe; 25% of these investments were in Western Germany, followed by Great Britain (21%), Italy (18%), France (17%), and the Netherlands (10%).[1] The development of large petrochemical companies that dominated the industry usually followed one of three paths. The majority of oil companies integrated forward by constructing their own chemical plants. A second path was for chemical companies to integrate backward to secure supplies of raw materials. For example, to secure deliveries of naphtha, ICI joined forces with the American oil company, Phillips Petroleum, to build and operate a refinery in Britain. Chemical companies such as DuPont, Monsanto, Dow, Pechiney, St. Gobain, Rhône-Poulenc, Bayer, and Hoechst all signed long-term contracts for feedstock deliveries with multinational oil companies.[2]

---

[1] J. E. Walker, "Utviklingstendenser i den petrokjemiske industri i Europa," *Teknisk Ukeblad*, September 8, 1966, pp. 577–81.

[2] Fred Aftalion, *A History of the International Chemical Industry* (Philadelphia: University of Pennsylvania Press, 1991), pp. 214–25.

The third path was for oil companies and chemical companies to form joint ventures in which, as a general rule, the oil company supplied the feedstock, the partners cooperated in basic chemical manufacture, and the chemical company took the chemical products for further processing and sale. In Germany, the chemical company BASF and the oil company Shell founded the company Rheinische OlefinWerke.[3] In Britain, British Petroleum entered into an alliance with the whiskey firm Distillers Company, which had taken up the production of organic chemicals in the interwar period. Distillers integrated forward into plastics manufacture shortly before the outbreak of the war. The joint venture, British Petroleum Chemicals, was incorporated in October 1947. The first plant, built adjacent to the BP Grangemouth refinery at the Firth of Forth, came on stream in 1951. It had a nameplate capacity of 30,000 tons ethylene, 25,000 tons isopropanol, 33,600 tons ethanol. The joint company was renamed British Hydrocarbon Chemicals in 1956, the same year that production at Grangemouth was more than doubled.[4] During the 1950s, BP also became interested in ventures outside Britain. Shortly after the war, the two French chemicals firms, Péchiney and Kuhlmann, invited BP's French subsidiary to discuss a joint petrochemical venture. In 1949, the three companies founded Naphtachimie, where each company had a third of the shares.[5] The first plant, much smaller than the one at Grangemouth, began operations in 1953. In Germany, BP signed an agreement with the chemical company Bayer, and in the fall of 1957, the two companies incorporated Erdölchemie GmbH. Their first ethylene cracker at Dormagen became operational in 1958.[6]

Many of the large oil companies integrated forward and built their own petrochemical plants. Royal Dutch Shell was the leader in this respect among the majors. In 1955, concerned that the company had virtually no foreign chemical activities whereas Shell was already firmly entrenched throughout Europe, Standard Oil (New Jersey) urged its subsidiaries to invest in petrochemical plants. By 1959, more than one-third of the company's total petrochemical investment lay outside the United States. From a negligible portion in 1956, foreign sales

[3] Raymond G. Stokes, *Opting for Oil: The Political Economy of Technological Change in the West German Chemical Industry, 1945-1961* (Cambridge: Cambridge University Press, 1994), pp. 137-53.
[4] James Bamberg, *British Petroleum and Global Oil, 1950-1975: The Challenge of Nationalism* (Cambridge: Cambridge University Press, 2000), pp. 350, 352-53.
[5] Ibid., p. 358.
[6] Ibid., pp. 360-61.

in 1959 represented 39 percent of the company's total chemical sales of $256 million.[7]

PETROCHEMICAL PLANTS IN SCANDINAVIA BEFORE 1970

In 1956, the Esso subsidiary in Sweden, Svenska Esso AB, conducted an analysis of the Swedish petrochemical market in order to find out when the consumption of plastics would reach a level where it might be profitable to build a petrochemical plant in the country. The production costs per ton for a new petrochemical plant in Sweden would have to be on the same level as competing factories elsewhere in Europe. The analysis showed that this level of consumption of ethylene would be reached in 1962 or 1963. Consumption grew faster than anticipated in 1956, and in 1961, Svenska Esso AB decided to build a petrochemical cracker in Stenungsund, 40 km north of Gothenburg. The American engineering company Fluor was chosen as the main contractor.[8]

The Esso ethylene cracker in Stenungsund was officially brought on line on June 16, 1964. The cracker had a yearly production capacity of sixty thousand tons ethylene, sixty thousand tons propylene, twenty thousand tons butylenes, and twelve thousand tons butadiene. The naphtha feedstock was imported from the Esso refineries at Slagentangen, Norway, and Kalundborg, Denmark, supplemented by more sporadic deliveries from other Esso refineries in Western Europe.

The Swedish paper group, Mo och Domsjø (MoDo) produced ethanol from the sugar in the residues from their sulphite cellulose production. The ethanol was further processed into ethyleneoxide and ethyleneglycol in a factory in Örnsköldsvik. The production of these chemicals was moved to Stenungsund, and ethylene substituted for ethanol as the raw material. The new subsidiary of MoDo, Berol, began production in the new plant in 1964 with a nameplate capacity of fifteen thousand tons ethyleneoxide and ten thousand tons of ethyleneglycol.

The largest user of ethylene in Stenungsund, however, was the new polyethylene factory, Unifos Kemi AB. In 1961, the American chemical concern, Union Carbide, signed a contract with the Swedish company, Stockholms Superfosfat Fabriks AB, concerning the building of a

---

[7] Bennett H. Wall, *Growth in a Changing Environment: A History of Standard Oil Company (New Jersey) 1950-1975* (New York: McGraw-Hill, 1988), pp. 185, 190-91.

[8] J. Gunnar Amnéus, "Krackningsanläggningen i Stenungsund," *Teknisk Tidsskrift*, 1961, p. 1031 ff.

polyethylene plant.[9] At that time, Union Carbide was the world's largest polyethylene producer with factories in ten countries, four of them in Europe. The Swedish company, Fosfatbolaget, had been a producer of plastics raw materials since World War II. The joint company, Unifos Kemi AB, had a share capital of SEK 40 million. The new plant would have a yearly capacity of fifteen thousand tons. When the plant began operations in the fall of 1963, the growth in demand had already reached a level that made it profitable for Unifos to enlarge production capacity to twenty-three thousand tons.[10] Unifos sold most of its polyethylene in the Scandinavian market. The packaging industry, producing plastic film and bottles, was the largest customer. Polyethylene was also widely used for tubes and electric insulation.

In 1965, Svenska Esso AB decided to almost double the capacity of the ethylene cracker, to 110,000 tons. Unifos Kemi AB followed up on this by expanding the capacity for LDPE from twenty-three thousand to fifty thousand tons. The first production line for HDPE, with a capacity of fifteen thousand tons, became operational in 1972.[11] That same year, it was decided to expand the capacity of HDPE by another thirty-five thousand tons.

In the mid-1960s, Fosfatbolaget decided to build a PVC plant in Stenungsund with yearly capacity of seventy-five thousand tons. Since the 1940s, a subsidiary of Fosfatbolaget had been producing VCM and PVC with acetylene as feedstock in a factory at Stockvik, outside Sundsvall. The consumption of PVC for floors, pipes, electrical insulating, and film increased steadily during the 1950s and 1960s. Part of the VCM production in Stenungsund would be transported in bulk to Stockvik to be further processed to PVC there, and PVC-capacity was extended from thirty-five thousand to fifty thousand tons.[12] Later, a new PVC factory was constructed at Stenungssund with a capacity of seventy thousand tons.

In Copenhagen, Denmark, the shipping firm, A. P. Møller, built a polyethylene plant in 1960, A/S Danbritkem Polyethylenfabrik. This was a joint venture with ICI. In the beginning, this factory manufactured eight

[9] "Stenungsund. Et tyngdepunkt for skandinavisk petro-kjemisk industri," *Teknisk Ukeblad*, November 14, 1963, pp. 1022–24.
[10] "De petrokemiska anläggningarna i Stenungsund," *Teknisk Tidsskrift*, 1964, p. 689ff.
[11] "Unifos med ny HD-fabrikk," *Plastnytt*, no. 6, 1972, p. 19.
[12] "De petrokemiska anläggningarna i Stenungsund byggs ut," *Teknisk Tidsskrift*, 1965, p. 899.

thousand tons yearly. The year before the factory was closed in 1978 production was close to thirty thousand tons.[13]

THE FIRST PETROCHEMICAL PLANT IN FINLAND

In 1948, the Finnish state founded a wholly state-owned oil company, Neste Oy. A refinery was built by Neste at Porvoo (Borgå) and an expanded refinery put into operation in 1972. An ethylene cracker was built near the new extension using naphtha as feedstock. The new Porvoo ethylene cracker with a yearly capacity of 165,000 tons marked the beginning of ethylene production and the petrochemicals industry in Finland (Table 8.1). The volumes of ethylene were sold to Pekema Oy, a wholly owned subsidiary of Neste Oy, making LDPE and PVC. Yearly production capacity was eighty thousand tons of LDPE and fifty thousand tons of PVC. The large Finnish paper concern, Kymmene Oy, built a factory for the production of polyester plastics and phthalat softener. A third company, Stymer, took up the production of polystyrene and had a capacity of twenty thousand tons.

All the new plants came on line in 1972.[14] It was expected that the plants would be able to cover the domestic demand for the three basic plastics materials within a short time, as well as export some of the volume produced. It was further hoped that products from the new petrochemical industry would help add value to the production of other Finnish industries. Much of the polyethylene produced would be used as paper coating and in other products from the wood processing industries.

NORTH SEA OIL AND THE EMERGENCE OF THE NORWEGIAN PETROCHEMICALS INDUSTRY

The consumption of plastics generally, and of polyethylene especially, was higher in the Nordic countries around 1970 than in most other European countries.[15] In 1972, approximately two hundred thousand

---

[13] Povl A. Hansen and Görin Serin, *Plast: Fra galanterivarer til "high-tech." Om innovationsudviklingen i plastindustrien*, København 1989, pp. 158–61.

[14] "The petrochemical industry and Finland's balance of trade," *Tidsskrift for kjemi, bergvesen og metallurgi*, vol. 32, no. 6, 1972, p. 17.

[15] Johan Lothe, "Polyetylener. Særlig høyt forbruk i Norden," *Kjemi*, June 14, 1973, p. 21–23.

Table 8.1. Import of Polyolefins, PVC, and Polystyrene to Finland in Tons, 1965–1972

	Polyolefins	PVC	Polystyrene
1965	17,070	12,255	3,985
1966	28,684	13,672	5,085
1967	32,650	14,299	6,142
1968	48,620	17,240	6,708
1969	61,000	18,800	9,400
1970	80,400	28,592	12,456
1971	80,060	26,600	13,300
1972	69,000	28,500	20,900

Table 8.2. Plastics Consumption in Kilos per Inhabitant in Some Industrialized Countries

	1960	1965	1970	1974
Sweden	14	25	56	79
Norway	9	19	35	48
Finland	8	–	59	62
USA	11	25	37	56
Belgium	8	24	37	47
BRD	15	27	54	75
Japan	6	14	37	54
France	7	14	30	48
Great Britain	9	15	25	42
Italy	5	12	28	38

Source: PVC i KemaNord 1944-1984, p. 27.

Table 8.3. Consumption of LDPE, HDPE, and PP in Kilos per Inhabitant in Some Industrialized Countries in 1970

	LDPE	HDPE	Polypropylene
Sweden	13.0	2.0	0.5
Norway	10.0	4.5	1.0
Finland	12.5	3.5	0.5
Denmark	13.0	3.5	1.0
USA	8.5	3.5	2.0
Belgium	8.0	2.5	0.5
BRD	7.0	3.0	0.5
Japan	6.0	2.5	4.0
Canada	6.5	2.0	1.0
France	5.5	1.5	0.5
Great Britain	4.5	1.0	1.5
Italy	5.0	1.0	0.5

Source: Plastnytt, no. 6, 1972, p. 20.

Table 8.4. Norwegian Import of Polyethylenes, PVC and
Polystyrene in Tons, 1962–1974

	Polyethylene	PVC	Polystyrene
1962	9,431	3,833	2,495
1963	11,312	3,788	3,026
1964	15,607	3,106	5,114
1965	19,832	3,151	5,490
1966	21,924	3,306	7,496
1967	25,876	4,844	9,022
1968	35,364	6,294	10,922
1969	49,464	8,269	14,604
1970	56,221	8,834	13,603
1971	59,054	11,101	13,273
1972	71,414	10,816	17,191
1973	82,109	15,096	19,657
1974	75,117	17,255	19,828

tons of polyethylene were produced in the Nordic countries, whereas consumption was 310,000 tons. The consumption of polypropylene was still in its infancy; total consumption in the Nordic countries was only twenty thousand tons.

By the early 1970s, Norway was the only Nordic country that did not have a petrochemicals factory. After the discovery of the giant Ekofisk petroleum field in the Norwegian sector of the North Sea in late 1969, it became clear that Norway would get access to large quantities of NGL. Proposals to use this NGL to found a Norwegian petrochemicals industry were put forward in 1971.

The Ekofisk discovery came to change the course of Norwegian history. A considerable income would flow into government coffers through royalties and taxes, whereas exploration and petroleum production would lead to the establishment of a new offshore-related industry. Norwegian society and industry wanted to participate fully in its development. Stimulated by the heated political debate about Norway joining the EU in 1972 and strong red-green political currents, the Norwegian parliament, Stortinget, voted in favor of strong national management and control of the petroleum resources on the Norwegian shelf. Stortinget laid down the ten commandments for the future Norwegian oil policy in June 1971. In 1972, it established the Norwegian Petroleum Directorate and a state-owned oil company, Statoil. The new, wholly state-owned company became the nation's major political instrument in safeguarding a fair percentage of the economic rent for the state.[16]

The seventh of the ten oil commandments stated that all petroleum from the Norwegian shelf should, in principle, be landed in Norway. There was some room for exceptions, however. At this time, it was considered very difficult technologically to cross the deep Norwegian Trench with a large diameter pipeline. The Ekofisk gas was sold to a consortium of European buyers, and the gas was transported in a pipeline from Ekofisk to Emden in Germany.[17] Two terminal points were considered for the transport of oil and NGL, Teesside in Great Britain and Egersund in Norway. Phillips Petroleum Co., as operator for the Phillips Group of companies favored Teesside. The Norwegian government forced the Group to do a feasibility study on the technological possibilities of crossing the Norwegian Trench with a large diameter pipeline. In the end, the Norwegian authorities let the Phillips Group build an oil and NGL pipeline from Ekofisk to Teesside, on the condition that the Group be willing to sell back NGL cheaply to the Norwegian state.

The Phillips Group was to supply the Norwegian petrochemicals industry with sufficient volume of NGL for a yearly production of 250,000 tons of ethylene in fifteen years at a heavily subsidized price.[18] At the same time as Stortinget voted in favor of landing oil and NGL from Ekofisk at Teesside in April 1973, it accepted the NGL deal with the Phillips Group. Before the vote and in the months after, three Norwegian companies or groups of companies lobbied hard to get all or most of the big gift of cheap NGL that the state would give away.[19]

Norsk Hydro, one of the largest chemical companies in Norway, wanted the whole package. Hydro felt that it was their right, considering that the state owned more than 50 percent of the Hydro shares. Plans were made to build a large ethylene cracker at Bamble, close to Hydro's main factories at Herøya. Most of the ethylene produced would be further processed into VCM and PVC, with some of the volume being used in a new LDPE-plant. The remaining ethylene and all the propylene would be sold on the open market.

---

[16] Tore Jørgen Hanisch and Gunnar Nerheim, *Norsk oljehistorie: Fra vantro til overmot*, vol. I (Oslo: Leseselskapet 1992).

[17] Gunnar Nerheim, *Norsk oljehistorie. En gassnasjon blir til*, vol. II, Oslo 1996, pp. 29-32.

[18] *Ilandføring av petroleum*, NOU 1972: 15, p. 19.

[19] St. prp. no. 79 (1973-74), "Kontrakt om levering av våtgass til en norsk petrokjemisk industri," p. 4.

Norsk Hydro was already well established in the production and marketing of polyvinylchloride (PVC). The company had produced PVC based on German acetylene technology since 1951 at a factory at Herøya. Norsk Hydro manufactured 75% of the PVC consumed in Norway in the 1960s.[20] In 1971, the old VCM technology was no longer profitable, and production was closed down. Instead, Norsk Hydro began importing the necessary volumes of VCM from Britain. Norsk Hydro wanted to build a new VCM-plant with a capacity of 300,000 tons, which would require 150,000 tons of ethylene.[21]

The second suitor for the NGL from Ekofisk was Saga Petrokjemi. In the spring of 1972, a large number of Norwegian shipping and industrial companies joined forces and established the oil company, Saga Petroleum, which they hoped would be favored in the future concession rounds on the Norwegian continental shelf. The new oil company immediately showed interest in engaging in the petrochemicals industry. In the summer of 1972, Saga Petroleum cooperated with five major Norwegian industrial companies – Aker, Dyno Industrier, Elkem-Spigerverket, Hafslund, and Årdal og Sunndal Verk – in the establishment of Saga Petrokjemi, where Saga Petroleum acquired 50 percent of the share capital and the five industrial companies 10 percent each.[22] Saga Petrokjemi wanted to produce LDPE, HDPE, and polypropylene, and wanted half of the NGL from Ekofisk.

The third suitor was Statoil. The newly established and wholly owned state company joined the competition in the summer of 1973.[23] From the beginning, Statoil had ambitions of becoming not only a downstream company, but a fully integrated oil company with upstream activities as well. This was the first opportunity to get a foothold in the petrochemicals industry. The board had great political clout and used this for all it was worth. Statoil argued that it would be fair to give each of the suitors a third of the gift and ministers and major ministry officials showed sympathy for thinking along these lines. The change of government in the fall of 1973, when Labor won the election, helped Statoil get a higher allotment than many had expected.

Discussions between the three companies to find a solution that all three could accept came to nothing because Norsk Hydro insisted that

---

[20] "Norvinyl-fabrikken ved Norsk Hydro," *Plastnytt*, no. 2, February 1958, pp. 31–32.

[21] "Rafnes-prosjektet," report from Norsk Hydro A.S. and Borregaard A.S., Oslo, September 27, 1973. Copy Statoil archive.

[22] Minutes from board meeting Saga Petroleum, May 16, 1973. Board archive Saga Petroleum.

[23] Minutes from board meeting Statoil, June 1, 1973. Board archive Statoil.

it was in its right to demand most of the NGL gift. The final decision
was made by the government and supported by a vote in Stortinget in
June 1974.[24] Two new companies would be incorporated, one for the
ethylene cracker and one for the polyolefin factory. Norsk Hydro was
appointed operator for the ethylene cracker and Saga Petrokjemi for the
polyolefin factory. Norsk Hydro got 51% of the shares in the cracker,
Statoil 33%, and Saga 16%. Each company was awarded a third of the
shares in the polyolefin factory.

BARRIERS TO ENTRY

Barriers to entry are obstacles that face companies attempting to break
into a market, and are typically erected and sustained by companies
already in the market. One of the fundamental features of the petrochemical industry was economy of scale. Petrochemical producers built
very large plants in an attempt to maintain high output levels to reduce
the unit cost of production. Larger plants tended to be more efficient
than smaller plants up to a certain level of output because fixed costs
were spread over a higher number of units produced. There also was
the possibility of cost synergies between internal processes. The need
for economies of scale was an effective barrier to entry because it forced
new entrants to come in on a large scale or risk being uncompetitive,
especially in a market where there were already a number of major
competitors.

It could be argued that one of the weaknesses of the petrochemicals
producers in the Nordic countries in the early 1970s, before the Norwegian plants were built, was that the plants in both Sweden and Finland
were on the edge of being uncompetitive because of their small scale.
The return on capital in petrochemical activities had been very satisfying
for most European companies throughout the 1950s but declined somewhat in the early 1960s. If companies wanted to remain competitive,
they had to build larger plants. During the 1960s, petrochemical companies in Britain leapfrogged one another with new, larger plants. In 1963
and 1964, Shell and ICI announced that they were going to construct
crackers on a scale not previously seen in Europe. Shell was building
a cracker with a capacity of 150,000 tons, and ICI a still larger one.

[24] St. prp. no. 79 (1973-74), "Kontrakt om levering av våtgass til en norsk petrokjemisk
industri," Innst. S. no. 333 (1973-74), "Innstilling fra industrikomitéen om kontrakt om
levering av våtgass til norsk petrokjemisk industri."

In 1965, British Hydrocarbon Chemicals announced that the company would build a fourth ethylene cracker at Grangemouth with an ethylene capacity of 250,000 tons a year. Not to be outdone, ICI began planning a 450,000-ton ethylene cracker.[25]

The combination of large plants and the need to operate at high throughputs tended to create a glut of supply when new plants came on stream. This, in turn, put pressure on prices and profits, leading to a downturn in investment until rising demand put pressure on supply, triggering new investment and a repetition of the cycle. This cycle, familiar to the shipping and shipbuilding businesses since the late nineteenth century, was evident in the petrochemicals industry even before the oil crisis in 1973. When the Norwegian three-hundred-thousand-ton ethylene cracker was constructed at Bamble in the late 1970s, competitors argued that it was already too small to be profitable. The Norwegian companies involved, however, felt that the competitive advantage of having available a cheap, heavily subsidized NGL feedstock, would be more than enough to stay competitive in the petrochemicals and plastics raw material markets in the Nordic countries. The depressed state of the petrochemical markets in the late 1970s, proved that cheap feedstock was not sufficient.

During the 1950s, proprietary technology was an effective barrier to entry into the petrochemicals market. A number of firms created products and processes that they protected with patents or secrecy. In a study of nine petrochemical products from the beginning of their product lives through 1974, Robert Stobaugh found that a total of 537 manufacturing units had been built worldwide for the production of these chemicals. Each time a new factory was built, a technology transfer occurred. Three types of firms were involved in the technology transfers: firms that had initially commercialized a product and exploited the technology in a company owned facility; firms that had developed a new commercial process for manufacturing a product that was already being produced by an existing process; and firms that purchased the technology.

Almost all of the principal product innovators and most of the process innovators were large oil or chemical companies. Some process innovators were engineering contractors, which sold a package of design, purchasing, and construction services along with the technology, but rarely produced petrochemicals themselves. Firms such as Bechtel, Fluor, Lummus, and Scientific Design were well known throughout the world

[25] Bamberg, *British Petroleum and Global Oil*, pp. 374–75.

in the 1970s. Three different channels were used for the transfer of technology: the sale of technology to an unrelated firm (licensing); investments in facilities partly owned by the technology owner (joint ventures);
and investments in facilities wholly owned by the technology owner.
Stobaugh found that more than half of the 586 technology transfers in his
study took place by licensing technology to unrelated firms. Investments
in wholly owned subsidiaries were more common than joint ventures.[26]
With the increase in the number of petrochemical engineering contractors in the 1960s, technological barriers to entry were considerably
lowered.

Scale and economies of scale were greater barriers to entry for the new
Norwegian companies in the 1970s than were access to technology. The
German engineering company, Linde AG, was chosen as main contractor
for the construction of the ethylene cracker, with a yearly production of
three hundred thousand tons of ethylene and fifty-five thousand tons of
propylene. Half of the ethylene production would be further processed
into three hundred thousand tons of VCM in a new Hydro plant. Saga
Petrokjemi would be supplied with 150,000 tons of ethylene and all
the propylene. Planning was underway to build an LDPE-plant with a
capacity of one hundred thousand tons, a HDPE plant with a capacity of
forty thousand tons, and a polypropylene plant with a capacity of fifty
thousand tons.

When Saga Petrokjemi was to choose technology for its polyolefin
plants, a number of potential suppliers were willing to license their
technology to the Norwegians.[27] The list of potential suppliers of LDPE-
technology initially had seven applicants. In the end, it was a choice
between two bidders, the French company Ethylène Plastique, owned
by the state company CdF Chimie, and the American National Distillers and Chemical Corporation (USI). The contract was awarded to
the American company. The lower language barrier played a large role
in this decision. Norwegian engineers and operators had to be trained
to master the use of the technology. It was felt that the transfer process would be much easier with English-speaking instructors than with
French.

[26] Robert Stobaugh, "Channels for Technology Transfer: The Petrochemical Industry," in
Robert Stobaugh and Louis T. Wells Jr., eds., *Technology Crossing Borders: The Choice,
Transfer and Management of International Technology Flows* (Boston: Harvard Business
School Press, 1984), pp. 158–60.
[27] "Valg av lisensgiver for myk polyetylen," memorandum from Saga Petrokjemi to the board
of I/S Norpolefin, September 26, 1974. Statoil archive.

A license for polypropylene technology was bought from the American company, Dart. In this instance, Shell lost in the last round because it wanted to be part owner in the plant.[28] In the fall of 1974, Saga Petrokjemi had talks with Solvay, USI, Phillips Petroleum Co., DuPont, and Mitsui concerning the licensing of HDPE technology, and Phillips Petroleum was preferred.

The Phillips Group was not able to open the NGL plant at Teesside in the summer of 1977 as planned. The first deliveries of NGL from Teesside to Bamble did not commence until the spring of 1979. This created problems for the ethylene cracker at Bamble, which had been put on stream in July 1978, more than one year behind schedule, and with a total cost almost 30% higher than calculated. Noretyl had to buy propane as feedstock on the open market, and this meant much higher raw material costs than planned. The Norwegian companies, Norsk Hydro, Saga Petrokjemi, and Statoil, demanded that the Phillips Group give them economic compensation for the delays in NGL deliveries. After some discussion, the Phillips Group signed an agreement worth 130 million NOK with the three companies in the spring of 1978.[29]

The American engineering company, Lummus, had problems completing the first polyethylene plant and the propylene plant on time. They came on stream in the summer of 1978. Because of weak demand, the second plant for LDPE was delayed until January 1979, whereas the HDPE plant was completed in March 1979.[30] The new factory had a capacity of 125,000 tons LDPE, 60,000 tons HDPE and 65,000 tons of polypropylene. Total investments were approximately 1.7 billion NOK.

POOR ECONOMIC PERFORMANCE IN SPITE OF CHEAP FEEDSTOCK

Norwegian polyolefins entered the market in the late 1970s. The mid-1970s proved to be difficult years for most of the chemicals industry, with volume growth in world production and consumption falling from some 10% annually in the decade to 1973, to about 3% in 1974. In 1975, the industry experienced a worldwide recession, with consumption falling 10%. Although figures for 1976 showed a good recovery, in 1977 and

---

[28] Minutes from board meeting I/S Norpolefin, October 3, 1974. Statoil archive.
[29] "Forhandlinger med Phillips-gruppen om kompensasjon for uteblitte våtgass-leveranser," memorandum to Statoil board, May 18,1978. Board archive Statoil.
[30] Annual report I/S Norpolefin 1978.

Table 8.5. Turnover and Operating Profit I/S Norpolefin, Million NOK

	1978	1979	1980	1981	1982
Turnover	108	610	821	914	947
Operating profit	(87)	(38)	(90)	(189)	(240)

Source: Annual reports I/S Norpolefin.

1978 it became evident that the chemicals industry was still adversely affected by rising fuel and feedstock prices and by plant overcapacity. There were clear signs that chemicals growth was slowing because substitution of traditional materials by synthetics had reached a plateau in some industries, whereas plant sizes had increased to the point at which the benefits of scale became less apparent. The economic recession of 1974–75 brought declining markets worldwide, forcing manufacturers to operate large plants at low throughputs. Low growth rates and greater competition in export led the OECD to predict that the surplus in capacity might last until 1985. European companies were hit hardest, and their problems might have been exacerbated by the growth of heavy chemicals capacity in the Eastern bloc countries and the Middle East.[31]

How did the recession in Western European petrochemicals affect the new Norwegian companies? Were cheap feedstocks from the North Sea a sufficient bulwark against the depressed prices for polyolefins? It became evident that even though the new Norwegian plants had cheap feedstock, low product prices made it very difficult to earn enough money to pay operating costs, interest on loans and depreciation. The new polyolefin factory, Norpolefin, had a negative operating profit each year from 1978 to 1982 (Table 8.5).

In the Nordic countries, Norpolefin was able to secure a market share of only between 12 and 14 percent; 60 percent of production was sold in the Nordic countries, whereas the rest was exported to other countries in Western Europe. The market share in the Nordic countries was much lower than Saga Petrokjemi had expected during the planning stage.

Because of overcapacity, fierce competition between European petrochemicals manufacturers continued in 1980 and 1981. As profit margins

---

[31] "Hopes rise after 1976 recovery," *Petroleum Economist*, June 1977, p. 233; "Concern over Middle East potential, *Petroleum Economist*, August 1976, p. 301; "Slow Recovery by Major Companies," *Petroleum Economist*, July 1978, p. 281; "Gloom in Western Europe," *Petroleum Economist*, July 1979, p. 233. See also Robert Stobaugh, *Innovation and Competition: The Global Management of Petrochemical Products* (Boston: Harvard Business School Press, 1988), pp. 128–34.

became slimmer during the late 1970s, most manufacturers had tried to trim down surplus capacity. Despite closures, demand and production for the five major plastics (LDPE, HDPE, polypropylene, polystyrene, and PVC) was running around 30% below capacity. It seemed clear that only massive closures would help improve profitability in the longer run.

In 1979, Union Carbide sold petrochemical plants or their shares in plants in Europe to British Petroleum, with the exception of Unifos in Sweden. BASF cut its LDPE capacity at its Antwerpen and Rheinische Olefinwerke plants by 20%. Hoechst intended to reduce its West German polyethylene capacity by 15%, while the Dutch company, DSM, moved to scrap 20% of its capacity. Gulf Oil pulled out of petrochemicals in Europe, closing its Europoort and Milford Haven facilities. In the spring of 1981, ICI announced an asset swapping deal with BP Chemicals in which seven of the two companies' plastics plants would be closed. BP would withdraw from the PVC market, whereas ICI would withdraw from the LDPE market in Britain, but not elsewhere in Europe.[32]

The 1982 economic results for chemical companies were the worst in the history of petrochemicals. Despite plant closures during the last years, world capacity for high-volume petrochemicals was still in excess of expected future demand. Profits continued to decline, reaching unacceptable levels for return on capital. Five of the seven major oil companies were showing serious losses in their chemicals operations. Despite the closure of fifteen ethylene crackers in Western Europe in the previous two years, capacity was still close to fifteen million tons, whereas output was just over ten million tons a year. When crackers in the planning or building stage came on stream in Europe as well as in OPEC-countries and developing countries overproduction could get even worse. Esso and Shell were, for instance, building a new five-hundred-thousand-ton ethylene cracker at Mossmorran in Scotland, using ethane from the North Sea as feedstock.[33]

STATOIL ACQUIRES SAGA PETROKJEMI

Future prospects for the industry did not look rosy. Companies that wanted to position themselves for an upturn in demand had to be willing to finance the losses in the meantime. Among the three Norwegian companies that entered the field in the 1970s, Norsk Hydro and Statoil

---

[32] "In search of profits," *Petroleum Economist*, July 1982, p. 286.
[33] "Profits continue to decline," *Petroleum Economist*, July 1983, p. 261.

could weather the financial losses on petrochemicals easier than Saga Petrokjemi.

Saga Petroleum owned 56% of Saga Petrokjemi, and smaller owners, the remainder. In October 1980, Saga Petroleum offered the smaller owners the chance to exchange their shares in Saga Petrokjemi for 10% of the share capital in Saga Petroleum. If the offer was accepted, it would mean that the 44% owners accepted that 75% of their capital in Saga Petrokjemi was lost.[34] The owners felt that Saga Petrokjemi was worth more, and turned down the offer. There was no increase in the prices for polyethylene and polypropylene in 1981. In October, Saga Petroleum expected that Saga Petrokjemi would have to sustain more than 115 million NOK in losses including depreciation for 1981. In January 1982, the 44% owners had to realize that their shares had become worthless. Saga Petroleum agreed to take over their shares free of charge.[35]

In the spring of 1982, Saga Petroleum wrote a letter to Norsk Hydro and Statoil, signaling that it wanted out of petrochemicals, provided that one of the two companies would buy their interest at "a fair price."[36] With profitability problems with its own VCM and PVC production, Norsk Hydro was only interested in buying out Saga Petroleum if it could get Saga Petrokjemi for a very cheap price. Statoil, by contrast, wanted to expand its involvement in petrochemicals. The company wanted to buy Saga Petrokjemi for a price equal to its debt, at that time 560 million NOK. Saga Petroleum wanted to sell for 1.5 billion NOK.

Statoil had time to wait. The European petrochemicals industry had had a very bad year in 1982, and Saga Petroleum had to book mounting losses. To avoid share capital turning negative, Saga Petroleum had to agree to a loan of USD 20 million; and, if Saga Petrokjemi wanted to be able to continue operating in 1983, another 150 million NOK had to be secured. By 1983, Saga Petroleum was desperate to sell to Statoil or Norsk Hydro.

In the summer of 1983, Statoil stated that the company was willing to buy Saga Petrokjemi from Saga Petroleum for 550 million NOK.[37] Even though the profitability at Bamble was very poor at the time, Statoil

[34] Minutes board meeting Saga Petroleum A.S. October 16, 1980. Board archive Saga Petroleum.
[35] Gunnar Nerheim, *Norsk oljehistorie. En gassnasjon blir til*, Vol. II, Oslo 1996, pp. 193–200.
[36] Letter from Saga Petroleum A.S. to Norsk Hydro A.S. and Statoil, April 23, 1982. Statoil archive.
[37] Minutes of board meeting Saga Petroleum, July 6, 1983. Board archive Saga Petroleum.

was of the opinion that it would be possible to get a decent return on capital in the longer run. It was of great strategic importance for the company, however, that this acquisition enable Statoil to become an operating company in the field of petrochemicals. The dream of becoming a fully integrated oil company could be realized sooner than expected.[38] Statoil would take over operations at Bamble on January 1, 1984. The sale had to be sanctioned by Stortinget but went through without a hitch late in 1983.

RESTRUCTURING OF THE NORDIC PETROCHEMICALS INDUSTRY

In the first part of the 1980s, major restructuring took place within the Nordic petrochemicals industry. During the 1970s, the Finnish company Neste had built up a dominating position for its plastics products in Finland (Table 8.6). The main products were plastics such as polyethylene, PVC, and polystyrene, but Neste also produced and marketed polyester, resins, plasticizers, phthalic acid anhydride, and solvents. The bulk of plastic materials were marketed in Finland and the Scandinavian countries, with Finland accounting for 60% of volume delivered, and another 25% of volume being sold in the other Scandinavian countries.

Although the Finnish market for plastics was more sheltered from outside competition than markets in many other European countries, the downward trend in demand and prices was also strongly felt by Neste. Even though the turnover of the chemicals division increased by 28% from 1979 to 1980, the operating profit remained on the same level in 1980 as in 1979. In 1982, turnover was 3% lower than the previous year because of falling prices. "As the starting point was already low, the profit center's financial result was not satisfactory."[39] Neste considered the 70% average utilization of capacity very satisfactory when compared with the situation elsewhere in Western Europe.

When other companies withdrew from petrochemicals and volume plastics production, Neste decided to expand in these fields. From its foundation, Neste had been a downstream company. High oil prices and declining domestic consumption of oil products led the company to believe that it had to grow in sectors other than the traditional ones. Neste had already established itself as a manufacturer of petrochemicals

---

[38] "Overtagelse av Saga Petroleums eierinteresser i Saga Petrokjemi," PM to the board of Statoil, September 5, 1983. Board archive Statoil.

[39] Neste. Annual report 1982, p. 42.

Table 8.6. Neste Oy, 1972–1981

	Plastics Production 1,000 tons	Oil Products 1,000 tons	Total Turnover Mill FIM
1972	63	8,660	1,376
1973	96	8,255	1,918
1974	122	8,212	3,891
1975	102	7,710	3,886
1976	159	10,509	5,242
1977	143	11,128	6,189
1978	173	10,573	6,567
1979	193	11,613	8,592
1980	195	11,932	13,376
1981	195	10,547	16,330

Source: Neste Annual Report 1981.

and plastics, and, in the early 1980s, the company chose a strategy of aggressive forward integration and internationalization. In 1981, close to 50% of Neste investments were made by the chemical division.

In 1981, Neste acquired one-third, and Asko Oy two-thirds, of the shares in Oy Uponor AS. The new company, which manufactured PVC pipes for sewage and water systems, began operating in 1982. Already during the first year, Uponor acquired further capacity by buying existing companies in Sweden, Norway, and Denmark. After several years, Uponor had acquired almost 50% of the market for PVC pipes in Scandinavia. In 1981, Neste also bought a 24% holding in AB Celloplast, the largest manufacturer of LDPE membrane products in Sweden. Neste felt that minority shares in the plastics processing industry would help the company gather important market information needed for further raw material development and help the company acquire important customers for its products.[40] The Neste chemical division acquired several firms in 1983. The aim of these acquisitions was certainly to strengthen Neste's market standing in its product sectors, but the company also presented it as their contribution to the rationalization of the Scandinavian plastics industry.[41] The purchase of the Finnish polystyrene manufacturer, Suomen Polystyreeni Tehdas Oy, and the Swedish company, Beolit Plast AB, which used polystyrene as a raw material in the manufacture of insulating boards, strengthened the polystyrene unit.

[40] Neste. Annual report 1982, p. 10.
[41] Neste. Annual Report 1983, p. 8.

Table 8.7. Key Figures. Statoil's Petrochemical Operations (million NOK)

	Turnover	Operating Profit	Operating Profit	In % of Turnover
1982	881	−12	−1.4	
1983	939	39	4.1	
1984	1,425	171	12.0	
1985	4,441	107	2.4	
1986	3,687	123	3.3	
1987	3,718	574	15.4	
1988	4,533	1,299	28.6	
1989	4,350	796	18.3	
1990	5,333	718	13.5	
1991	6,002	−30	−0.5	
1992	5,508	−446	−8.1	
1993	5,524	−423	−7.6	

Source: Statoil, annual reports.

By far the largest project, however, was the signing of the preliminary agreement to buy Unifos Kemi in Stenungssund. It was believed that successful completion of this deal would make Neste one of Europe's leading manufacturers of plastics in terms of its production capacity and highly advanced grades. In 1984, the internationalization and expansion of Neste Chemicals led to a doubling of net sales. A crucial factor in this growth was the purchase of Unifos Kemi. Neste felt that Unifos had a highly developed and diversified portfolio of products that supplemented the products Neste already had. Through the Unifos deal, Neste also gained an extensive European marketing network.[42]

Although Neste was a downstream oil company that diversified into the chemicals business, the Norwegian state-owned company, Statoil, was becoming a crude heavy upstream oil company eager to integrate forward (Table 8.7). When Statoil bought Saga Petrokjemi, it stopped being a passive investor in petrochemicals, and became an active operator. Petrochemicals were singled out as an area for further growth.

In the fall of 1984, Statoil began discussions with Esso, which wanted to sell its business in Sweden, above all its network of gasoline stations, but the ethylene cracker in Stenungsund as well.[43] When the Statoil wet gas terminal at Kårstø in Western Norway came on stream in the fall of 1985, the company would have large volumes of NGL on its hands. In

---

[42] Neste. Annual report 1984, p. 4.

[43] Gunnar Nerheim, *Norsk oljehistorie. En gassnasjon blir til*, Vol. II, Oslo 1996, pp. 202–05.

1984, the Esso plant at Stenungsund produced 335,000 tons of ethylene, 155,000 tons of propylene, and some 200,000 tons of other products. Most of the volume produced was sold on long-term contracts to the other chemicals factories at Stenungssund; Unifos was by far the largest customer. Half of the propylene volume and all other products were exported.[44]

If Statoil bought the Esso cracker in Stenungsund, the company would strengthen its market position in the field of petrochemicals and ensure that the polyolefin plants at Bamble would have sufficient volumes of ethylene and propane at all times. At the same time, it would have a new and secure customer for eight hundred thousand tons of petrochemical feedstocks each year. Statoil could supply the Stenungsund cracker with naphtha, gasoil, natural gasoline, ethane, propane, and butane. In the spring of 1985, the Statoil board decided to buy all of Exxon's assets in Sweden for a price of USD 260 million.[45] Both the Norwegian government and Stortinget later sanctioned the decision.

From 1980 to 1984, yearly production at Statoil's plant at Bamble increased from 172,000 to 253,000 tons. This growth was a result of learning by doing. Statoil was still the only manufacturer of polypropylene in the Nordic countries, and polypropylene production increased from thirty-five thousand to fifty-eight thousand tons. If Statoil wanted to hold on to its market share in polypropylene, however, it was necessary to change to catalytic technology as had the major competitors. Statoil signed a licensing agreement with the American company, Himont, the world's leading polypropylene manufacturer and the owner of important proprietary technology.[46] Statoil invested fifty-five million NOK in new technology; of this, twelve million NOK was paid to Himont as a licensing fee. When the new product line came on stream, Statoil could produce ninety thousand tons of polypropylene yearly.

After the takeover of Norpolefin, Statoil improved its operating profit from petrochemicals operations. In 1984, operating profit was 12% of turnover. The next two years, operating profit was around 3 percent. In its 1987 annual report, Statoil proudly wrote that: "Effective use of the company's own feedstocks, high and even production and a positive

---

[44] "Kjøp av Exxons petroleumsvirksomhet i Sverige," memorandum to Statoil board, March 15, 1985. Board archive Statoil.

[45] Minutes from board meeeting Statoil, March 22, 1985. Board archive Statoil.

[46] "Ny teknologi ved I/S Norpolefin. Polypropylen," memorandum to the Statoil board, August 20, 1985. Board archive Statoil.

business cycle has led to a record high production."[47] The operating result was the best in this sector because the company entered the petrochemicals field. Strong demand for petrochemicals in 1988 led to even better results than the previous year. The operating result as a percentage of sales was 28.6% in 1988 compared with 15.4% in 1987.

THE TEMPTATION OF PROFITS AND FURTHER EXPANSION

Because of its favored position on the Norwegian continental shelf, Statoil would have at its disposal increasing volumes of NGL in the 1990s. Encouraged by the good economic results, Statoil decided to expand. In 1988, Statoil entered into a fifty-fifty joint venture with Himont to build a new propane and polypropylene factory in Antwerp under the name of North Sea Petrochemicals.[48] The new propylene plant, with a yearly capacity of 180,000 tons, came on stream in 1990, while the propane plant was ready in 1991. Utilization rates and profits were good in both 1989 and 1990, and Statoil decided to expand total production capacity at Bamble from 320,000 to 480,000 tons of polyolefins.

Throughout the 1980s, Statoil had managed to resist the temptation to integrate forward. It was company policy not to compete with the company's customers for plastics raw materials. In 1990, Statoil was no longer able to resist. The company became an end producer of plastics when it bought the Swedish producer of plastics compounds, AB Nobel Plast. This company was Sweden's leading supplier of plastics components to the automobile industry. Nobel Plast owned four factories in southern Sweden with a total of two thousand employees. Deliveries to the automobile industry accounted for 75% of turnover.

By the late 1980s, Neste in Finland and Statoil in Norway had emerged as major producers of petrochemicals, not only in the Nordic countries, but also in Europe. Neste strengthened its marketing position in 1986, when it bought a polypropylene plant in Beringen, Belgium, from Himont, Inc. It also decided to build a new polypropylene plant at Porvoo in Finland with a capacity of 120,000 tons. The technology was bought from Himont, and the new plant came on stream in 1988. Neste signed a long-term contract with Statoil Petrokemi AB in 1986, in which Statoil agreed to supply Neste with most of its demand for ethylene feedstock.

---

[47] Statoil. Annual report 1987, p. 36.
[48] Statoil. Annual report 1990, pp. 36–37.

Neste Chemicals had an exceptionally good year in 1987, followed by a record year in 1988. The good results whetted the company's appetite for further expansion. In January 1988, Neste and the Belgian company, Petrofina, set up an enterprise called Finaneste S.A., with Neste owning 35% of share capital. The company was to build a new ethylene plant, and through this, Neste hoped to "safeguard a supply of raw materials for its plastics factories."[49] In 1988, the Neste Group invested 3.5 billion FIM; 60% of this was invested in chemicals. Construction of a new polypropylene production line with a capacity of 120,000 tons began in Beringen in 1988 and came on stream in the summer of 1989. In July of that year, Neste signed an agreement with the Portuguese government regarding the purchase of Empresa de Polimeros de Sines S.A. The three polyethylene plants and a polypropylene plant had a combined capacity of 250,000 tons.

A new polyethylene plant with a capacity of 120,000 tons, based on Neste proprietary technology, was built in Beringen. When it came on stream in 1991, Neste decided to build an equivalent plant in Finland, designed to increase overall production of specialty products.[50] In 1991, a third cracker was completed in Antwerp as part of Neste's joint venture with Petrofina. From now on, Finaneste could guarantee sufficient feedstock supplies for all the Neste polyolefin units in Belgium.

At the beginning of the 1990s, Neste Chemicals was one of the largest manufacturers of polyolefins, petrochemicals, adhesive resins, and polyester gelcoats in Europe. The chemicals division had ambitions of further strengthening its position by increasing the proportion of specialty products in its output, diversifying its product range, and expanding operations outside Western Europe. The ambitious managers soon had to shelve their expansion plans however. The crisis in the Gulf, the overall economic downturn, and the completion of a number of new plants, resulted in oversupply in the market for petrochemicals and plastics. This negatively affected Neste's profitability. The same was true for Statoil, which had a negative operating profit of a half percent in 1991. Demand for new cars slowed considerably, and this strongly affected the plastics compound activities of Statoil EuroParts AB in Sweden. Prices for polyolefins in Europe continued to fall in 1992. In the fourth quarter of that year, prices were the lowest in ten years. The North Sea Petrochemicals plant in Antwerp was closed down both at the beginning and end

[49] Neste. Annual report 1987, p. 8.
[50] Neste. Annual Report 1991, p. 13.

of 1993 because of unprofitable product prices. Statoil recorded large losses in 1992 and 1993. Neste Chemicals did not fare any better. Consumption of polyethylene and polypropylene grew between 5% and 10% yearly in the late 1980s. Growth in 1993 was only 1%.

BOREALIS – THE LARGEST EVER MERGER BETWEEN TWO NORDIC INDUSTRIAL COMPANIES

The world chemicals industry was familiar with boom or bust business cycles, but that did not prevent managers from taking a particularly gloomy view of the future prospects. The industry was in the middle of a lengthy recession, and 1992 was the worst year for the chemicals sector since 1982. In the last three years of the 1980s, demand had been strong and chemical firms had been flush with cash. New plant expansions came on stream in the early 1990s, at the same time as demand weakened. Surplus capacity led to price cuts. In 1992, the major oil companies made a negative return on their chemical operations. This stood in sharp contrast to the peak of the cycle in 1988, when return on assets in the chemical sector averaged 13.6%.[51]

The major operators had in the 1980s already shut down their older, less efficient capacity in sectors with a surplus. The emphasis in the early 1990s was on putting together deals with other producers, where both sides could benefit from the exchange of business. ICI created the European Vinyls joint venture with Enichem. The company later agreed to swap its nylon fiber businesses for DuPont's acrylics business. In the spring of 1993, ICI announced that it would exchange its polypropylene business in Western Europe for BASF's acrylics operations in Germany and Spain.

The Nordic petrochemicals producers were part of these general trends. Top managers in the petrochemicals divisions of the Finnish company, Neste, and the Norwegian company, Statoil, came together several times in 1992 to discuss the possibility of merging the assets of the two companies in basic petrochemicals and polyolefins. The discussions led to the signing of a letter of understanding in June 1993, covering the merger of the polyolefin activities of the two companies. The assets in the new company, to be named Borealis, were to be owned 50/50 by Neste and Statoil. Beginning in September 1993, task forces drawn from the Neste and Statoil organizations were set up to plan the integration

---

[51] "Still no green shoots," *Petroleum Economist*, July 1993, pp. 4–5.

Table 8.8. Borealis Economic Performance

Borealis (EUR million)	1995	1996	1997	1998	1999
Sales of products	2,404	2,132	2,516	2,739	2,987
Operating profit	403	113	235	177	216
Operating profit as percent of sales	17%	5%	9%	6%	7%

Source: Borealis, Annual Report 1999.

of key areas and functions in the new company. The integration process went smoothly, and the final agreement between Neste and Statoil was signed in January 1994. Neste and Statoil presented the establishment of Borealis officially as an important contribution toward the restructuring of the European petrochemicals industry (Table 8.8).[52]

Borealis became operational on March 1, 1994. The company operated on three continents, produced over two million tons of polyethylene and polypropylene plastics a year, and employed some sixty-five hundred people, two-thirds of whom worked in the Nordic countries. This meant that Borealis had become the largest polyolefin producer in Europe and the fifth largest worldwide.

The company produced petrochemicals such as ethylene, propylene, and aromatics, and polyolefins such as polyethylene and propylene. The manufacturing plants were located at six main sites: Beringen and Antwerpen in Belgium, Porvoo in Finland, Rønningen in Norway, Sines in Portugal, and Stenungsund in Sweden. Some smaller plants were located in France, Germany, Sweden, and the United States. Borealis Industrier AB in Sweden produced polymer compounds that were to be found in everything from food packaging to hygiene and medical applications, sports goods, housewares, carpets and cars, construction materials, pipes, and cables.

The new company had wholly owned crackers in Finland, Portugal, and Sweden. The nameplate capacity in Porvoo, Finland, was 230,000 tons ethylene, and 130,000 tons of propylene; in Sines, Portugal, 330,000 tons ethylene and 165,000 tons propylene; and at the plant in Stenungsund, Sweden, 400,000 tons ethylene and 200,000 tons of propylene. Borealis owned 35% of the Fina-Borealis cracker in Antwerp, Belgium where the Borealis share of nameplate capacity was 350,000 tons of

[52] Borealis. Annual report 1994.

ethylene and 175,000 tons of propylene. It owned 50% of North Sea Chemicals in Antwerp and a corresponding nameplate capacity of 200,000 tons of propylene. Borealis also owned 49% of the Noretyl cracker at Bamble, corresponding to a nameplate capacity of 205,000 tons ethylene and 75,000 tons propylene. Added up, this meant that the new Borealis company could produce 1,515,000 tons of ethylene and 945,000 tons of propylene.

Total Borealis nameplate capacity for polyethylene was 1,460,000 tons. The plants in Antwerp and Beringen in Belgium could produce 250,000 tons. The capacity in Porvoo, Finland, was 200,000 tons, and at Bamble, Norway, 275,000 tons, followed by 270,000 tons in Sines, Portugal. By far the largest polyethylene plant was to be found at Stenungsund, Sweden, with a yearly capacity of 465,000 tons.

The merger brought Borealis to the forefront in the European polypropylene business. The company got an enhanced product profile and better market coverage in geographical terms. This, in turn, resulted in valuable savings. Significant synergy benefits were achieved in research, development, and sales. After the merger, total polypropylene capacity was 680,000 tons. The Beringen-plant had a nameplate capacity of 300,000 tons, followed by 160,000 at Porvoo and 105,000 at Bamble. Through the joint venture in North Sea Petrochemicals, Borealis had an additional 105,000 tons of polypropylene.

During its first year of operations, Borealis was able to meet the demands of an expanding market and increase production accordingly. Operating profit as a percentage of sales was 7%. The market demand for both polyethylene and polypropylene was much stronger in 1994 than the years before. Utilization rates climbed by as much as ten percentage points. Borealis economic performance was even better in 1995. Most of the profit was generated during the first half of the year, when capacity utilization and profit margins were still very high. Toward the end of the year, demand had fallen back substantially and prices decreased sharply. The trend was reminiscent of developments typical of the slump three years earlier. In sum, the production of petrochemicals in the Nordic countries started around 1960 when consumption had reached a level where the foreign multinationals felt it could be profitable, and not only economically, to construct local plants. The production substituted volume that had so far been imported from elsewhere in Europe or the United States. Customs barriers, scarcity of dollars, and politics all played an important role in the decision processes. Governments in the 1950s had urged foreign multinationals to invest more in the Scandinavian

countries. At the same time that Esso built the ethylene cracker in Stenungsund, it also built the first large refinery in Norway, at Slagentangen
in the Oslo fjord. Politics were all over the place when it was decided
to build a petrochemicals complex in Norway with cheap NGL from the
North Sea. The emergence of Neste as a petrochemical producer had
very much to do with Finnish politics and Finland's position between
east and west.

Markets and company strategies came to play a larger role as the 1980s
progressed. One may argue that without the economic recession in the
early 1990s, the Borealis merger would never have taken place. The
decisions of business managers in the Nordic petrochemicals companies
in the 1990s were much more influenced by economic results in the
short run than they had been in the early 1980s.

By contrast, top management at both Neste and Statoil began regarding
Borealis as an important investment but not part of their daily operations.
Less than four years after the merger, in the fall of 1997, Neste announced
that it would sell its 50% share in Borealis; 25% to the Austrian gas company OMV and 25% to the International Petroleum Investment Company
of Abu Dhabi. Still, in the course of the last thirty years, the Nordic
latecomers to the international petrochemicals industry have built up
Borealis as the largest polyolefin manufacturer in Europe and one of the
largest in the world.

9

Repositioning of European Chemical Groups and Changes in Innovation Management

The Case of the French Chemical Industry

FLORENCE CHARUE-DUBOC

Germany, Great Britain, and Switzerland are usually considered to be the dominant powers in the European chemical industry, as they are associated with companies of international stature such as Bayer, BASF, ICI, Ciba-Geigy, and Sandoz. Founded decades ago, these firms have marked the sector dynamics with their strategies regarding academic relations, the importance of patents, diversification within a multidivisional structure, and internationalization. At first sight, France would seem to be a less important player in the European chemical industry. However, today it is ranked second in Europe, just behind Germany, and with a greater annual growth rate, as shown in Table 9.1. The current position of the French chemical industry is a result of two simultaneous factors: a change in the relative weight of sectors of activity and the repositioning of companies. This chapter will examine the strategies used by the four French firms that today are the major players in achieving France's unexpected success: Atofina, the chemicals division of Total-Fina-Elf; Aventis,[1] the company formed by a merger between Hoechst Life Sciences and Rhône-Poulenc; L'Oréal; and Air Liquide.[2] The companies

---

[1] Continuing the concentration in the pharmaceutical industry, Aventis finally merged with Sanofi-Synthélabo in 2004.

[2] The choice of these companies obviously reflects a number of assumptions about what constitutes the chemical industry. Are the pharmaceutical and cosmetics industries part

Table 9.1. Breakdown of Turnover by Country in 2000 and Mean Annual Growth Rate by Volume from 1990 to 2000

Country	Turnover as % EU Turnover in 2000	Mean Growth 1990–2000
Germany	22.1%	2.3%
France	16.7%, i.e., 82 Geuros	3.3%
Great Britain	11.3%	3%
Italy	10.6%	1.3%
Belgium	8.1%	
Spain	7.5%	
Netherlands	6.7%	
Switzerland	5.4%	9.2%
Ireland	4%	
Others	7.6%	
EU	490 Geuros	3.3%

Sources: CEFIC-UIC.

have differing histories with regards to the characteristics of the firm at its creation, the consequences of the oil crisis, and recent mergers and acquisitions.

Table 9.2 shows the change in the chemical industry over the past twenty-five years by presenting the respective weights of various sectors of activity in Europe and France in 2000, and the mean growth of each one over the last decade. Pharmaceutical activities and some specialty sectors (perfumes and toiletries) have grown, often at the expense of more traditional sectors of activity on which the growth of the chemical industry has historically been based since World War II.

Companies in the sector have been transformed, first by internationalization, and then by intensive concentration and specialization. Three different path of development can be characterized.

The case of Rhône-Poulenc in France, demonstrates the first one. Like other major diversified European groups, such as ICI in England, Hoechst

of it? Does it include the oil industry? At the European level, the OECD consolidates statistics regrouping sectors of activity mentioned in Table 9.2. The oil industry is not included, whereas pharmaceutical products and cosmetics are. In contrast, in the United States, the conventions are different; the pharmaceutical industry is separate as is the cosmetics industry while the oil industry is part of the chemical industry. I favored the European conventions, feeling it was important to use a definition consistent with the scope of most major French and European companies. The major European chemical groups historically have developed a pharmaceutical activity that is of increasing importance and has only recently become autonomous. Oil companies have developed a petrochemical activity since the 1960s; in the case of France today, these are grouped in Atofina.

Table 9.2. Breakdown by Activity Sector in 2000 for the European Union and France

European Union	% Turnover 2000	Growth 1990–2000	France	% Turnover 2000	Growth 1990–2000
Petrochemical: plastics and polymers	29.4%	3.6%	Organic chemicals	25.6%	3.2%
Specialties: consumer-oriented products	21.6%		Parachemicals*	17%	2.3%
Agriculture	4.2%	1.1%			
Inorganic chemical	5.1%		Inorganic chemicals	7.5%	0.6%
Oleochemicals and derivatives	12.5%	1.6%	Soaps and perfumes	16%	4.5%
Pharmaceuticals	25.2%	4.9%	Pharmaceuticals	34%	4.5%

* Parachemicals include paint, glue, varnish, ink, cleaning products, cosmetics, and phytosanitary products used directly by clients (www.sarpindustries.fr/anglais/metier_chimie_centre2.htm).

and Bayer in Germany, Ciba-Geigy and Sandoz in Switzerland, it was directly affected by the oil crisis, then by market stagnation and a drop in commodity prices. In the 1990s, these companies engaged in specialization and concentration, leading to the creation of separate companies for pharmaceutical activities on one hand, and for other chemical activities on the other. Rhône-Poulenc, a one-hundred-year-old chemical group, was nationalized in 1982 at the time of reorganization of the industry by the government,[3] expanded its activity in the life-sciences continuously from the middle of the 1980s, made its specialty chemicals sector independent by creating Rhodia in 1998, and finally merged with Hoechst Life Sciences, creating Aventis[4] at the end of 1999.

Air Liquide and L'Oréal demonstrate a second possible trajectory. These historically specialized companies have grown continuously because of strategies of alliance, internationalization, innovation, and developing uses and applications for chemical products with their customers. Today, they appear as national champions. I will examine these companies and the characteristics of their growth process.

[3] With this restructuring, two large French companies, Pechiney and Saint-Gobain, having historically developed chemical activities, had to give away their assets in this sector to specialize on raw material.

[4] In 2004, Sanofi-Synthélabo, a French pharmaceutical company, merged with Aventis to become the first European pharmaceutical group, Sanofi-Aventis.

A third path covers activities in the oil industry. In France, the role of the national government has been extremely important in the organization and development of these activities. I will examine this path based on the key events that have marked the history of the oil industry in France and that led to the formation of Atofina.

Specialization, internationalization, and merger and acquisition are the common characteristics of the three trajectories. Nonetheless, I should stress that differences are also important and each of the trajectories explains part of the evolution in the chemical industry.

RHÔNE-POULENC: FROM A DIVERSIFIED GROUP TO A STRATEGY OF SPECIALIZATION AND CONCENTRATION

In this section, I describe the process that led the diversified group, Rhône-Poulenc, to reposition itself in the life sciences and finally cede its chemical activities with the creation of Rhodia. To a certain extent, this evolution is the exact opposite of that of Total-Fina-Elf, which kept all its chemical activities in the petroleum group and made the pharmaceutical branch autonomous with the creation of Sanofi-Synthélabo. The trajectory of Rhône-Poulenc is interesting for several reasons. First, as stated in the introduction, many European chemical groups present diversified product portfolios similar to that of Rhône-Poulenc and have followed parallel strategies, separating chemical and pharmaceutical activities into independent companies. Second, the pharmaceutical activities in the national chemical industry figures are important and show strong growth. Determining the specific features of this activity, as spotlighted by the gradual separation process that occurred in Rhône-Poulenc, helps to explain the sector dynamics. Finally, the redefinition of the company's field of activity is recent, but it is so radical and irreversible that tracing the process that led to this point would seem to be a most robust analytical strategy.

Rhône-Poulenc, the mother company of Aventis, was created on a portfolio of diversified products. Its growth was driven by the development of chemical products such as synthetic fibers and mastery of major processes to synthesize commodities. In the early 1990s, the firm was still producing a broad spectrum of products and was organized into five major sectors: fibers and polymers, major intermediates, fine chemicals, agrochemicals, and pharmaceuticals.

The history of Rhône-Poulenc is generally traced back to 1895, with the creation of the "Société Chimique des Usines du Rhône," which produced dyes and raw materials for perfumes, and the creation in 1900 of the "établissements des Frères Poulenc," which produced substances used in pharmacy and textile dye. In 1928, the two companies merged and became the "Société des usines chimiques Rhône-Poulenc." At that time, the company synthesized and marketed products with therapeutic properties, as well as products for other uses (packaging, detergents). Distribution via hardware stores, the principal network for chemical products in the first half of the century, reflected this mix of products of varied uses. From the start, the company based its growth on both chemical and pharmaceutical products. The historic factories at Saint-Fons and Vitry produced vanillin, aspirin and synthetic colorants at the former, and stovain (one of the first synthetic anesthetics) and photography products at the latter. This positioning is similar to that of other large European chemical companies.

After World War II, development of synthetic fibers drove the firm's growth and allowed various acquisitions. Savings from large-scale production of products, basic chemicals, and innovative processes were the dominant characteristics of the firm, although its product portfolio remained diversified with fine chemicals and pharmaceuticals.

However, several years of deficit, the slowdown in the chemical sector, and the oil crisis led to the reorganization of the French chemical industry by the state in 1982. Rhône-Poulenc had to cede its activities in fertilizers and petrochemicals, and the company was nationalized. Elf Atochem (entirely owned by Elf, itself state-owned) took over several of Rhône-Poulenc's commodity assets.

The subsequent period was one of continued reorganization and development of more profitable activities. Acquisitions and divestitures followed one after another, and internal rationalization also was conducted. Since the mid-1980s, there has been continual reinforcement of life-sciences, with the purchase of divisions of American companies, Union Carbide and RTZ in Agrochemicals, Rorer in 1986, and Fison in 1994 in pharmaceuticals. The company, which was once again profitable, was privatized in 1993.

For a century, Rhône-Poulenc favored first agrochemicals, then chemicals and petrochemicals, then fibers, and then again agrochemicals and pharmaceuticals. In this way, in spite of sometimes divergent strategies and antagonisms between branches of activity, Rhône-Poulenc grew, gradually organizing its varied but synergistic activities (Figures 9.1, 9.2). This heterogeneity allowed the financing of some external acquisitions (in agrochemicals, for example) using cash generated by years of high

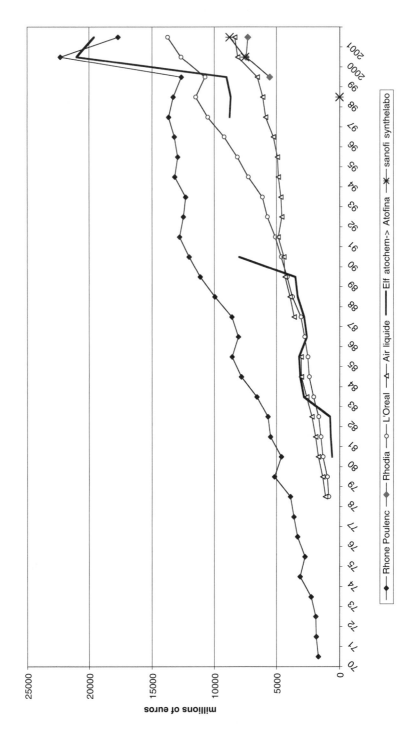

Figure 9.1. Evolution of the turnover of major French chemical companies.

Legend: —♦— Rhone Poulenc —●— Rhodia —○— L'Oreal —△— Air liquide —— Elf atochem-> Atofina —✳— sanofi synthelabo

256

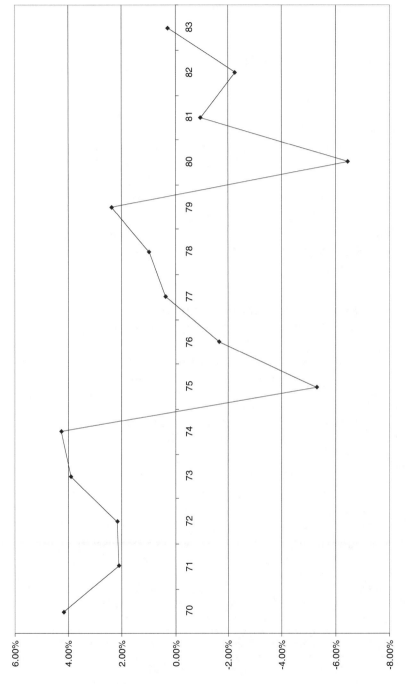

Figure 9.2. Evolution of Rhône-Poulenc net income as a percentage of its sales.

257

profit in chemical activities. Another constant during this century of development was the important part played by external growth – acquisitions, partnerships and other alliances – marked by periods of major reorganization and divestitures that redefined the frontiers of the company and its domain of specialization. From this history, in which mergers were important, two specialized companies were born: Aventis and Rhodia.

The creation of Aventis is very recent, at the end of 1999. Rhodia, which groups the chemical activities of Rhône-Poulenc (fibers, intermediates and specialties), was created and placed on the stock exchange in 1998. Rhône-Poulenc reduced its share of Rhodia's capital to less than 28%. During this same time period, Rhône-Poulenc acquired the Health division of Hoechst, becoming Aventis. With a turnover of 21,000 million euros in 2000, Aventis, together with Novartis, has become one of the leading European pharmaceutical enterprises. Its headquarters is located in Strasbourg, and former executives of the two merged companies form the top management of this European firm.

FROM RHÔNE-POULENC TO RHODIA AND AVENTIS: A FOCUS ON PRODUCT INNOVATION

The Move toward Specialties of the Chemicals Division and the Specialization of the Pharmaceuticals Division

Although various explanations can be offered as to why chemical activities were separated from pharmaceutical activities, this chapter concentrates on management of innovation and its changes during the last twenty years.[5] The place of innovation in company strategy is increasingly critical in the competitive business world. In Rhône-Poulenc, transformation of the project management mode in the chemicals division reveals the key place of innovation in the strategy and type of innovations. These changes accentuate the differences between the chemical and pharmaceutical branch as regards the innovation process. To understand

---

[5] The materials used as the basis for the analysis were gathered by three methods: examination of a course for project leaders (access to about one hundred ongoing projects in the company), more detailed analysis of over a dozen projects based on interviews with the project teams, and, finally, a longitudinal approach monitoring a project throughout its life (two years). Florence Charue-Duboc and Christophe Midler, "Le développement du management de projet chez Rhône-Poulenc," Rapport de recherche Rhône-Poulenc, (1994); "Le développement du management de projet chez Rhône-Poulenc – II," Rapport de recherche Rhône-Poulenc (1995); and "Le développement du management de projet chez Rhône-Poulenc – III," Rapport de recherche Rhône-Poulenc (1998).

and analyze this trend, we will distinguish three periods. The first period is characterized by company growth based on economies of scale and the preponderance of heavy chemical activities. This results in a project management model, principally dedicated to industrial production capacity construction projects. The second period was a time of transition for the group: increasing the weight of life-science activities, and strategic repositioning of the chemical branch to specialty chemicals. During this period, a new project management model emerged related to the emphasis on product innovation throughout the company. Finally, the third period corresponds to the anchoring of this innovation based strategy and highlights the increasing differences between the life-sciences and chemical activities.

PROJECT MANAGEMENT FOR ECONOMIES OF SCALE

The Strategic Context: A Growth Driven by "Major" Products

Although Rhône-Poulenc's growth was based on a group of diverse products, ranging from therapeutic and crop protection products (copper sulfate), to specialty chemicals (colorings, flavorings), intermediates (phenol), and fibers (cellulose acetate), the 1960s and 1970s saw spectacular growth in the chemicals activities. This was partly due to the remarkable development of synthetic fibers (such as nylon, polyester, and Terylene), which represented 60% of turnover by the beginning of the 1960s. It was also partly because of the crucial decision taken by Rhône-Poulenc with regard to what was then its leading product – phenol – at a time when the chemical industry was shifting to synthesizing processes using petrochemical-based material. Finally, the beginning of the 1970s was marked by acquisitions in heavy-chemical, organic-chemical, and inorganic-chemical sectors with the purchase of Pechiney, Saint-Gobain, and Progil.[6]

Project management at the beginning of the 1980s was related to the strategic model, which was an underlying factor in the growth experienced by the company up until 1975, and which was based on the production and sale of major products such as Terylene, nylon, phosphates, and phenol. In this context, projects that justified a specific management

---

[6] L. Bibard, et al., "Recherche et développement et stratégie: Rhône-Poulenc Agrochimie et Rhône-Poulenc Santé," in *Stratégie technologique et avantage concurrentiel, rapport de recherche IREPD* (Grenoble: 1993).

method were the construction projects for new production facilities. A typical example would be an increase in the capacity of a phenol factory, which represented an investment of several hundred million francs: the product is known, as is the process. The project is restricted to the construction of a production unit. A process, which has been outlined on paper or tested on a laboratory pilot, is developed into a full-scale industrial unit. The project involves finalizing process engineering studies in order to define the unit in detail, signing supply contracts, optimizing the installation of the main machinery and pipe-work, coordinating the construction site, and providing an interface with the host site.

Characteristics of the Project Management Model

The organization and method of undertaking projects, drawn up in the 1980s, were similar to those for large building sites (engineering model).[7] The client would stipulate their requirements with regard to the production facility (volumes, unitary cost, product quality, time-scale of start-up) and the prime contractor – the construction project manager – would coordinate the various means necessary to construct the facility that had been ordered. It would be the construction project manager who would organize site managers, instrumentation experts, purchasers, and draftspeople from the design office, and others. Such project organization is particularly appropriate to Rhône-Poulenc's matrix structure, which involves a partitioning of responsibilities according to activity or product group, and a hierarchical organization within each activity, particularly in engineering (Chart 9.1). The client belongs to the product structure, and is often the industrial director of the "enterprise" (strategic business unit). The main contractor is from engineering. The quasi-commercial relationship between these two is borne out by an internal contract similar to that between a client and a supplier.

The development of this type of project organization can be explained by the specific characteristics of the projects to be carried out; that is, size of the budgets, large numbers of people to be managed, and numerous suppliers. In such projects, the main uncertainty relates to time-scale and costs, two aspects brought under control by project monitoring tools. The planning tool makes it possible to plan the project's time frame and to coordinate the various actions, as well as to spot actual delays and

---

[7] Christophe Midler, "Modèles gestionnaires et régulations économiques de la conception," in Gilbert De Terssac et Ehrard Friedberg, eds., *Coopération et conception* (Toulouse: Octares, 1996).

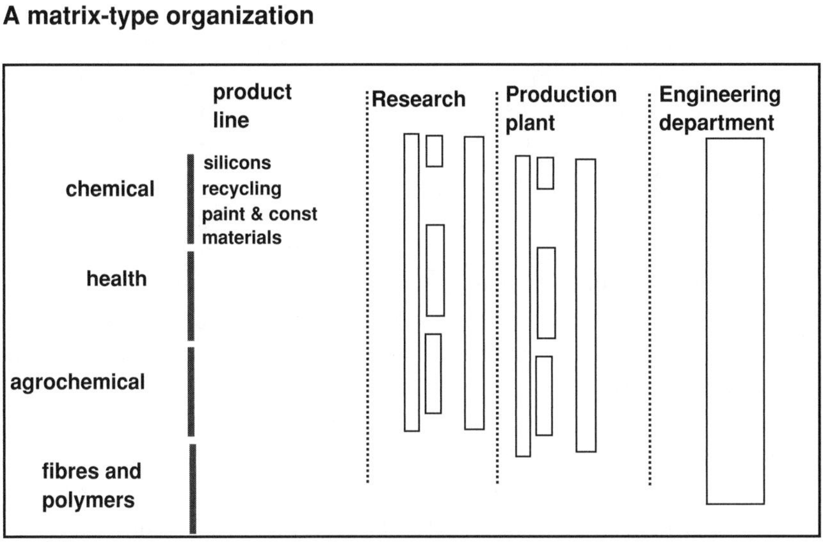

A matrix-type organization

Chart 9.1. Rhône Poulenc organization; a matrix structure.

anticipate their consequences. Expenditure control and the early identification of discrepancies are facilitated by generic profiles of planned
expenditure according to the project's progression. On such questions as
duration and cost, a great deal of experience has been capitalized across
projects in the engineering department.

THE RISE OF THE LIFE-SCIENCES SECTOR AND AN EMPHASIS
ON INNOVATION MANAGEMENT

1982: A Restructuring to Face the Economic Crisis
and the Emergence of a New Strategy

Following the restructuring of the chemical industry by the state in 1982
and the nationalization of Rhône-Poulenc, steps taken during the 1980s
to enable the group's recovery later redefined the company. A number of
divestitures were made in the textile, heavy chemical, and petrochemical
areas.[8] Some were decided during the reorganization conducted by
the French government, whereas others were concluded directly with

---

[8] S. J. Lane, "Corporate Restructuring in the Chemical Industry," in Margaret Blair, ed., *The
Deal Decade, What Takeovers and Leveraged Buyouts Mean for Corporate Governance*
(Washington, DC: The Brookings Institution, 1993).

private companies. However, the second half of the 1980s was marked by acquisitions in agrochemicals (the agrochemical division of Union Carbide in 1986), pharmaceuticals (Pasteur vaccines, Connaught, Virogenetics, Rorer in 1990), and specialties (RTZ). Thus, between 1986 and 1991, forty billion French francs were invested in acquisitions. These acquisitions also had a significant role in the internationalization of the French group.

In the early 1990s, cyclical effects within the chemical industry again challenged the strategy of mass economy and economies of scale. Manufacturing overcapacity and economic underperformance in the downstream sectors, together with an increase in production costs resulted in a drop in world prices for standardized products. Competition from countries with cheap labor also proved to have a particularly negative impact on this type of strategy.

The agrochemical division, with the acquisition of the agrochemical division of Union Carbide, saw its turnover rise from 6.5 billion francs in 1986, to 10.5 in 1989, whereas the operational margin increased from 7.9% to 11.4% over the same period. At the beginning of the 1990s, the agrochemical and pharmaceutical activities already represented 48% of the turnover in 1991, whereas in the 1960s, the chemicals division and fibers represented 80% of turnover. These numbers can serve as a reference for fixing objectives for other sectors: 15% of operational margin with regard to turnover. The company was cited for its exemplary innovation strategy, and its capacity – thanks to the production and marketing of innovative products – to maintain high margins. The sectors in the life-sciences came to exemplify the new models of economic excellence. Jean René Fourtou, CEO of Rhône-Poulenc from 1986 to 2002 probably had that vision early on and led the group to Rhodia and Aventis (Figures 9.3, 9.4).

Search for Synergies: Innovation Management as Means of Integration

Following this period of divestitures and acquisition, the 1990s saw a focus on the internal organization of the different sectors of activity in order to consolidate their positions, bring together teams with differing histories from various backgrounds, and gain maximum benefit from the synergies between these recently united entities. Amid the internal rationalization at the beginning of the 1990s, an initiative was launched relating to innovation project management. In the 1990s, project management

Figure 9.3. Evolution of Rhône-Poulenc net income in the 1980s.

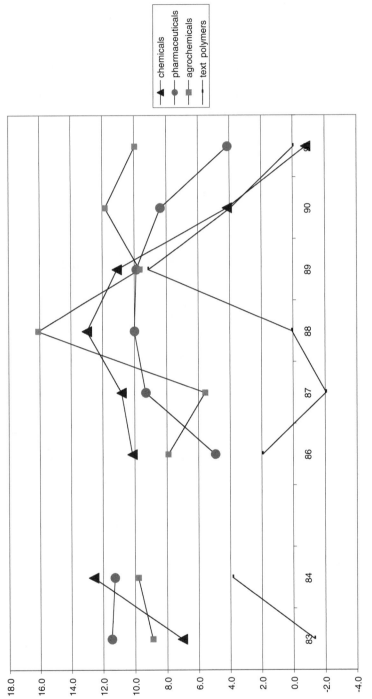

Figure 9.4. Rhône-Poulenc operating margin by sector in the 1980s.

264

was a managerial fashion that came from the automobile industry.[9] It was hardly surprising that Rhône-Poulenc headquarters noted its importance for the company as it was being implemented in several industrial sectors. Project management requires integration between services and the various specialty areas, and also relates to the company's internal team management.

From 1992 onward, innovation project management was referred to in communications from the general management as one of the five key areas to be developed. The aim was to provide the best possible prospects for growth while at the same time achieving double-figure margin levels in terms of percentage of turnover. It therefore played a pivotal role as far as the company's competitive position was concerned, and affected all of the group's business activities and sectors.

Innovation project management at Rhône-Poulenc began initially with a benchmarking initiative between the various divisions. At the beginning of the 1990s, there were five such divisions: organic and inorganic intermediates, specialty chemicals, fibers and polymers, agrochemicals, and health. This type of approach demonstrated a desire to draw on the group's diversity and to exchange the techniques and expertise developed in the various divisions in order to capitalize on, unite and improve project management in each of the divisions. R&D managers from the various divisions formed a working group for exchanging their various related experiences, which operated for one year. The end result was a white paper on project management setting out best practices, which was in fact a list of recommendations. The white paper also included a self-assessment guide, designed to help each sector identify any necessary action for improvement. Two points need to be emphasized. The first is that the life-science activities, in which several major innovative products were being developed, rapidly proved to be the model to follow. At management level, the project management initiative was led by the former agrochemicals director. The second is that, at a time when operational management was being decentralized, it quickly became clear that there was a need to develop a project management "doctrine" for each division. The second stage was then to ask each division to put forward action plans, with a view to improving the implementation of project innovation.

---

[9] Kim B. Clark and Takahiro Fujimoto, *Product Development Performance: Strategy, Organization and Management in the Auto Industry* (Boston: Harvard Business School Press, 1991).

STRATEGIC REDIRECTION TOWARD CHEMICAL SPECIALTIES AND EMERGENCE OF A NEW PROJECT MANAGEMENT MODEL

Innovation Projects in the Chemical Division

For the chemical division, implementing a strategy similar to that of the life sciences division meant exploring a greater variety of products and their properties, instead of being mainly geared toward developing new synthesizing processes for products that were already well known. Because, in the 1980s, the division had concentrated on major products and on economies of scale, innovation was confined to improving processes, and development of innovative products was not at the forefront. The chemicals division began structuring the development of innovative products/services/processes referred to as "innovation projects" in the white paper. Greater emphasis was placed on product innovation. Innovation projects were set to develop specialty products (pigments, solvent-free paints, additives for recycling and so on) for niche markets. Basic products were reworked in order to achieve the properties required for specific applications (silica for tires, phosphates for the treatment of salmonella). These projects marked a change from a strategy based on major standardized products to a strategy of offering innovative products that had been optimized for specific uses.[10] This emphasis on innovation project was both a sign and an integral part of the strategic redirection toward specialty chemicals that the chemicals division had taken. This decision was, no doubt, based on a comparison with the life-sciences sectors. This redirection led to a progressive yet profound change in the style of project management. A new method of project management emerged that was significantly different from what had been previously used (as described earlier). The project scope changed: research activities (exploring the technical characteristics of a new product and conditions for synthesis), marketing activities, an understanding of product use, market analysis and industrialization were incorporated into a unified and integrated project. Projects were no longer limited to the phase of industrializing a process and building a unit. A multidisciplinary team was created. A project manager was dedicated to the project for its whole course. He was in charge of the overall success of the project. This

---

[10] Patrick Cohendet, ed., *La chimie en Europe: innovations, mutations et perspectives*, (Paris: Economica, 1984); Patrick Cohendet, J. A. Herault, and M. Ledoux, "Quelle chimie pour l'an 2000?" *La Recherche*, 166 (1989): 1254–57; and U. Colombo, "A Viewpoint on Innovation and the Chemical Industry," *Research Policy*, 9 (1980): 204–31.

project organization had many similarities with the heavyweight project management model.[11]

Some Specificities of Innovation Project Management

Concurrent engineering[12] was developed as the result of various factors: there was an overall project responsibility; the project scope incorporated the various aspects; shortening the time-to-market for new products was very important from a competitive point of view; and uncertainty was very high. Uncertainty lay with the process, as well as with the suitability of the product vis-à-vis the clients' uses, its acceptability, and with comprehension of its most sought-after properties, and therefore those most capable of generating value for the product and, potentially, profit. Managing uncertainty required the ability to integrate new information at any step of the project and to react quickly because of the virtual impossibility of foreseeing properly on all these dimensions. A client analysis approach was adopted to study product usage and the various processing intermediaries up until the product reached the end user. Expertise within the firm had to be developed with regard to product properties for the processor, on the one hand, and with regard to the transformation processes, and particularly required properties, on the other. Application laboratories and applicability teams were set up. Application laboratories studied the end-use properties of the finished product in the composition of which the chemical product was used. For example, the antifoam properties of a silicone to be used in the composition of a washing powder would be measured and characterized. The role of applicability laboratories was to establish correlations between physicochemical properties of molecules and their application properties. Such understandings were key at the product development stage in order to direct the development process and to fine-tune marketing strategies.

The relationship with clients was also transformed; from price and volume negotiations for a given quality, partnerships began to develop with clients to explore jointly the properties of the product under development and decide which one was the more valuable for the consumer. This

---

[11] Clark and Fujimoto, *Product Development Performance.*
[12] C. Navarre, "Pilotage stratégique de la firme et gestion des projets: de Ford et Taylor à Agile et IMS," in V. Giard and C. Midler, eds., *Pilotage de Projet et Entreprises, diversités et convergences* (Paris: Economica, 1993).

trend of development of win-win partnership is similar to that described in the automotive industry[13] between the car manufacturer and their main suppliers. The chemicals division therefore underwent a series of fundamental transformations:

- transformation as far as product strategies were concerned, from standardized products sold for their physicochemical specifications to diversified products targeted in relation to the client for their properties in specific applications;
- transformation as far as the methods of innovation and project management were concerned, leading to the development of a closer relationship between researchers and markets, and the establishment of integrated teams (research, process, marketing), whereas in the past these had been sequential and compartmentalized;
- transformation with regard to external relations leading to the establishment of development partnerships based on complementing the expertise of each of the partners and not on direct pursuit of external growth.

This transformation took several years: in 1991–92 the need for a strategic repositioning was identified; in 1995 the specialty and raw chemicals branches merged; and in 1998 Rhodia was created and progressively separated from the rest of Rhône-Poulenc (Figure 9.5).

For such a strategic evolution to be successful, several learning processes needed to occur. These learning processes had to take place at different levels and in various areas of the organization. New skills had to be built within the application and applicability labs; new relationships had to be set up with clients based on a larger market knowledge and better understanding of the client constraints and stakes; new project management methods had to be implemented involving the learning of new behavior such as working and coordinating actors under uncertainty; and, finally, the company had to learn how to select projects in a portfolio in order to balance different kinds of risk and different timing. Today, the results of this new strategy still appear fragile, as can be seen in Figure 9.6. The evolution, however, is irreversible because of the divestiture undertaken in the commodities. Learning processes have already developed but remain to be completed. The challenge for the company

---

[13] G. Garel and A. Kesseler, "New Car Development Projects and Supplier Partnership," in Rolf A. Lundin and Christophe Midler, eds., *Projects as Arenas for Renewal and Learning Processes* (Dordrecht: Kluwer, 1998).

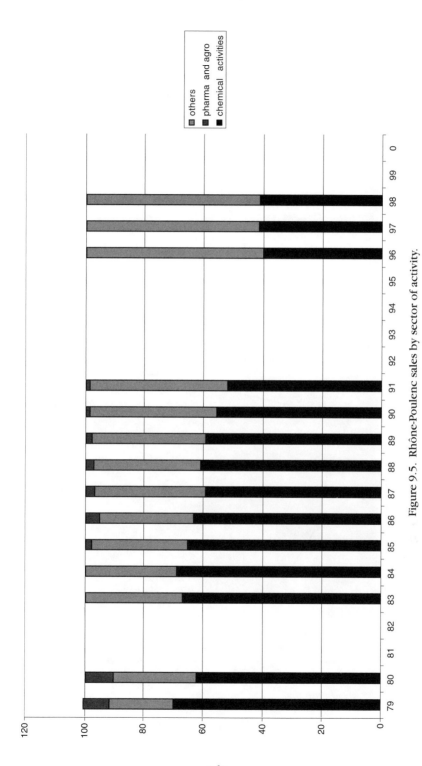

Figure 9.5. Rhône-Poulenc sales by sector of activity.

others
pharma and agro
chemical activities

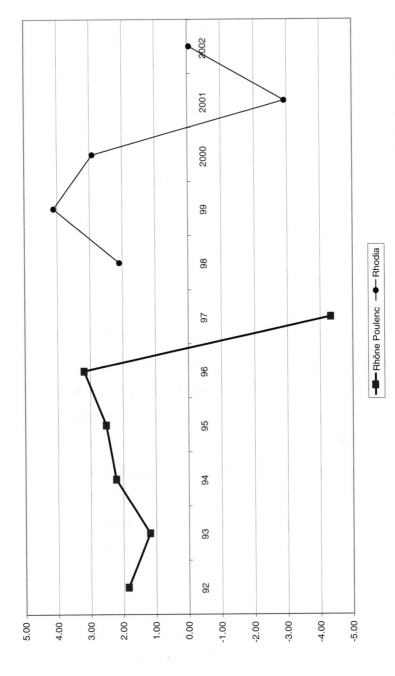

Figure 9.6. Comparative evolution of Rhône-Poulenc and Rhodia net income as a percentage of sales during the 1990s.

is to reinforce its position as an innovative company that can survive "bad" years (currency exchange rate, economic cycles) with reduced benefits but no deficits. This innovative dynamic has already proven fruitful with 14% of new products contributing to the group sales and an increase in the group's market share in eight strategic markets. Nevertheless, today, debt related to previous deficits and acquisition is negatively impacting the results. In addition, bad economic conditions have slowed learning processes because cost cuts often strongly affect R&D expenses, and, as a consequence, new product development.

This fundamental transformation, in which the life-sciences sector was to have such an instigative and referential role, might have brought the sectors closer together. In fact, to a certain extent, that came about as a result of the replication of the same strategic model, thus leaving the way open for product innovation. The aim was to establish competitive ranking by introducing new products, which improved on existing products, and to optimize profitability by adding value to the client system. Contrary to the premise that the chemicals division would imitate the strategic model of the life-sciences division, it would seem that the chemicals division's strategic redirection actually intensified differences. It could even be said that these differences were more limited when the chemicals division followed a strategy of economies of scale, and that they paradoxically became more pronounced after the strategic move toward a competitive model through innovation.

INNOVATION PROCESSES: DIFFERENCES BETWEEN PHARMACEUTICALS AND CHEMICALS

In the preceding sections, emphasis has been placed on the manner in which project management was transformed in the chemicals division. Indeed, it was in this division that the changes were the most spectacular. In the pharmaceutical sector, innovation projects also were introduced, but these projects have distinctive characteristics that determine the way they are managed. Consequently, innovation projects actually increased differences compared with what had come before.

The first distinctive characteristic relates to the regulatory requirements, which have become increasingly stringent. Any new drug or agrochemical must undergo an extremely detailed examination by regulatory authorities before receiving the authorization to be introduced on the market. The approval dossiers must show the effectiveness of a new product, as well as demonstrate its safety (i.e., for the environment and patient health). The dossiers contain the results of numerous tests; they describe the product and the manufacturing process in great detail.

The second characteristic is the importance of patents. As in any industry, patents protect companies that have incurred heavy research and development costs from competitors, who would otherwise simply need to develop a production process for a product that had already been approved. In the drug industry, the patent system is further strengthened because public health authorities will grant market authorization if a therapeutic benefit can be shown over existing products. As this benefit increases, so does the price the public health authority will allow. It is therefore very important to be the first to file for a patent. In addition, the major companies tend to focus on a few potential flagship products and on exploring the same therapeutic targets or the same crops. As a result, there is a veritable race to apply for patents. These two specificities have had important implications on the evolution of innovation processes in the pharmaceutical industry.

Pharmaceutical and agrochemical groups have tended to concentrate their research and development efforts on products that are likely to generate a high turnover, so as to compensate for the development costs. In the past decade, the amount of study required for the approval of a product has continuously increased and development costs have become extremely high. Companies are thus focusing on high-volume products that are often marketed on a worldwide scale, and/or products with high added value, often corresponding to therapeutic or preventive products that previously did not exist. Niche products are not viable in terms of development costs, given the narrow market targeted. Isolated or breakthrough innovation strategies[14] are followed and enabled by patent protection. On the contrary, repeated innovation strategies as observed in the appliances[15] and IT industries[16] are not required to stay ahead of the competition.

Within the development process, concurrent engineering is almost impossible because of regulatory requirements. Efficacy tests are not

[14] M. Tushman and P. Andersen "Technological Discontinuities and Organizational Environments," *Administrative Science Quarterly*, 31 (1986): 439-65.
[15] V. Chapel, "La croissance par l'innovation: de la dynamique d'apprentissage à la révélation d'un modèle industriel. Le cas Tefal," Thèse de doctorat de l'Ecole des Mines de Paris (1997).
[16] S. Brown, and K. Eisenhardt, "The Art of Continuous Change: Linking Complexity Theory and Time-Paced Evolution in Relentlessly Shifting Organizations," *Administrative Science Quarterly*, 42 (1997): 1-34.

allowed unless safety tests have proved conclusive. Efficacy tests must be carried out on a synthesized product according to an established manufacturing process. For any new product development, the major issue at stake is obtaining approval. But the regulatory authorities may quite easily delay market entry by requiring additional tests. Such a delay would leave the way open for a competitor with a completed approval dossier to be the first on the market. Thus, any late modification to the process or the product that might optimize its efficiency and the global profitability of the project is regarded very carefully as at risk for not getting the authorization on time and having to undertake additional tests. I can say that development planning is governed by this regulatory system.

Pharmaceutical companies prepare the ground for patent application at the earliest possible opportunity. The protection of intellectual property is no longer just a defensive strategy: it also can be seen as an offensive strategy, by using patents to protect a wide field, thus limiting competitors' ability to explore promising avenues. Greater cooperation with public research establishments is developed because of the importance of patents. Early identification of the most promising new avenues can lead to a speedy investigation of potential drugs. The approval process also gives an ever-greater importance to relationship with the public research center. Scientific recognition of the company research labs by the research community and regulatory authority add to the credibility of the approval dossier. This provides an additional opportunity for closer relationships between industrial and public research centers through conferences, publications, research partnerships, and recruitment.

Because of regulatory authorities' requirements for approval, concurrent engineering is very limited, even though time constraints as regards competition for patenting and approval are important. Within this context of increased development costs the focus is more and more on "big products" to be sold on a worldwide basis and niche strategies are not attractive from an economic point of view. Because the patent system gives such an advantage to the first on the market, alliances, partnerships and mergers are increasing. This trend for pharmaceutical groups to join forces is also designed to provide companies with sufficient financial means to engage in increasingly costly development programs. By choosing alliances or partnerships with companies focused on the same therapeutic targets and on molecular compounds, groups are ensured access to the global market and to the most promising areas of new development. Development partnerships are also set up with research laboratories upstream. The most important ones are definitely not with clients downstream or for issues concerning the chemical process.

AFFIRMATION OF DIVERGENCE IN PRODUCT INNOVATION STRATEGY

In the current context, the common competence on which the chemical and pharmaceutical branches were based – competence in chemical processes and their industrialization – seems only to be of secondary importance. The competences involving the products and their properties and an understanding of the mechanisms that underlie their efficacy have become key elements. The development of these competences draws the chemical and pharmaceutical branches in opposite directions.

The chemical branch has moved away from standardized products manufactured in large quantities to concentrate on diversification in specialty products with high margins. Niche and specialty strategies have therefore been developed. In this type of strategy, patents offer less protection from competition. Indeed, different products may have quite similar properties. Competition by innovation therefore implies that new innovations be brought onto the market at regular intervals. This is the strategic model of repeated innovation.[17] Short development time is a condition of the competitiveness of this strategy. The firm must be quick to offer a "me-too" product, or to respond to identified client expectation. Project organization must allow for concurrent engineering. Implementing this specialty strategy requires an understanding of client expectation, the transformation processes they employ, the properties they require and how they are measured. Client partnerships, and even alliances, have been set up. The importance of application and applicability laboratories also has increased.

In contrast to the chemical branch, where downstream development partnerships with customers have been developed, in the health sector upstream partnerships with academic research centers have been set up. Their goal is to improve understanding of the mechanism of action of medicinal drugs, with a view to registration and patents. In the chemical branch, highly integrated development processes have been instituted. Concurrent engineering was privileged to reduce development delays as

---

[17] A. Hatchuel, et al., 1998, "Innovation répétée et croissance de la firme," Rapport du programme CNRS "Enjeu économiques de l'innovation," 1998.

far as possible and test products on customers. In the health sector, on the other hand, processes are highly sequential given the regulatory constraints, each project targets a major market and a product with a long life span, conditions needed to offset development costs. In all, it can be said that if innovation is crucial to both branches of activity, it nonetheless presents virtually opposite characteristics for each. Each branch has adopted an innovation strategy consistent with its characteristics. The chemicals division has turned to a repeated innovation strategy while the pharmaceutical branch tries to protect its position building barriers to entry that are as durable as possible.

AIR LIQUIDE AND L'ORÉAL: SPECIALIZED COMPANIES THAT EXPERIENCED A TREMENDOUS GROWTH IN EXPANDING THEIR MARKET

These companies targeted niche markets when they were created almost a century ago. In inventing new products matching the needs of their customers, they enabled a continuous growth of their specific markets. It is only very recently, at the end of the 1990s, that these companies joined the leaders in the sector. I will try to stress a few outstanding features of their growth strategy using a macroscopic view.

Air Liquide, today, is number one worldwide in industrial gases, with a turnover of 8,100 million euros in 2000. One-third of its sales are made in the United States, half in Europe, and 15% in Asia Pacific. The company was created in 1902 by a chemist, Georges Claude, who had developed an acetylene liquefaction process and a process to separate and liquefy oxygen and nitrogen in the air. Innovation, patents, and international development have been characteristics of the company since its creation. As early as 1906, the company acquired installations in Belgium and Brazil. In 1907, a patent for neon tubes was obtained.

After a period of growth, the 1960s saw a reduction in demand. Air Liquide then decided to specialize in the industrial gas market. Targeted acquisitions allowed it to extend its customer base. In 1969, the firm developed its activities in the United States. In 1986, Air Liquide held 14% of the American market, and with the purchase of Big Three Industry, became the second most important producer of industrial gas in the United States.

Innovation allowed the company to reinforce its competitive position and to adapt to changes in the market. When the market for oxygen (for the steel industry) and acetylene (welding) declined, new products were developed for the electronics and food sectors (certain inert gases improve food conservation).

Finally, since its creation, the company has needed to raise capital and has distributed dividends. Today, it is still a private company quoted on the Paris stock exchange. The net profit in 2000 was 8% of turnover. Specialization, innovation, and internationalization were virtually continual growth vectors.

L'Oréal is the world cosmetics leader, with a turnover in 2000 of 12,671 million euros, and sales in Europe (51%), the United States (30%), and other countries (19%). The company traces its history back to the chemist Eugène Schueller. Having developed a colorant manufacturing process, he developed a market for hair dyes. An eccentric visionary, in 1907 he proposed his product to hairdressers and created the *Société des Teintures Inoffensives pour Cheveux* (Society of Dyes Safe for Hair) in 1909.

The company's strategy of diversification started in 1928 with the purchase of Monsavon. In 1933, a shampoo for the general public – Dop – was launched. In 1936, a sun cream was marketed for the first paid holidays granted by the Popular Front government. From this time on, the company has been remarkable for its publicity strategy using all possible media and designs by the best artists.

The 1960s and 1970s were marked by a diversification in distribution methods. Hair products reserved for professionals (dyes, hairsprays) were introduced onto the public market. Several acquisitions in perfumes, beauty creams and cosmetics gave access to the perfume distribution network. In 2000, the company was organized into four branches: professionals (hairdressers), the general public (skin and hair care, cosmetics), luxury (cosmetics and perfumes), and active cosmetics (skin care). Sales were 55% on the public market, 27% in perfume shops, 12% to hairdressers, and 5% for skin care.

Finally, international development started long ago. In 1953, L'Oréal sold its products in the United States via Cosmair. In 1994, L'Oréal acquired control of Cosmair, which in 2000 became L'Oréal USA. Multiple modes of distribution, an original publicity strategy, and early international development have always been characteristics of the company.

Although it is positioned in general public products, L'Oréal has always based its new products on technical innovations. Research and patents have played a key part in the company's development. Hence, when L'Oréal decided to develop antiaging creams at the end of the 1950s, the company formed a research team specializing in these areas. In 1973, L'Oréal even invested large sums in the pharmaceutical sector with the acquisition of 53% of Synthélabo. However, this majority investment did not lead to integration of the companies' activities. Today, L'Oréal holds 20% of Sanofi-Synthélabo, an autonomous company in the pharmaceutical sector.

Still in private hands, the company's capital was entirely held by the family until 1963. At that time, Eugène Schueller's daughter, Lilliane de Bettencourt, sold part of the shares. In 1974, a second batch was sold, and the Swiss group, Nestlé, became one of the shareholders. Today, 54% of the group is owned by a holding company, Gesparal, of which 51% is held by L. de Bettencourt and 49% by Nestlé.

With similar starting points, Air Liquide and L'Oréal occupy very different market sectors. Air Liquide specializes in services to industrial companies, and customized services with a strong technical component. L'Oréal has developed innovative products for the general public and supported their development with sophisticated research and marketing actions. The two companies have had similar growth processes: created by chemists through mastery of novel chemical processes, relatively unconnected to oil chemicals, they stressed development of their competence in their products/services and anticipated customer needs. Growth and internationalization are based on alliances and targeted acquisitions, the most fruitful of these today date back to the 1960s.

A CHEMICAL LEADER BACKED ON A OIL GIANT: THE CREATION OF TOTAL-FINA-ELF

Turning to oil exploitation, the growth of a chemical branch in oil companies, we find that Total-Fina-Elf was created following Total's acquisition of the Belgian company, Petrofina, in 1998, and the long state-owned and recently privatized French company, Elf Aquitaine, in 1999. Turnover in 2000 was 114 billion euros. It is the fourth largest petroleum group worldwide. Sales are split between Europe (54%, with only 31% in France), North America (9%), Africa (4%), and the rest of the world (33%). Atofina was created in April 2000 to consolidate the petrochemical activities of the three companies and is wholly owned by the oil group. Turnover in 2001 was 19.6 billion euros, with 38% for commodity polymers, 25% for performance polymers, and 36% for specialties. The markets are divided, with 63% in Europe, 27% in North America, and 10% in the rest of the world.

The history of Total-Fina-Elf is very different from that of Air Liquide, L'Oréal, or Rhône-Poulenc. The company was created around a raw material, and growth in chemicals was a consequence of positions upstream in the industry. It is much more recent. Describing the origin of the company means going back to the creation of its three components: Total, Fina, and Elf. All three were created to exploit petroleum, and national interests were critical in the creation of Total and Elf. Governmental involvement in the strategy of these companies was important, as shown by their capital holdings.

Elf resulted from the consolidation of three entities whose capital was entirely held by the French government: the RAP (Autonomous Petrol Board) created in 1939, the SNPA (National Society for Aquitaine Oils) in 1941, and the BRP (Oil Research Bureau) in 1945. The aim of these three "companies" was to explore and exploit oil and gas resources in France and its colonies. The politicians' goal was to make France independent in energy.

The CFP (French Petroleum Company), which became Total, was created in 1924 to develop a petrol industry in France. The involvement of the French government was considerable here, too, providing 25% of the capital on creation, and 35% from 1931. Each of these companies based its development on exploitation of oil resources, refining, and distributing fuel. Elf used the gas resources in Lacq in southwestern France, and then Algeria. Total historically developed from the oil fields in Turkey and Iraq.

Petrofina was created in 1920. Belgian financial groups took over Rumanian oil exploitation installations taken from the Germans during World War I. These Soviet oil fields, and then American fields in Texas, were exploited before the discovery of North Sea oil.

The diversification of these companies into the chemical industry occurred during a second phase. The development of chemicals in Total occurred conjointly with Elf with the creation in 1971 of Atochimie and Chloé (chlorine and ethylene) in 1980. In 1981, the development of Elf in the United States was boosted by the acquisition of Texas Gulf, which added very significant industrial holdings in phosphates and fertilizers. However, the place of Elf in the chemical industry is linked to French industrial policy in the years 1970–80. The oil crisis in the 1970s led to disastrous results for chemical companies confronted with an increase in the price of raw materials and a reduction in the price of chemical products related to overcapacity, internationalization of markets, and low growth of downstream consumers of chemical products who also were affected by the oil crisis. The French government then decided to reorganize the chemical industry by redistributing assets between companies and nationalizing Rhône-Poulenc. The industrial policy was intended to limit competition between French firms at the national level, and to form large companies capable of competing with their rivals at an international level. In 1983, Atochem was created. This company is wholly owned by Elf, which is itself entirely owned by the State. Atochem regroups the industrial assets of Rhône-Poulenc in chlorine and ethylene derivatives, those of Ugine Kuhlman in chlorine and fluorine derivatives, and in phosphates. Atochem then organized its activities into three product lines: bulk products for plastic and chemical materials (ethylene, propylene, benzene styrene); plastics and technical polymers; and specialties with chlorinated, fluorinated, and sulphur products. In 1990, Atochem took another step, acquiring Orkem, the chemical division of Charbonnages de France.

For Fina, diversification in chemicals goes back to the early 1960s, with investment in Petrochim (fifty-four) and development in the production of ethylene, polyethylene, paint polymers (sixty-three), and then styrene and polystyrene. At much the same time, Fina positioned itself in the soap segment with Oléochim (1957). In 1972, Sigma Coatings was created, grouping together Fina's paint interests. In 1998, Sigma Coating incorporated the activities of Lafarge in paints and is now the third-ranked company in Europe in the paint domain. Fina holds 80% of the company, and Lafarge, one of the largest French building companies, the remaining 20%. In 1983, Fina was one of the major European producers of high-density polyethylene. In 1998, agreements were concluded with the Belgian company, Solvay, to double the production capacity in the next ten years.

The French government did play an important role but has been withdrawing since the mid-1990s. First, there was a reduction in direct participation in the capital of Total in 1992 from 31.7% to 5.4%, then in 1996 to 0.97%. Elf was privatized in 1996. These changes in the structure of the capital opened the way to the concentration seen afterward. Total-Fina-Elf is a private company quoted on the Paris and New York stock exchanges with 13% of stable shareholders.

Atofina is, today, divided into three branches: petrochemicals and commodity polymers (polyethylene, polypropylene, styrenes, PVC, and others), intermediates and performance polymers (chlorine, fluorine, oxygen derivatives, functional polymers, etc.), and specialties (such as paint, adhesives, and rubber). Commodity polymers account for 38%, performance polymers for 25%, and specialties for 36% of activity.

Sanofi-Synthélabo

Similar to the creation of Atochem in 1973, Sanofi was created by merging several French pharmaceutical companies to form a company of international stature. Initially, the firm was entirely held by Elf. However, in 1979, the capital was opened up, with Elf keeping the majority. The company grew by multiple alliances, notably with Japanese companies. The merger with Synthélabo in 1999 accounts for its current stature. It is the second (largest) French pharmaceutical company and the seventh in Europe, with a turnover of 7,508 million euros in 2000. The reference shareholders are Total-Fina-Elf (33%) and L'Oréal (20%) in 2000.

CONCLUSION

Since World War II, the chemical industry has gone through several evolutions. My analysis has specifically focused on the period covering the last twenty-five years of the twentieth century. During this period, the French chemical industry, which has historically based its growth on the mastering of high-volume production processes of well-known chemicals, progressively changed its position. Growth was increasingly driven by the use of chemicals in various applications: specialty goods with high value added, diversified products dedicated to end-users (in the cosmetics, for example) and pharmaceuticals. This evolution also can be traced in society at large: health expenditures have been continuously growing and consumers are looking for personalized and frequently renewed goods. The demand for these products is as much a result of the strategies of the companies as an opportunity they have taken. The place of the French chemical industry today is a result of the strategy that the main companies have deployed to face this global evolution.

In the area of specialty goods, two typical trajectories can be characterized. The first is the continuous growth of companies focused from their creation on a type of specialized products, having developed relationships with their customer, gained market knowledge, structured new product development, and proactively offered new products to their customer. Air Liquide and L'Oréal exemplify this strategy. These companies were founded on the mastering of a chemical process, but the exploration and invention of new usages for their products and of products with new use value for their customers propelled their rapid growth. Air Liquide has diversified the use of industrial gases from traditional customer to electronics and food industry for example. L'Oréal has expanded its product portfolio from toilet products to cosmetics and skin care.

The second trajectory is the strategic repositioning of diversified chemical groups that have based their growth on commodities and economies of scale and have recently focused their activity on specialty products instead. I have analyzed this strategic change using the case of Rhône-Poulenc and its chemical division. I have underlined the impact of this evolution on the organization of product development: coordination among researchers, process engineers, and marketing people; development of new technical knowledge; and new relationships with customers. This evolution led to a new firm: Rhodia.

Historically intertwined with the chemical industry in France, the pharmaceutical branches of large companies have become autonomous and specialized in pharmaceuticals as a result of acquisitions and concentration of pharmaceutical activities to gain a critical size. This evolution has taken place in France as well as in other European countries in the last decade. The case of Rhône-Poulenc and its transformation into Aventis is typical of this trend. The specific nature of the regulatory system and the patenting regime as well as the necessity of commercializing products on a worldwide basis have induced concentration and specialization in this sector. So despite their common focus on innovation, the pharmaceutical and chemical division evolved in diverging direction as far as scientific and industrial partnerships are concerned.

The production of commodities has remained an important component of the chemical industry. Globalization and merger and acquisition have characterized the recent period in this activity also. In France, only companies having assets in oil exploitation have been able to remain competitive in raw chemicals. The intervention of the French government, which has been massive at different point in time in the past but that has today become more and more limited, is still imprinted on Total-Fina-Elf. The French government was part of the origin of Elf and Total because of the importance of oil for industrial development and economic independence. The state as the prime shareholder (indeed, the only one for Elf until the mid-1990s) has long taken part in the strategic decisions of these companies. The restructuring undertaken by the French government after the oil crisis in 1982 has led to the division of assets in the raw chemicals between Elf Atochem (created as a subsidiary of Elf totally owned by the company) and Rhône-Poulenc, which was nationalized. Today, there is no more state-owned company in this industry. This evolution of the French state from a "hands-on" to a "hands-off" position was a precondition to the mergers that occurred in the late 1990s. It had led to a dramatic change of the industrial policy in this sector in France.

BIBLIOGRAPHY

Abescat, B., 2002, *La saga des Bettencourt. L'Oréal: Une fortune française*, Plon, Paris.

Beltran, A., and S. Chauveau, 1999, *Elf, des origines à 1989*, Fayard, Paris.

Benghozi, P. J., F. Charue-Duboc, and C. Midler (ed.), 2000, *Innovation Based Competition and Design Systems Dynamics*, l'Harmattan, Paris.

Ben Mahmoud-Jouini, S., and C. Midler, 1999, "Compétition par l'innovation et dynamique des systèmes de conception dans les entreprises françaises. Réflexions à partir de la confrontation de trois secteurs," *Entreprises et Histoire*, 23, pp. 36–62.

Bibard, L. et al., 1993, "Recherche et développement et stratégie: Rhône-Poulenc Agrochimie et Rhône-Poulenc Santé" in *Stratégie technologique et avantage concurrentiel*, rapport de recherche IREPD, Grenoble.

Bonin, H., 2002, "'The French Touch': International Beauty and Health Care at L'Oréal," in Bonin et al. *Transnational Companies*, pp. 91–101, Editions PLAGE, Paris.

Bram, G. et al., 1995, *La chimie dans la société, son rôle son image*, l'Harmattan, Paris.

Brown, S., and Eisenhardt K., 1997, "The Art of Continuous Change: Linking Complexity Theory and Time-Paced Evolution in Relentlessly Shifting Organizations," *Administrative Science Quarterly*, 42, pp. 1–34.

Cayez, P., 1988, *Rhône-Poulenc, 1895–1975*, Colin et Masson, Paris.

Chapel, V., 1997, "La croissance par l'innovation: de la dynamique d'apprentissage à la révélation d'un modèle industriel. Le cas Tefal," Thèse de doctorat de l'Ecole des Mines de Paris.

Charue, F., and C. Midler, 1994, "Le développement du management de projet chez Rhône-Poulenc," Rapport de recherche Rhône-Poulenc.

Charue-Duboc, F., 1997, "Maîtrise d'oeuvre, maîtrise d'ouvrage et direction de projet, pour comprendre l'évolution des projets chez Rhône-Poulenc," *Gérer et Comprendre*, 49, pp. 54–64.

Charue-Duboc, F., and C. Midler, 1995, "Le développement du management de projet chez Rhône-Poulenc – II," Rapport de recherche Rhône-Poulenc.

Charue-Duboc, F., and C. Midler, 1998, "Le développement du management de projet chez Rhône-Poulenc – III," Rapport de recherche Rhône-Poulenc.

Clark, K., and T. Fujimoto, 1991, *Product Development Performance: Strategy, Organization and Management in the Auto Industry*, Harvard Business Press, Cambridge.

Cockburn, I., and R. Henderson, 1999, "The Economics of Drug Discovery," in R. Landau, B. Achiadellis, and A. Scriabine (ed.), *Pharmaceutical Innovation*, The Chemical Heritage Foundation, Philadelphia.

Cohendet, P. (ed.), 1984, *La chimie en Europe*, Economica, Paris.

Cohendet, P., J. A. Herault, and M. Ledoux, 1989, "Quelle chimie pour l'an 2000?" *La Recherche*, pp. 1254–57.

Colombo, U., 1980, "A Viewpoint on Innovation and the Chemical Industry," *Research Policy*, 9, pp. 204–31.

Dalle, F., 2001, *L'Aventure l'Oréal*, Editions Odile Jacob, Paris.

Eisenhardt, K., and B. Tabrizi, 1995, "Accelerating Adaptive Processes: Product Innovation in the Global Computer Industry," *Administrative Science Quaterly*, 40, pp. 84–110.

Fridenson, P., 1997, "France: The Relatively Slow Development of Big Business in the Twentieth Century," in A. Chandler, F. Amatori, and T. Hikino, (eds.), *Big Business and the Wealth of Nations,* pp. 207–45, Cambridge University Press, Cambridge.

Gaffard, J. L., et al., 1993, *Cohérence et diversité des systèmes d'innovation en Europe*, rapport de synthèse du FAST vol. 19, CEE, Bruxelles.

Garel, G., and A. Kesseler, 1998, "New Car Development Projects and Supplier Partnership," in R. A. Lundin, and C. Midler, (eds.), *Projects as Arenas for Renewal and Learning Processes*, Kluwer, Dordrecht.

Hatchuel, A., and B. Weil, 1999, "Design Oriented Organizations, toward a Unified Theory of Design Activity," communication, 6th international product development management conference, Cambridge, UK.

Hatchuel, A., V. Chapel, X. Deroy, and P. Le Masson, 1998, "Innovation répétée et croissance de la firme," Rapport du programme CNRS, "Enjeu économiques de l'innovation."

Internal document, Rhône-Poulenc, 1995, *100 ans d'innovations.*

Jemain, A., 2002, *Les conquérants de l'invisible, Air liquide 100 ans d'histoire*, Fayard, Paris.

Lane, S. J., 1993, "Corporate Restructuring in the Chemical Industry," in M. Blair, (ed.), *The Deal Decade, What Takeovers and Leveraged Buyouts Mean for Corporate Governance*, The Brookings Institution, Washington DC.

Longhi, C., 1993, "Stratégies organisationnelles et système d'innovation: le cas du groupe Rhône-Poulenc et de la SBU silicones," in J. L. Gaffard et al., *Cohérence et diversité des systèmes d'innovation en Europe*, Rapport de recherche du FAST, CEE, Bruxelles.

Lynn, G., J. Morone, and A. Paulson, 1996, "Marketing Discontinuous Innovation: The Probe and Learn Process," *California Management Review*, 38, pp. 8–37.

Midler, C., 1996, "Modèles gestionnaires et régulations économiques de la conception," in G. De Terssac et E. Friedberg (ed.), *Coopération et conception,* Octares, Toulouse.

Midler, C., and Charue-Duboc, F., 1998, "Beyond Advanced Project Management: Renewing Engineering Practices and Organizations," in R. Lundin and C. Midler (eds.), *Projects as Arenas for Renewal and Learning Processes*, Kluwer, Dordrecht.

Navarre, C., 1993, "Pilotage stratégique de la firme et gestion des projets: de Ford et Taylor à Agile et IMS," in V. Giard et C. Midler (eds.), *Pilotage de Projet et Entreprises, diversités et convergences*, Economica, Paris.

Nouschi, A., 2001, *La France et le pétrole*, Picard, Paris.

Porter, M. E., 1980, *Competitive Strategy: Techniques for Analyzing Industries and Competitors*, The Free Press, New York.

Quéré, M., 1997, "Le paradoxe de la fonction recherche-développement dans la dynamique des firmes industrielles," *Cahier de recherche du GIP Mutations Industrielles* no. 71.

Rhône-Poulenc document, activity report.

Ruffat, M., 1996, *175 ans d'industrie pharmaceutique française. Histoire de Synthélabo*, La Découverte, Paris.

Smith, J. G., 1979, *The Origins and Early Development of the Heavy Chemical Industry in France*, Clarendon Press, Oxford.

Tushman, M., and P. Andersen, 1986, "Technological Discontinuities and Organizational Environments," *Administrative Science Quarterly*, 31, pp. 439-65.

10

The United Kingdom

WYN GRANT

In a country whose manufacturing industry has been blighted by relative decline and poor productivity, the chemical industry has been a relative success. However, like the rest of the British manufacturing, the chemical industry has experienced a decline in competitiveness. The watershed decade for the industry was the 1970s. Before that, as in other countries, the story was one of rapid and continuing expansion as a growing economy and processes of substitution for more traditional materials provided apparently limitless demand for the industry's products. A rapid rate of innovation in both process and product technologies helped to stimulate this growth in demand. However, the two oil shocks not only blunted demand, they also revealed underlying weaknesses in the industry.

These trends are illustrated by the example of polyvinyl chloride, a key plastic used extensively in the construction industry among other applications. Production took some time to develop after the war, but more than doubled between 1955 and 1960. It almost doubled again between 1960 and 1965, and increased by 62 percent between 1965 and 1970. However, in the fifteen years from 1970 to 1985, production increased by just over 11 percent.

Production of ethylene, the basic building block of the petrochemical industry, displayed a long-running upward trend but also has been susceptible to cyclical variations in the economy. The statistical series peaked in 1988 at the time of the "Lawson boom." Production almost doubled between 1965 and 1970 but actually fell back by 1975. There was a very modest increase up until 1980, and then an increase of over 30 percent between 1980 and 1985. By the time the statistical series came to an end in 1992, the volume of production had increased 3.7 times since 1965.

Looking at the industry as a whole, the index of industrial production for chemicals and allied trades increased from 100 in 1948 to 192.4 in 1957, with a particularly sharp rise between 1950 (124) and 1955 (178.5). The index was rebased at 100 in 1958 and showed a further marked increase to 158.5 in 1967. A new index level of 100 was again set in 1963, and this series showed continuing growth to 157.9 in 1970. The rebased index of 100 in 1970 grew to only 116.5 in 1975. When the index was rebased at 100 in 1975, the index numbers for 1973 and 1974 were shown to be below 100, the first decline in the postwar history of the industry. The rebased index struggled to reach 109.7 in 1980.

As a capital-intensive industry, the chemical industry has never been a leading employer compared with, for example, the motor industry. Long-term data on employment in the industry are difficult to provide as a narrower basis of calculation was used after 1980, leading to some seventy thousand workers no longer being classified as falling within the industry. However, it is evident that employment in the industry increased in the immediate postwar period, reaching a peak of 531,000 in 1961. Employment then started to fall away steadily, although the decline in the 1990s was particularly marked (from 330,000 in 1990 to 251,000 in 2000). Allowing for differences in the basis of calculation, it can be said that employment in the industry has been approximately halved from its peak, from around half a million to around quarter of a million.

The industry's loss of its flagship status is graphically illustrated by the rise in import penetration after 1970. At that time, import penetration was only 18 percent, marginally above the level for the manufacturing industry as a whole. By the end of the decade, the level of import penetration for chemicals was well above that for the manufacturing industry, and the last available figure is an import penetration ratio of 57 percent. In the meantime, however, the deterioration in performance in other sectors means that the figure for manufacturing as a whole is once again marginally below that for chemicals (see Table 10.1). These trends reflect, in part, the increasing internationalization of the industry, but they also suggest the presence of some competitive weaknesses, although the position relative to British industry as a whole has improved recently.

Table 10.1. Import Penetration in the Chemical Manufacturing Industries

Year	Import Penetration in the Chemical Industry (Imports as a Percentage of Home Demand)	Import Penetration in Manufacturing Industry (Imports as a Percentage of Home Demand)
1970	18 percent	17 percent
1975	23 percent	22 percent
1980	29 percent	25 percent
1985	41 percent	34 percent
1988	42 percent	36 percent
1996	57 percent	56 percent

Source: Annual Abstract of Statistics, various volumes. There was a break in the statistical series in the late 1980s and early 1990s. 1996 is the last year for which data is available.

In a country that has had chronic balance of payments problems, the industry has made a positive contribution to the balance of payments throughout the postwar period. However, the size of that contribution has diminished substantially (see Table 10.2). In the early to mid-1950s, exports were more than double the level of imports in terms of value. By 1975, the gap had narrowed, but exports still exceeded imports in value by over 50 percent. By 1985, the gap was somewhat over a third. By 1999, the gap was below a quarter, although the ratio of exports to imports had stabilized at around 1.2 in the 1990s. Moreover, it was still highly positive compared with most manufacturing industries, but poor in comparison to the industry's historical record. The first sharp surge in imports occurred in the late 1950s, and there was a particularly marked increase in the early 1970s. In the five years from 1970 to 1975, the imports index increased from 100 to 229, whereas that for exports increased to 209.

Table 10.2. Balance of Imports and Exports in the U.K. Chemical Industry

Year	Surplus of Exports Over Imports (£m)	Ratio of Exports to Imports
1950	74.9	2.1
1960	141.2	1.8
1970	241.6	1.5
1980	2141.7	1.7
1990	2347.4	1.2
1999	4367.0	1.2

Source: Calculations from *Annual Abstract of Statistics*, various volumes.

The U.K. chemical industry is characterized by deterioration in some indicators of performance but has been much more successful in terms of international competitiveness than most British industries. How can this record of relative success be explained?

A central theme in any history of the chemical industry in Britain has to be that of Imperial Chemical Industries (ICI). That does not mean that other companies in the industry, such as the chemical divisions of the oil companies or the various specialist firms, were unimportant. However, ICI had a significance that went beyond the chemical industry itself. It was regarded as an industrial flagship company in Britain.

In a country where many manufacturing industries were seen to be relatively unsuccessful, at least in terms of international competitiveness, ICI was seen as a remarkable success. It helped to transform a straitened British chemical industry at the time of World War I into one that was technologically innovative and well managed. The way in which this was done has a broader significance. After World War II, the Labor Government relied on nationalization to secure a variety of (not always clearly specified) objectives. As far as making a contribution to economic planning and national competitiveness was concerned, the nationalized industries were not a success. As Dell puts it, "A series of baronies was created which, while not immune from government influence, was quite difficult to influence against their own predilections."[1] Far more successful was the ICI model of setting up the company as a "chosen instrument" that enjoyed a special, privileged relationship with government.

THE SECOND INDUSTRIAL REVOLUTION

Any account of the British chemical industry since World War II would be incomplete without a consideration of earlier events that helped to shape the industry. It is generally accepted that Britain was not successful in the "second Industrial Revolution" that took place in the late nineteenth and early twentieth century. One of the characteristics of this industrial revolution was that it was science-based: it involved the systematic application of scientific knowledge to products and processes. Rather than the talented inventors who had played an important role in the first Industrial Revolution, it required innovators who could design effective large-scale production plants and find economic uses for by-products.

[1] Edmund Dell, *A Strange Eventful History: Democratic Socialism in Britain* (London: HarperCollins, 2000), p. 142.

It also required sufficient numbers of technically qualified individuals to operate the plants effectively. They, in turn, required a systematic technical education that would enable them to improve existing processes, rather than learning how to do the job by observing those who had the skills already, as might happen, for example, in the pottery industry.

It was generally recognized both at the time, and in subsequent analyses, that British technical education was deficient when compared with what was available in, for example, Germany. The problem received the attention of the Royal Commission on the Depression in Trade and Industry in 1886, in what was one of the earliest attempts to investigate the underlying causes of poor industrial performance in Britain. In a book that attracted wide attention when it was published at the end of the nineteenth century, Williams wrote: "The attention paid by the State in Germany to Education – and particularly to Scientific and Technical Education – is a matter of common knowledge the world over. . . . The Technical Education to be obtained in Germany is thorough, and thoroughly scientific; *but it is meant for application.*"[2] Williams insisted that the system of technical education "is an integral factor in Germany's industrial success, and which, compared with anything in the nature of technical education to be found in England, is as an electric lamp to a rush-light."[3] These deficiencies did not provide a good basis for the development of the chemical industry.

Nevertheless, the chemical industry in Britain got off to a good start, in part because of demand for soda ash from industries such as textiles, glass, and paper. "In the 1870s, Britain had the world's largest chemical industry, centered on Merseyside, which was close to the Cheshire salt deposits for making soda ash and chlorine, to supplies of limestone and coal and to the busy port of Liverpool."[4] Yet, by 1896, the industry was in dire straits:

. . . it is no exaggeration to say that Germany is a more formidable rival, and has already given us a sounder beating in Chemicals than in any other field of Trade, not even excepting Iron and Steel. . . . In England the industry in several branches is little better than a Bottomless Pit for capital: in proof whereof may I instance the disappearance of many of the smaller manufac-

---

[2] E. E., Williams, *Made in Germany* (London: William Heinemann, 1896), pp. 151-52.
[3] Ibid., p. 156.
[4] Carol Kennedy, *ICI: The Company that Changed Our Lives* (London: Hutchinson, 1986), p. 10.

turers and the sore straits of those syndicates in which some businesses are merged.[5]

The underlying problem was the failure of the industry to adapt to a major technological change, that is, the replacement of the Leblanc process by the Solvay process. Not only was the Leblanc process very wasteful, it was also environmentally damaging, to the extent that it prompted government intervention in the form of what was probably the first pollution control agency in the world, the Alkalai Inspectorate. One of the reasons the Leblanc process survived at all was that it was possible to extract chlorine as a by-product that could be used to produce bleaching powder, much in demand from the textile industry.

The Solvay process produced purer soda ash more economically, but Brunner, Mond (a forerunner of ICI) was the only company to adopt it in Britain. As Williams observed: "The fact that Brunner, Mond, and Company, the one really successful firm in England, is practically an international concern, puts it in a separate category from the genuine local English industry."[6] One of its founders was a German immigrant to England.

The alkali companies decided that the way out of their difficulties was to suppress the operation of market forces through consolidation, and in 1891 they came together as the United Alkalai Company. This new company had a rather elderly and overly large board of directors "openly dedicated . . . to preserve, if they could, a system of technology threatened by obsolescence and very efficient competition."[7] This was hardly a recipe for responding positively to new technological changes, and so it proved with the electrolysis process.

This offered the possibility of commercially producing chlorine out of the Leblanc process by the electrolysis of brine. "Almost the whole of UAC's capital investment . . . was threatened with obsolescence within four or five years of being made."[8] Brunner, Mond quickly took advantage of an electrolysis process known as the Castner-Kellner process. The only thing that saved UAC from immediate disaster was that it took a while to sort out some of the problems associated with getting the process into

---

[5] Williams, *Made in Germany*, p. 90.

[6] Ibid., 96n.

[7] W. J. Reader, *Imperial Chemical Industries, A History. Volume 1: The Forerunners, 1970–1926* (London: Oxford University Press, 1970), p. 107.

[8] Ibid., p. 114.

production. Although UAC managed to stagger on to become one of the founding companies of ICI, their attachment to outmoded technology exemplified many of the problems British industry faced in coming to terms with the second Industrial Revolution. Reader comments, "It is hard to see anything that the hapless Company's management did right. . . . The management displayed nepotism, amateurism, lack of technical knowledge and scientific training, and its policy was based on all things right-minded people abhor: restriction of output, price-fixing, market-sharing, instead of good, clean, ruthless competition."[9]

Another branch of the industry that faced difficulties was dyestuffs. At the outbreak of World War I, 88 percent of all synthetic dyes on the market were produced in Germany.[10] As Williams discussed, this was an extraordinary record of failure, given that aniline dyes were discovered in England and that they were extracted from coal-tar products, the country having an abundant supply of coal. Williams attributed this failure to the lack of investment in research and development by the British companies in the way that had been done by their German counterparts.[11] Reader emphasizes the way in which dyestuffs lived under the shadow of the textile trade and therefore failed to diversify into new and promising areas of production such as photographic chemicals.[12]

WORLD WAR I

Before World War I, there had been no systematic government interest in the plight of the industry. The Royal Commission on the Depression in Trade and Industry did not even discuss it, preferring to concentrate on sectors such as lace and silk. The war, however, brought about a transformation in the industry and its relations with government. Progress in a science-based industry such as chemicals was essential to the success of modern warfare. The importance of the explosives sector of the industry to the war effort needs no emphasis, but shortages of dyestuffs also caused problems that led to government intervention. Haber notes: "The structural changes resulting from the emergency measures created the framework of the contemporary chemical industry . . . the large factories,

[9] Ibid., p. 122.
[10] Kennedy, *ICI*, p. 10.
[11] Williams, *Made in Germany*, pp. 105–06.
[12] Reader, *Imperial Chemical Industries*, pp. 259–60.

the big concerns that control them, and their enormous capital requirements."[13]

The plight of the dyestuffs industry was such that khaki dye for troops'
uniforms had to be procured from Germany through neutral countries. In
1915, the government set up British Dyes Limited with a Treasury majority shareholding. However, this company had a narrow base as one of the
most important and successful companies, Levinsteins of Manchester,
remained outside it. "By 1917 the Board of Trade was convinced that
only a merger could establish a viable industry."[14] It was not until 1919
that the British Dyestuffs Corporation was set up, controlling 75 percent
of U.K. production and benefiting from nearly a half million pounds in
government aid.

These government interventions were relatively modest and somewhat confused, but what is significant was that they took place at all.
The chemical industry was now an important agenda item for government. Reader refers to "a revolution in British ideas on the national importance of having a complete chemical industry. The Government would
never again be indifferent to the chemical companies' development as
it affected the country's self-sufficiency."[15] One consequence was that
the government became more involved in the development of scientific research through the formation of the Department of Scientific and
Industrial Research. It took a series of initiatives that led eventually to
the establishment of the National Chemical Laboratory in 1927.

Nevertheless, Germany's technological lead in some areas remained
substantial. A key problem that had faced the industry for many years
was how to "fix" nitrogen from the air so that it could be used to make
fertilizers. The Haber-Bosch process for synthesizing ammonia offered
the way forward, representing "what was generally regarded as the most
important advance for many years in the heavy chemical industry."[16]
In order to acquire the technology, a strategy of appropriating it was
followed. A Chemicals Mission was attached to the advancing British
Army with the objective, as the major in charge put it, of "pinching
everything they've got."[17] BASF did everything they could to frustrate the

[13] L. F. Haber, *The Chemical Industry, 1900–1930: International Growth and Technological Change* (Oxford: Clarendon Press, 1971), p. 216.

[14] Wyn Grant, William Paterson, and Colin Whitston, *Government and the Chemical Industry* (Oxford: Clarendon Press, 1988), p. 20.

[15] Reader, *Imperial Chemical Industries*, p. 351.

[16] Ibid., p. 320.

[17] Kennedy, *ICI*, p. 18.

plan, including painting out dials on the machinery, and removing stairs between floors. Brunner, Mond obtained more reliable data by in effect bribing two chemical engineers who had worked for BASF (generally referred to as K and A) to provide plans, price lists, and other technical information. "The K and A transaction, it seems, rather that the Oppau mission, was the really successful burglary."[18]

An important development, given the research deficiencies of the industry before the war, was the decision to equip an excellent new laboratory at Billingham. Determined to see the synthesis of ammonia produced in the country, the British government persuaded an initially reluctant Brunner, Mond to take on the Billingham site, a deal sweetened by help in securing confiscated BASF patents and a £2 million loan guarantee. Brunner, Mond provided themselves "with a first-class research centre under a scientist of academic distinction who was also a member of the general management."[19] They were, thus, moving closer to the German model of not only emphasizing research expertise but also integrating it as a central part of the overall management of the company rather than having a group of "boffins" shut away in a collection of sheds.

THE AFTERMATH OF WAR

At the end of the war, the productioneers movement, associated with such figures as Dudley Docker, the founder of the Federation of British Industries, and Christopher Addison, the Minister for Reconstruction, sought to take British industrial policy down a new path, aimed at creating a high-wage, high-output economy with a measure of government intervention and an emphasis on cooperation between employers and labor.[20] Later in the twentieth century, it might have been described as a "corporatist" approach to policy making. For a variety of reasons, the productioneers were defeated by industrial politicians of a more laissez-faire orientation. However, it was not a case of a simple reversion to "business as usual" as it had been conducted before the war.

One of the changes brought about by the war had been the direct involvement of businessmen in the operations of government. This brought government and business closer together both personally and

---

[18] Reader, *Imperial Chemical Industries*, p. 365.
[19] Ibid., p. 366.
[20] R. P. T. Davenport-Hines, *Dudley Docker: The Life and Times of a Trade Warrior* (Cambridge: Cambridge University Press, 1984).

institutionally. A network of "industrial politicians" formed who had experience of both business and government. As Turner notes, the establishment of ICI "was made possible by a network of businessmen, politicians and civil servants who co-operated to promote, but also control, state intervention in the economy."[21]

New structures were being set up to mediate between government and industry. The formation of the Federation of British Industries in 1916 was itself indicative of a closer relationship between business and government. Wartime controls encouraged manufacturers to come together at a sectoral level. The Association of British Chemical Manufacturers (ABCM) was formed in 1917 and quickly won praise from government for its representativeness and willingness to cooperate with government. The ABCM established a new hegemony as the representative of the British chemical industry after the Federation of the British Dyestuffs Industry was forced into dissolution.[22] In a country noted for its relatively ineffective and fragmented associations, the ABCM (later the Chemical Industries Association) provided a united voice that enjoyed a good working relationship with government.

THE FORMATION OF ICI

The immediate stimulus for the formation of ICI was the establishment in 1925 in Germany of IG Farben by the fusion of the six largest chemical firms. "The idea of what was at first called a "British IG" began to take shape, not quite on government initiative, but certainly with government blessing."[23] Apart from events in Germany, another imperative was the continuing difficulties of the British dyestuffs industry and the BDC.

One of the central figures in the formation of ICI was Reginald McKenna, a typical example of the new breed of industrial politician. Chancellor of the Exchequer during the first part of the war, he subsequently became chair of Midland Bank but retained his connections in political circles. Late in January 1926, he had lunch with Sir Harry McGowan, chair of Nobel Industries, the explosives company that had

---

[21] John Turner, "The Politics of Business," in John Turner, ed., *Businessmen and Politics: Studies of Business Activity in British Politics, 1900–1945* (London: Heinemann, 1984), p. 13.

[22] Grant et al., *Government and the Chemical Industry*, p. 23.

[23] W. J. Reader, "Imperial Chemical Industries and the State, 1926–1945," in B. J. Supple, ed., *Essays in British Business History* (Oxford: Clarendon Press, 1987), p. 230.

absorbed many other leading companies during the war. McKenna let it be known over the lunch table "that it would be acceptable in the highest circles of government if a coalition of British chemical companies were to rescue British Dyestuffs."[24] McGowan was not interested in the idea that Nobel Industries should take over BDC but suggested a more ambitious plan for a British IG that "would provide a mighty counter-balance to the German combine."[25] The broad outline of the agreement to set up ICI was worked out by Mond and McGowan on board the *Aquitania* while it was sailing from New York to Southampton. The deal was really a takeover rather than a merger with the stronger businesses (Brunner, Mond and Nobel) acquiring the weaker ones (United Alkalai and BDC). From the beginning, ICI enjoyed an unusual and special status as a company. It "appeared to be regarded more as a public service than a powerful new private enterprise."[26] ICI's Board could not make decisions simply on commercial criteria. The ICI merger was essentially a matter of public policy "making the Board conscious from the start that they would have to take the public interest into account, even, if necessary, ahead of the interest of shareholders."[27] As Reader emphasizes in his account, there was nothing inevitable about the formation or ICI or its success. Difficulties over the Billingham fertilizer plant, which came on stream just as the slump hit the demand for fertilizer, could have brought the company down. In general, however, the company was seen as a successful example of the benefits of indirect government intervention. Its management was highly professional, if sometimes autocratic, at a time when that was not the norm.

One of the strengths of the company was clearly technological. "By the time ICI had been in existence for ten or twelve years it had a very high concentration of scientific and technological manpower – much the highest in British industry."[28] There was sometimes suspicion within the company that there was too much encouragement of expensive, "ivory tower" research. However, it was research that brought results. Among the key inventions made in ICI laboratories in the 1930s were polythene,

[24] Kennedy, *ICI*, p. 22.
[25] *Ibid.*
[26] Kennedy, *ICI*, p. 50.
[27] W. J. Reader, *Imperial Chemical Industries, A History, Volume 2: The First Quarter Century, 1926-1952* (London: Oxford University Press, 1975), p. 473.
[28] Ibid., p. 77.

"Perspex," and various dyestuffs. In the 1940s, ICI was able to exploit the discovery of terylene.

ICI had a "good war" in terms of the expansion of its business, the development of new products and processes, and an increase in profits. The good performance of the industry compared with World War I validated the decision to set up ICI.

After the war, ICI assumed a new importance in terms of government efforts to redress regional imbalances in the economy through its willingness to start businesses in "depressed areas." ICI also calculated that such areas would receive priority when buildings and materials licenses were being allocated. One practical consequence was the enlargement of the proposed North-East development area to encompass ICI's new Wilton works, which was near its existing plant at Billingham. Teesside became one of the major geographical concentrations of the British chemical industry.

THE ANGLO-AMERICAN PRODUCTIVITY COUNCIL REPORT

Corelli Barnett has lambasted the postwar performance of British industries in general but argues "there was in fact only one branch of second-industrial-revolution technology where the Second World War did not find Britain to a greater or lesser extent wanting – the chemical industry and, in particular, ICI, the huge combine which dominated it."[29]

It is interesting to contrast this judgment with the report of the Anglo-American Productivity Council on heavy chemicals that appeared in 1953. The Productivity Council set up sixty-six sectoral teams that attempted to learn from the achievement of high levels of productivity in the United States. The sectoral teams were drawn from both managers and workers. The heavy chemicals team set sail for the United States on 6 March 1952 on the *Queen Mary*, returning on 23 April from New York on the *Queen Elizabeth*. The party leader was, not surprisingly, from ICI, the only company to have two representatives from both the managerial and "workshop" levels.

The report that emerged was a striking indictment of what was supposed to be one of Britain's successful industries. It calculated that the productivity of many American chemical plants was at least three times

[29] Correlli Barnett, *The Audit of War: The Illusion and Reality of Britain as a Great Nation* (London: Macmillan, 1986), p. 181.

that of their British counterparts.[30] A wide range of deficiencies was identified:

Many of the American factories visited were superior to the British in such diverse matters as the selection of recruits, the *continuous* training of employees at all levels, the delegation of authority, the control of costs, the flexibility of the apprenticeship schemes, the organization of maintenance work, the use of promotion as an incentive, the poor use of foremen, and the cultivation of effective attitudes towards production, advertising and the use of mechanical aids.[31]

When one looks at the specific problems identified by the investigating team they can be seen to be an echo of the problems identified in the British manufacturing industry in general during the postwar period. Assets per employee in the United States were calculated at $18,800 in 1951 compared with £3,250 in Britain.[32] The report found that the American plants were more modern, reflecting the more recent development of the industry, and used more automatic controls than did the British plants. The instrumentation and controls were not, in general, different from that available in Britain, they were just used more extensively. This gave steadier plant operation, although the British plants were more likely to use continuous as against batch production methods. The committee thought that the American adoption of instrumentation and automatic control was in part driven by the higher cost of labor, although it also reflected a shortage of instrument maintenance engineers and designers in Britain. Their American hosts told the committee that in the British industry there was a greater reluctance to scrap and renew old and inefficient plant and equipment, a phenomenon noted in studies of other industries, although the committee itself seemed reluctant to accept this conclusion.

Maintenance practices were found to be more effective in the United States. The committee believed that the British industry "was not making sufficient use of planned maintenance and this was no doubt due, in part, to a shortage of engineering staff and, for the rest, to an inadequate faith in its advantages."[33] These dismissive attitudes toward the value of maintenance also were found in relation to preventive maintenance, which was "too often confined to somewhat sporadic studies carried out on pieces

---

[30] British Productivity Council, *Heavy Chemicals* (London: British Productivity Council, 1953, p. 1.
[31] Ibid., p. 2.
[32] Ibid., p. 52.
[33] Ibid., p. 34.

of equipment which were being exceptionally troublesome, rather than being directed towards a systematic study of all equipment."[34] In the committee's view, the single most important difference between the two industries was that there were fewer technically qualified personnel in the British plants. In the United States, there was a "very generous use of technical graduates to overcome plant difficulties."[35] The report noted a continuous interaction between these technically qualified staff and the process and maintenance workers, what we would perhaps today call a continuous search for higher quality. As far as the committee was concerned, their most "fundamental observation" was that: "The American companies had, on average, one technically qualified man to every six hourly-paid workers, whereas in Britain the ratio was about one to sixteen."[36] Given all these difficulties, why did the chemical industry have such a positive image in the postwar period? One reason was that even if its productivity record was poorer than that of its overseas competitors it was still good when compared with other British industries. The committee calculated, admittedly on the basis of rather inadequate data, that productivity in the industry was growing at around 8 percent a year in the immediate postwar years. Another reason was that it, or more particularly ICI, handled its political relations very well. One minister in the 1945 Labor government suggested that it was easier to get cooperation from ICI than from the British Electricity Authority.[37] Reader notes that ICI showed "sufficient political skill and sense of responsibility to respond to movements of power and opinion in the world at large and to adjust policy continually to the mood of the times."[38]

THE RISE OF THE PETROCHEMICAL INDUSTRY

The displacement of coal and fermentation technology by oil as a feedstock changed the shape of the industry after World War II. The major oil companies set up chemicals divisions, increasing the number of players in the market. One of the longer-run consequences of the shift from an internationally cartelized industry to one in which there was much more intensive competition on price was the eventual recurrent overcapacity crises.

---

[34] Ibid., p. 35.
[35] Ibid., p. 10.
[36] Ibid., p. 2.
[37] Philip M. Williams, *Hugh Gaitskell* (Oxford: Oxford University Press, 1982), p. 133.
[38] Reader, *Imperial Chemical Industries, Volume 2*, p. 479.

The shift to petrochemicals happened relatively early in Britain in comparison to other countries. This was "partly as a result of the infusion of American technology, but also due to important British technology developed in the early 1940s."[39] Work was carried out at Billingham by ICI, while Shell started producing a detergent at Stanlow in 1942. Between 1948 and 1958, $280 million was spent on petrochemicals development, "which was then receiving a higher proportion of total industrial investment than any other British industry."[40] Overall production of petrochemicals in the United Kingdom tripled between 1953 and 1959.

ICI made a strategic decision to construct a cracking plant at Wilton "for themselves rather than relying on the oil industry to do it for them. These decisions were profoundly important to the development of ICI over the next quarter-century and more."[41] Shell, BP (at first as a joint venture with Distillers called British Hydrocarbon Chemicals), and, at a later stage, Esso, were all able to challenge ICI in particular product markets, but they were never really able to erode its dominance of bulk chemicals. By the 1980s, there were four producers of ethylene (the basic building block of the petrochemicals industry) in Britain: ICI, BP Chemicals, Shell Chemicals, and Esso Chemicals. "However, even BP Chemicals had a turnover in 1985 of less [than] one-fifth of that of ICI (£1922 million against £10725 million)."[42] When the Board of Trade conducted its investigation into investment in the chemical industry in the mid 1960s, it initially found it sufficient to talk to just six firms: ICI, DCL, Shell, Laporte, Albright, and Wilson; and British Titan. Later, BP Chemicals, Esso Chemicals, Fisons, and Monsanto were added to the list.[43] The number of significant players in the industry was seen by government to be limited to ten firms.

CI REMAINS DOMINANT

ICI was still very much seen, however, as the leader of the industry whose cooperation had to be obtained for any government initiatives in

---

[39] Peter H. Spitz, *Petrochemicals: The Rise of an Industry* (New York: Wiley, 1988), p. 364.

[40] Ibid., p. 365.

[41] Reader, *Imperial Chemical Industries, Volume 2*, pp. 394-95.

[42] Wyn Grant, "Government-Industry Relationships in the British Chemical Industry," in Martin Chick, ed., *Governments, Industries, and Markets: Aspects of Government-Industry Relations in the UK, Japan, West Germany, and the USA since 1945* (Cheltenham: Edward Elgar, 1990), p. 144.

[43] PRO: BT 258/2498, "Future Investment Plans: Approaches to Individual Firms in the Chemical Industry, 1966-67."

advance of consultations with other firms. This is evident in a memorandum written by a civil servant when the Board of Trade was launching
its investigation into investment in the industry in 1966:

> If we are going to get on in this industry, it is really very important that we
> should get moving with ICI . . . [an informant] was quite sure that Allbright
> and Wilson – or for that matter anyone else in the chemical industry – would
> be likely to ask ICI whether they were cooperating with us or not. If ICI's
> reply was that they had heard nothing about us, I imagine the others would
> be rather flummoxed.[44]

   This, in part, reflected the fact that ICI could be seen as standing
"midway between a company in private business, properly so-called and
a public corporation."[45] In an earlier paper (1984), the author characterized ICI as a typical example of a "tripartite" as opposed to a "capitalist
aggressive" firm in a categorization of corporate political philosophies.
Tripartite firms were characterized by such features as a relatively bureaucratized relationship with government, being strong supporters of business associations, taking a moderate stance on employee relations issues,
and favoring moderate social policies generally.
   Not every ICI senior executive fell into this mold. Sir Paul Chambers,
ICI's chairman in the 1960s, was an advocate of neoliberal policies and an
outspoken critic of the Labor government. "However, even as independent a figure as he was constrained by corporate considerations and by
ICI collegial conventions not to break its consensus on industrial relations
policies."[46] A more typical figure was Michael Clapham, CBI president
from 1972 to 1974. The characterization of what they call a "revision-
ist" position by Boswell and Peters is somewhat similar to this author's
notion of a "tripartite" stance, under which they award Clapham the
accolade of becoming "intellectually the most comprehensive expositor
of advanced revisionism."[47] Finally, Harvey-Jones in the 1980s, combined
a strong attachment to competitive values and a commitment to restructuring ICI with the public development (most notably in a nationally
televised Dimbleby lecture) of an alternative economic and social vision
to that of Thatcherism.

---

[44] Ibid.
[45] Reader, *Imperial Chemical Industries, Volume 2*, p. 476.
[46] Jonathan Boswell and James Peters, *Capitalism in Contention: Business Leaders and
    Political Economy in Modern Britain* (Cambridge/New York: Cambridge University Press,
    1997), pp. 62–63.
[47] Ibid., p. 84.

ICI was not just the dominant company in the British chemical industry, it also became the exponent of a certain set of ideas about how the manufacturing industry should be run, and a defender of the central role of manufacturing in the economy.

WARNING SIGNS IN THE INDUSTRY

The combination of innovation and demand for its products meant that from the 1950s to the early 1970s the petrochemical industry was one of the fastest-growing sectors of the British economy. For example, in the period from 1963 to 1969, the ratio of the average annual growth rate of chemicals to manufacturing output was 1.7.[48] Expansion was facilitated by the maturation of chemical engineering, with the most striking achievement being the change from batch to continuous production in most chemical plants. Economies of scale were readily available. Nonetheless, most of the industry's major product innovations had been achieved by 1960.

However, even during this period of maximum growth, and in line with the general performance record of the British economy, the chemical industry in Britain was not doing as well as its major competitors:

There is evidence ... that during the 1950s and 1960s the UK chemical industry grew less substantially than its United States, German and French competitors, experienced a more pronounced decline of growth in the 1970s, and in the 1980s, for jointly economic, political and business reasons has been more active in reducing capacity and manpower than some of its West European counterparts.[49]

One of the underlying problems was that: "The UK chemical industry has had to try to keep up with the rapid advance of the world chemical industry without the assistance of a buoyant home economy."[50] It was also an economy susceptible to "stop-go" cycles in the 1950s and 1960s, as fiscal and monetary instruments were used to manage the economy in accordance with the precepts of neo-Keynesian demand management. These cycles were often shorter than the investment planning horizon of the chemical industry, but they were a complication nevertheless.

---

[48] Chemicals EDC, *Investment in the Chemical Industry* (London: National Economic Development Office, 1972), p. 13.

[49] Andrew M. Pettigrew, *The Awakening Giant: Continuity and Change in ICI* (Oxford: Basil Blackwell, 1985), p. 52.

[50] Chemicals EDC, 1972, p. 13.

The Board of Trade became increasingly concerned about the state of the industry and initiated a series of discussions with the leading firms about their future investment plans. In a summary memorandum on the subject, the Board of Trade noted, "The level of investment by the chemical industry is a matter of concern both because of its absolute size in relation to the whole of industrial investment, and because the industry is one of the major growth sectors of the economy."[51]

What the Board of Trade learned from its discussions with the companies gave grounds for concern: In sum, the general picture of the industry is one in which despite the likely fall in 1967, investment will continue for some time at a fairly high level, but in which profitability is too low and either cash resources are too strained to sustain previously planned investment programmes or, where cash is available, some projects of substantial size which might have gone ahead in 1967 and 1968 are now showing a rate of return which will make them less attractive.[52]

The matter became politicized when Dr. Jeremy Bray, a former ICI employee who had become a junior minister at Fuel and Power, wrote to the prime minister drawing attention to cutbacks in ICI's investment plans. The very fact that a junior minister was minuting the prime minister disturbed the Board of Trade. The prime minister took the view that these cutbacks in ICI's investment plans represented a variation from their usual policy of maintaining a steady level of investment and became concerned that the decision was politically motivated. The fact that ICI's investment decisions could be interpreted as having broader political consequences is of itself significant. The Board of Trade sent a robust reply to Downing Street on 15 November 1966, stating: "The President is satisfied that their decisions have been based entirely on their need to match their expenditure to their resources and that there has been no political motivation."[53] As the Board of Trade's summary memorandum made clear, "I.C.I., which is by far the biggest single investor in chemicals had undoubtedly overstretched itself in the earlier investment boom and is probably the hardest hit of all by sheer cash shortage."[54] It was evident that ICI had more fundamental problems than a cash shortage and some poor investment decisions. One of the civil servants involved in the

[51] PRO: BT 258/2498.
[52] Ibid.
[53] Ibid.
[54] Ibid.

investment review was asked by ICI to give his views on the company as it appeared to an outsider. This revealed some worrying concerns:

[ICI] was regarded as production rather than market orientated. . . . Secondly, it appeared to others that investment planning in ICI had not been as rigorous as it might have been particularly in 1964/65. Thirdly, as a production orientated company, it had appeared to take too optimistic a view of the level of price at which the market would buy.[55]

THE INDUSTRY IN DIFFICULTY

The concerns expressed by government in the 1960s turned out to be well founded. The continuous growth recorded in the 1950s and 1960s came to an end. The Chemicals EDC[56] found that "from 1969 onwards there appears to have been a significant break in a number of trends. Since about mid-1970, the rate of growth of output of the chemical industry has declined, growing less than [0.5] percent during the year ended September 1971." The Chemicals EDC also noted that "Investment is now being severely cut back. . . . We believe . . . that substantial and widespread overcapacity is the primary reason for the cut back."[57]

Decision makers in the industry were, however, accustomed to continuous steady growth and found it difficult to adjust to changed circumstances. In ICI's Plastics Division, there was a belief "in the trading cycle with its good and bad years. This belief tended to obscure the basic irreversible structural change in the industry and market underlying the movements of the trading cycle."[58] An executive in another company interviewed by the author commented:

I don't think anybody in 1970 expected that there wouldn't be continued growth. We weren't attuned to thinking that. The first oil shock was absorbed, petrodollars recycled, it was not until 1980 that people woke up to the fact that they were not in the same world as the 1960s.[59]

The second oil shock of 1979 hit the chemical industry hard and brought to the surface a structural overcapacity crisis that had been looming for some time. Petrochemicals was now a mature industry and the continuous increases in demand experienced in the 1950s and 1960s

[55] PRO: BT 258/2498.
[56] Chemicals EDC, 1972, p. 5.
[57] Chemicals EDC, 1972, p. 7.
[58] Pettigrew, *The Awakening Giant*, p. 269.
[59] Quoted in Grant et al., *Government and the Chemical Industry*, p. 208.

were now a thing of the past. The oil crisis hit at a time when a number of new plants in Europe were coming on stream, and countries such as Saudi Arabia were becoming involved in production.

The British industry reacted by cutting capacity quickly and sharply. It was, nevertheless, operating in an unfavorable political environment. The relationship between ICI and the government had been damaged by the election of Mrs. Thatcher. There was a feeling in government circles that ICI was "just as bureaucratic as the Civil Service."[60] The then head of ICI, the somewhat unorthodox Sir John Harvey-Jones, who had been chosen to shake up the company, was a Social Democrat in his personal life. When asked why the recession had been so much harder in the United Kingdom than elsewhere, he replied: "We've got Thatcher."[61] ICI was hit by cutbacks and changes in regional policy that it had used to fund a substantial portion of its investment program.

The industry as a whole was unhappy about a number of aspects of government policy. The Chemical Industries Association informed the House of Lords Committee on Overseas Trade: "The government's policies aim to create the conditions for profitable and noninflationary growth and for the encouragement of enterprise. From the chemical industry's point of view, this is epitomized by very high energy prices, the removal of regional grants, and the changes in the regulations on the collection of VAT on imports, and so on."[62]

Nevertheless, in its 1987 review of the industry, the Chemicals EDC painted an upbeat picture. They argued that "circumstances are favourable to the UK chemical industry. We have seen successful restructuring by companies, great improvements in the efficiency of plant operations and plant construction, and movements in exchange rates which have left its products more competitive in many markets."[63] This last advantage was lost once Britain started to pursue a policy of shadowing the D-mark and then joined the Exchange Rate Mechanism at a relatively high rate. It is also worth noting that the committee's own figures showed that Britain ranked third in West European chemicals production, just ahead of Italy, and with a share (13 percent) that was only just over half that of West Germany (25 percent). The committee's definition of the

---

[60] William Keegan, *Mrs. Thatcher's Economic Experiment* (Harmondsworth: Penguin, 1984), p. 148.

[61] *Financial Times*, 1 April 1982.

[62] Chemical Industries Association, "Evidence to the House of Lords Select Committee on Overseas Trade" (1985), p. 329.

[63] Chemicals EDC, 1987, p. v.

industry included pharmaceuticals that accounted for nearly a quarter of output, compared with 11 percent for petrochemicals and plastics.

However, since the trough of 1980, the industry as a whole had resumed a growth rate of 3.25 percent per annum compared with 0.75 percent for manufacturing industry as a whole. Nevertheless, "Import penetration in plastics has increased from 34 per cent to 46 percent in this period. In some products, particularly polyethylene, UK capacity is now substantially lower than domestic demand."[64] In some respects, the performance of the chemical industry in the United Kingdom looked good because so much of the rest of the manufacturing economy was in dire trouble.

CONCLUSIONS: THE INDUSTRY AT THE END OF THE TWENTIETH CENTURY

By the end of the twentieth century, the chemical industry was experiencing considerable restructuring. A number of major mergers were taking place and it seemed likely that the global production of bulk chemicals would be dominated by a relatively small number of companies. Within Britain, the competitiveness of bulk chemicals production was offset by transport costs to continental markets and, at the turn of the century, by the high value of the pound.

Much of this chapter has been concerned with ICI that has been so dominant economically and politically in the industry. Throughout the 1990s, however, ICI had gone through a process of much needed change. By listing itself on the New York and Tokyo stock exchanges, it signaled that it wished to be seen as a global company rather than as a British based multinational. It sold off its pharmaceuticals business (initially as Zeneca, later to merge with other businesses). It also tried to get out of commodity chemicals production and switch toward speciality chemicals and paints. In January 2001, it sold its remaining industrial chemicals businesses, fundamentally severing its historic link with bulk chemicals production. Its headquarters operation was downsized. There was even speculation that it might be dropped from the FT-100 list of leading companies. It was no longer the flagship of British manufacturing industry enjoying a special relationship with government. Indeed, in an increasingly service oriented economy industries such as chemicals were perceived as being part of the "old" economy.

[64] Chemicals EDC, 1987, p. 13.

Nevertheless, the overall story of the British chemical industry in the last three-quarters of the twentieth century is one of relative success. It made a remarkable recovery from a bad start and this was in large part due to ICI. ICI in turn benefited from a special relationship with government that characterized a type of industrial partnership rarely found in Britain. However, the chemical industry could not be entirely insulated from the performance problems of the British economy as a whole. ICI also became perhaps too self-satisfied, too proud of its past achievements, and insufficiently responsive to changing circumstances. Success in an innovative, expanding industry came easier than in a mature, contracting one.

The industry had always been highly internationalized, but necessarily became more so as trade barriers diminished and the international economy became more integrated. The beginning of the twenty-first century is perhaps the last point at which it will be possible to write a credible account of a distinctively British chemical industry.

REFERENCES

Barnett, C., *The Audit of War* (London: Macmillan, 1986).
Boswell, J., and Peters, J., *Capitalism in Contention* (Cambridge: Cambridge University Press, 1997).
British Productivity Council, *Heavy Chemicals* (London: British Productivity Council, 1953).
Chemical Industries Association, Evidence to the House of Lords Select Committee on Overseas Trade, 1985.
Chemicals EDC, *Investment in the Chemical Industry* (London: National Economic Development Office, 1972).
Davenport-Hines, R. P. T., *Dudley Docker: The Life and Times of a Trade Warrior* (Cambridge: Cambridge University Press, 1984).
Dell, E., *A Strange Eventful History: Democratic Socialism in Britain* (London: HarperCollins, 2000).
Grant, W., "Large Firms and Public Policy in Britain," *Journal of Public Policy*, 4: 1–17.
Grant, W., "Government-Industry Relationships in the British Chemical Industry," in M. Chick (ed.), *Governments, Industries and Markets* (Cheltenham: Edward Elgar, 1990), 142–56.
Grant, W., Paterson,W., and Whitston, C., *Government and the Chemical Industry* (Oxford: Clarendon Press, 1988).
Haber, L. F., *The Chemical Industry, 1900–1930: International Growth and Technological Change* (Oxford: Clarendon Press, 1971).
Keegan, W., *Mrs Thatcher's Economic Experiment* (Harmondsworth: Penguin, 1984).

Kennedy, C., *ICI: The Company That Changed Our Lives* (London: Hutchinson, 1986).

Pettigrew, A., *The Awakening Giant: Continuity and Change in ICI* (Oxford: Basil Blackwell, 1985).

PRO: BT 258/2498, "Future Investment Plans: Approaches to Individual Firms in the Chemical Industry, 1966-67."

Reader, W. J., *Imperial Chemical Industries, a History. Volume 1: The Forerunners, 1970-1926* (London: Oxford University Press, 1970).

Reader, W. J., *Imperial Chemical Industries, a History. Volume 2: The First Quarter Century, 1926-1952* (London: Oxford University Press, 1975).

Reader, W. J., "Imperial Chemical Industries and the State, 1926-1945," in B. J. Supple (ed.), *Essays in British Business History* (Oxford: Clarendon Press, 1987), 227-43.

Spitz, P. H., *Petrochemicals: The Rise of an Industry* (New York: John Wiley, 1988).

Turner, J., "The Politics of Business," in J. Turner (ed.), *Businessmen and Politics* (London: Heinemann, 1984), 1-9.

Williams, E. E., *Made in Germany* (London: William Heinemann, 1896).

Williams, P., *Hugh Gaitskell* (Oxford: Oxford University Press, 1982).

11

The Development and Struggle of Japanese Chemical Enterprises since the Petrochemical Revolution

TAKASHI HIKINO

Japan's chemical industry since the Petrochemical Revolution in the 1950s and 1960s represents a notable case in which a new industry in the high-growth economy rapidly caught up with the established front-runners in North America and Western Europe. Aided by MITI's industry targeting and the group structure of large enterprises (*kigyo shudan*), the Japanese chemical industry actually became one of the major players in the world chemical market. As seen in Table 11.1, representative large enterprises that exploited the opportunities created by the policies and the group organization developed similar technological capabilities that were suitable during the catch-up phase of the Japanese industry. The core of the capabilities lay in the capacity of those enterprises to import the latest technological developments in advanced chemical economies and then to make incremental process improvements on them. The resulting cost savings in manufacturing were the major source of the international competitiveness of Japan's industry. The nurtured capabilities, however, would not transform the firms into real innovators on the world technological frontier. Consequently, when the industry started to suffer from structural excess capacity after 1971, the continued reliance on conventional capabilities could not create new sources of growth for the enterprises. The Japanese chemical industry has thus struggled for more than a quarter of the century until the present moment. Compared

Table 11.1. The Largest Industries in the Japanese Manufacturing Economy: 1995

Industry	Value Added (Billion Yen)	Persons Engaged (1,000)	Value Added per Person (Million Yen)
Manufacturing total	117,204	10,321	11.4
Electrical machinery	19,643	1,750	11.2
Chemicals	16,194	841	19.3
Transportation equipment	12,494	914	13.7
Foods and beverages	12,373	1,259	9.8
Nonelectrical machinery	12,131	1,087	11.2
Fabricated metals	7,970	817	9.7
Primary metals	6,936	458	15.1

Source: Compiled and calculated from unpublished data in kogyo Tokei Chosa, 1995, courtesy of Japan Ministry of International Trade and Industry.

with other major modern industries such as automobiles, communication equipment, and consumer electronics, therefore, Japan's chemical industry remains a marginal and invisible player in the global industry.

BASIC CHARACTERISTICS OF THE JAPANESE CHEMICAL INDUSTRY

For the illustrative purpose of international comparison, the basic traits of Japan's chemical industry can be summarized in seven points. The first is the capability issue. The second and third are concerned with strategic choices of individual enterprises. The fourth relates to organizational and structural aspects, whereas the remaining three are performance attributes.

First, Japanese chemical companies in general have not yet exhibited their technological competencies in radical product or process innovation. Their strength lies in learning speed and incremental process innovation capabilities. In other words, firms possess high commercialization and customization competencies. Until recently, most companies concentrated their efforts in development at the cost of research.

Second, even the largest enterprises are narrowly diversified with limited product portfolios. Except for a few notable cases such as Showa Denko and Ube Industries, firms rarely operate in two industry categories. The limited extent of vertical specialization into upstream and downstream areas is also notable.

Table 11.2. Ten Largest Chemical-Trading Nations in the World: 1995

Country	Exports ($ Million)	Imports ($ Million)	Balance ($ Million)
Germany	70,477	43,263	27,214
United States	61,701	40,378	21,323
The Netherlands	32,327	20,988	11,339
Belgium-Luxembourg	35,889	25,339	10,550
Switzerland	20,332	10,978	9,354
France	40,821	32,673	8,148
Great Britain	33,585	27,343	6,242
Japan	30,077	24,548	5,529
Canada	8,882	12,405	−3,523
Italy	19,575	27,553	−7,978
World Total	436,810	436,810	—

Source: Compiled and calculated from "Facts and Figures for the Chemical Industry," *Chemical and Engineering News,* June 24, 1996, p. 69.

Third, internationalization takes different forms of product penetration depending on the developmental nature of specific foreign markets. To penetrate advanced economies Japanese companies usually rely on exporting. Manufacturing investment in such nations still remains limited. For emerging markets, companies often engage in aggressive direct investments by establishing subsidiaries or joint ventures.

Fourth, most major enterprises are members of diversified group businesses called *kigyo shudan*. All such groups have at least a few prominent chemical enterprises as their members. This is true not only for the three biggest groups, Mitsubishi, Mitsui, and Sumitomo, but also for other groups such as Fuyo, DKB, and Sanwa, which are organized around large commercial banks. The membership distribution of the entire Mitsubishi group and the position of chemical firms fall within this structure. (In the chemical industry, however, intergroup strategic alliances are more frequent compared to other industries.)

Fifth, the chemical industry as a whole has enjoyed relatively high productivity in the narrow technical and economic sense. This is evident in total factor calculations by David Dollar and Edward Wolff, the OECD, and Dale Jorgenson.

Sixth, prominent Japanese chemical companies are located on the downstream side, whereas upstream commodity, bulk-chemical producers perform badly. Although this characteristic can be found in most industrial economies, Japan's case is an extreme one in that all upstream

Table 11.3. Largest Chemical Companies Listed in Fortune Global 500: 1995

Rank	Company	Country	Sales ($ Million)	Profit ($ Million)	Profit Margin	Founding Year
58	Du Pont	United States	37,607	3,293	9%	1802
63	Hoechst	Germany	36,409	1,193	3%	1863
71	Procter & Gamble	United States	33,434	2,645	8%	1837
78	BASF	Germany	32,258	1,724	5%	1861
87	Bayer	Germany	31,108	1,671	5%	1863
156	Dow Chemical	United States	20,957	2,078	10%	1897
198	Ciba-Geigy	Switzerland	17,509	1,824	10%	1884
206	Mitsubishi Chemical	Japan	17,074	233	1%	1934
207	Rhφne-Poulenc	France	16,996	665	4%	1928
227	ICI	Great Britain	16,206	844	5%	1926
247	Kodak	United States	15,269	1,252	9%	1884
286	Akzo Nobel	Netherlands	13,383	818	6%	1899
327	Norsk Hydro	Norway	12,578	1,125	9%	1905
328	Asahi Chemical Industries	Japan	12,538	96	1%	1931
388	Fuji Photo & Film	Japan	11,241	755	7%	1934
413	L'Oreal	France	10,698	631	6%	1909
445	Henkel	Germany	9,907	302	3%	1876
448	Sumitomo Chemical	Japan	9,862	192	2%	1925
453	Toray Industries	Japan	9,753	189	2%	1926
475	Solvay	Belgium	9,268	417	4%	1863
489	Dainippon Ink & Chemical	Japan	8,996	72	1%	1908
492	Monsanto	United States	8,962	739	8%	1901

Source: Compiled and reorganized form "The Fortune Global 500," *Fortune*, August 5, 1996. Date on founding years comes from various company publications.

producers have been suffering from poor performance for a quarter of a century.

Seventh, as noted earlier, by world standards many Japanese chemical companies failed to become internationally known and respected. This is in spite of the fact that many play a large role in global operations. In terms of pure numbers, the Japanese big businesses in chemicals historically have had a prominent presence in the world industry. Even compared to other major industries, the relative position of Japan's chemical players is not inferior (Tables 11.2, 11.3). After all, then, invisibility seems to result

from the absence of breakthroughs that advance Japanese companies on
the world technological as well as commercial frontier.

GOVERNMENT POLICY

In part because of Japan's lateness in developing a chemical industry,
the Japanese government played a significant role, negative as well as
positive, in determining the speed and direction of the industry evolution. The intervention of the government has not uniformly affected
all branches of the chemical industry. Before World War II, the role of
the government was actually marginal and indirect. There was no systematic industrial policy so to speak. The most important contribution
of the government came when MITI guided the growth of petrochemicals beginning in the mid-1950s. It should be noted that MITI worked
closely with the Ministry of Finance in developing industrial policy instruments mostly because of their mutual concern about the balance of payments and foreign exchange. This concern came not only from the macroeconomic considerations for the overall economy. More immediately for
the chemical industry, as Japan started to reindustrialize after the destruction from World War II, chemical imports, particularly those of feedstock,
soared substantially, resulting in the acute necessity for import substitution. When these macroeconomic issues became less pressing around
1968, policies advocated by MITI and the MOF started to show some
conflict.

It is also noteworthy that, with the exception of petrochemicals, MITI's
influence was weak in forming the basic structure of industries. Synthetic
fibers, for instance, administered by the MITI textile division, had little in
common in its development with petrochemicals. On the other hand, the
Ministry of Welfare tightly regulated the pharmaceutical industry, whose
systems of price and safety regulations are much different from MITI's.
When many chemical companies tried to enter pharmaceuticals, therefore, they found out that working with the Welfare Ministry in addition
to MITI was very difficult.

The screening rule, which MITI introduced in the First Plan in 1955, is
significant in that it has performance criteria by which individual applications for large-scale petrochemical plant construction are evaluated.
When a bureaucratic organization tries to monitor the performance of
private businesses they often encounter the information barrier, because
a political entity does not typically possess technical capabilities equal
to those of private business. This asymmetry makes the application of

industrial policy difficult. MITI came up with a clear target of international prices that was a neutral measure for efficiency. If MITI contributed to the chemical industry's productivity improvement as well as the industry's overall growth, the imposition of performance standards was significant.

NATURE OF TECHNOLOGY ACQUISITION

A basic trait of the Japanese national system of innovation throughout the twentieth century has been the importation and incremental improvement of the latest foreign technology. This common pattern across modern industries has certainly been the case with the chemical industry. In this regard, Japan's history makes a sharp contrast to the United States, Germany, and Great Britain. Those three nations have acted as technology generators and innovators, whereas Japan (and many other late-industrializing nations) developed as a technology learner and a commercializing specialist. This difference in the nature of technology acquisition critically differentiated the growth strategies of chemical enterprises.

Japan's international competitiveness has thus been critically conditioned by the development of the world's technological frontier. The continuous success of the U.S. and West European chemical industries has made it difficult for a latecomer such as Japan to catch up. As Chandler emphasized, in many other industries in which Japan caught up with the world leadership, such as automobiles and consumer electronics, we find a failure to innovate and commercialize new processes on the part of established oligopolistic firms. By contrast, large chemical enterprises from the United States, Germany, and Great Britain continued to excel in radical and incremental product and process innovations, which made Japan's catch up difficult.

The general Japanese level of technological competence at the time of the development of the chemical industry resulted in the dual character of Japan's technology and enterprise evolution. For one thing, the level was high enough to learn and absorb advanced technical achievements of the United States and Western Europe. At the same time, it was not mature enough to compete directly with the leading large enterprises that made up the world oligopoly.

Because of Japan's immature technological competence, it was probably efficient and economical for each firm to concentrate in relatively narrow product areas, and to commit to incremental process innovations to lower cost and improve product and service quality. Given a sufficient level of technical accumulation, the enterprises have been successful in achieving these goals, whereas the technology learning process became localized, and the technical competencies for the entire industry stayed uneven. The outcome of this long-term strategy is a fragmented industry crowded with many similar firms possessing similar competitiveness with a similar product portfolio.

Thanks to the continuous developments of chemical technology and the enterprises embodying them, however, Japanese chemical firms have not achieved the prominence in international markets. At the same time, developing nations such as South Korea have been quickly catching up in basic, commodity chemicals. The resulting organization of the Japanese chemical industry now is many narrowly diversified and integrated firms struggling to reorganize themselves by developing new technologies in such similar areas as fine and specialty chemicals, pharmaceuticals, and new materials.

SIGNIFICANCE OF GROUP STRUCTURE AND MANAGERIAL CONTROL

The group enterprise organization developed after World War II affected the competitive behavior of large enterprises in two basic directions. Group membership, to a certain degree, guarantees the stable growth of individual constituent companies. Intragroup product sales and the mutual shareholdings among member firms created a comfortable setting in which managers could formulate long-term growth strategies. Information networks within a group lowered the transaction costs.

In order to enjoy these and other possible benefits, however, the firms have to pay opportunity costs. Given the comprehensive coverage of the major strategic industries by all the groups, the interindustry mobility barrier for individual companies became high. No matter which product areas they attempted to enter, another enterprise of the same group was already operating. This restrictive situation certainly conditioned the corporate growth by diversification.

The group structure often has acted as an obstacle for the external corporate growth by mergers and acquisitions. This is mostly because the intergroup rivalry made the mergers of companies across the groups almost impossible. Except for a severe case of depressed industries such as ocean shipping, the intergroup mergers and the resulting restructuring of the entire industry cannot be accomplished. Often, even within a group, or sometimes exactly because of the group membership, the merger of constituent enterprises is difficult for personal and company rivalry and other "path-dependent" reasons. The cases of Mitsubishi Kasei and Mitsubishi Petrochemical and of Mitsui Toatsu and Mitsui Petrochemical are two good examples.

The resistance of group structure and rivalry for restructuring is further compounded by the management control of all the constituent enterprises. As Robin Marris, John Kenneth Galbraith, William Baumol, and Oliver Williamson theorized, management-controlled firms exhibit a strong drive to grow regardless of the uncertainty and profitability of new investment outlets. Given little equity stake of the company's senior management, it can take even a high risk in investment, as Joseph Schumpeter suggested. By contrast, management tends to ignore and resist downsizing requirements and continues to invest the firm's resources in projects whose expected return is substantially low. This is because managers try to create more opportunities in terms of their own promotion and prestige, while profitability remains a secondary concern for them. The orientation toward sales growth and market share may work in the positive way as long as growth opportunities exist within the boundaries of a firm's capabilities. When core capabilities do not constitute the source of further growth, the momentum for expansion does not work in a positive manner for the firm.

Although this growth-oriented investment behavior of managerial enterprises is not unique to the Japanese firms, their management certainly showed this inclination for expansion. As long as opportunities for investments existed in petrochemicals as was the case of the 1950s and 1960s, this managerial drive worked in a positive manner for their companies and the chemical industry as a whole. Beginning in the 1970s, however, when restructuring and exit became necessary, growth orientation became dysfunctional and harmful. Compounded by the absence of active shareholders and other disciplinary mechanisms of capital markets, the management control of large chemical producers did not make the industry reorganization easy.

COMPETENCIES AND STRATEGIES

The Japanese chemical companies shared some of their functional capabilities with other stakeholders and participants. The chemical companies divided some capabilities with other parties or basically did not have them. Group banks and, to a lesser extent, government, took the burden of financing huge projects for chemical companies. Foreign firms, either

Table 11.4. Profitability of the Largest Chemical Companies of Japan and the United States: 1995

	Japan	Return on Equity (%)	Profit Margin (%)	United States	Return on Equity (%)	Profit Margin (%)
Diversified Firms	Mitsubishi Chemical	1.9	0.8	Du Pont	29.3	9.1
	Sumitomo Chemical	5.9	1.5	Dow Chemical	23.0	8.9
	Mitsui Toatsu	2.9	1.0	Monsanto	22.0	8.0
	Showa Denko	0.1	0.0	Union Carbide	59.3	15.4
	Ube Industries	5.3	1.4	FMC	49.4	4.9
Specialized Firms	Asahi Chemical Industries	3.6	1.4	Eastman Chemical	41.9	10.9
	Sekisui Chemical	4.8	2.2	W.R.Grace	16.7	6.0
	Kao	7.6	4.0	Air Products & Chemical	16.9	9.5
	Toray Industries	4.1	3.2	Morton International	19.3	8.8
	Dainippon Ink & Chemical	2.7	1.3	Sherwin-Williams	18.6	6.1

Note: The five largest companies, in terms of sales, are selected for both diversified and specialized firms of Japan and the United States. Return on equity is defined as net income divided by equity. Profit margin is defined as net income divided by sales.

Source: For Japanese companies, compiled and calculated from *Kaisha Shikiho*, 1997 Shunki; for U.S. Companies, compiled from "Annual Survey of American Industries," *Forbes*, January 1, 1996, pp. 94–95.

chemical firms, petroleum enterprises, or specialized engineering companies, generated usable technology and sold it to Japanese companies. Trading companies specialized in marketing and exporting the products of chemical companies, whereas other enterprises within a group bought their raw materials from group chemical companies.

Japanese chemical companies concentrated on two basic functions: project execution capabilities and operational capabilities. They excelled in finding an appropriate source of technical know-how and making the knowledge workable for production facilities. And the company particularly invested in improving various operational aspects for incremental productivity, performance, and costs. Given the general level of their accumulated technical knowledge, the specialization and concentration on a narrow function made perfect sense.

The weakness of technological capabilities, however, became significant when, after the early 1970s, accumulated knowledge related to project execution and operational capabilities was not what was needed. Japanese chemical companies continued to employ a part of their project execution capabilities for allocating foreign or domestic enterprises with appropriate technical excellence. Establishing a joint venture with those companies had been a significant way to diversify product portfolios of Japanese chemical companies. Or they could export plant engineering techniques to many extensive projects of emerging markets. However, they did not seem so far to have the genuine technological capabilities that would have led those companies out of long recession.

THREE EVOLUTIONARY PHASES OF THE JAPANESE CHEMICAL INDUSTRY

The First Phase

The first phase, up to the mid-1950s, was the period of coal chemicals and electrochemicals. New entrepreneurial groups as well as old *zaibatsu* groups played a significant role. Customer bases were the textile industry and agriculture.

In the 1920s and 1930s, the old *zaibatsu* groups, such as Mitsui, Mitsubishi, and Sumitomo, were generally reluctant to commit themselves to such emerging new fields as chemicals and heavy industries. This was mainly because of the entrepreneurial conservatism shared by the owning family members and senior managers. It was a small group of aggressive entrepreneurs who took this opportunity to advance their interest in chemicals. Industrial groups such as Nissan, Nichitsu, Nisso, and Mori became actually organized around large electrochemical firms. By successively importing the latest technology available, those enterprises became leaders of Japan's chemical industry. Because the war eventually stopped the inflow of new technical information, old *zaibatsu* chemical firms could catch up with those leaders by nurturing their coal-based chemical technology.

World War II, by contrast, had a destructive impact on the newly emerged groups in two ways. First, since the late 1920s, many of the groups, seeking a cheaper source of energy, aggressively transferred their operation to Korea and China. As the war ended, the groups not only lost a substantial part of their production facilities but also became a political target because of their colonial aggression. Second, because the organizational design created by those new groups was not coherent, the groups could not weather the turmoil that resulted from the war. As groups, they therefore, simply collapsed, although many constituent companies such as Asahi Chemical, Hitachi, and Nissan Automobile survived because of their technical and organizational competencies.

The Second Phase

Opportunities for petrochemical development in Japan were ripe in the early 1950s. The major factors working for petrochemicals are illustrated in Figure 11.1. The most important of all was that, although Japan was basically isolated from various developments in the world chemical industry, petrochemical technology was developed, tested, and accumulated in the United States and Western Europe. Another factor on the supply side was the reconstruction of petroleum-refining facilities by the end of the 1940s. On the demand side, in the meanwhile, many chemical user industries were growing rapidly in the early 1950s. Representative cases were plastics and synthetic fiber industries, with the automobile manufacturing and electronics industries coming slightly later.

This was a typical case of the possibility of a new Schumpeterian combination. Because the coal chemical industry could not meet the demand for price and other reasons, there was a vacuum between supply and demand. Many young Japanese managers were eager to take advantage of this disequilibrium by investing in the development of petrochemicals.

The rapid replacement of coal by petroleum as the key raw material thus characterized the development of the Japanese chemical industry since the 1950s. The timing of the reorganization of the petroleum refining industry after World War II was critical in its relationships to the petrochemical industry. By the time that the petrochemical industry in postwar Japan started around 1955, the petroleum-refining industry had been reorganized by 1951 under the overwhelming influence of occupation policies.

Although the coal mining industry became a significant target of the Japanese government's industrial policy, the petroleum-refining industry reconstructed itself along the structural plan that was originally made by the corporate members on the advisory panel for the General Headquarters. Five major international oil companies were represented on the panel: Standard-Vacuum, Shell, Caltex, Tidewater, and Union. Fearing the total loss of the vast markets in Mainland China because of political and economic uncertainties, the company representatives seriously sought to expand their own interest in the Japanese market. In place of tariff protection and industrial targeting, which were usual for other strategic industries including petrochemicals, therefore, the petroleum-refining industry relied on free trade and foreign capital for its postwar development.

Petrochemical technology was completely novel in Japan in the late 1940s, but some enterprises were eager to initiate petrochemical production in Japan. As the plastics and synthetic fiber industries grew starting around 1948, pressure for the domestic supply became acute in the early 1950s. In spite of the eagerness on the part of some industrial enterprises, the initial process through which both industrial enterprises and MITI and the Ministry of Finance tried to establish the petrochemical industry took several years. MITI established a consulting committee whose recommendations, formalized in the Petroleum Industry Development Plan (The First Plan) in July 1950, worked to evaluate and coordinate various schemes and possibilities.

The First Plan clarified the overall purpose of petrochemicals, according to which the new industry was supposed to supply basic petrochemical products to domestic users at internationally competitive prices. This criterion was significant for two reasons. First, the basic target was the domestic market, which resulted in import substitution, not export drive, at least at this stage. Second, the government schemes contained a key element of efficiency and price competitiveness, which is intriguing given the subsequent history of MITI's policies and the industry's troubles. The plan came up with three general directions and purposes of the foundation of the petrochemical industry in Japan. The MITI plan specified three directions for the introduction of petrochemical production. First, the immediate needs lay in securing basic domestic supplies

of benzol, organic acids, and acetone for the plastic and synthetic fiber
industries. Second, it aimed at introducing the domestic manufacturing
of ethylene and derivatives. Third, it aimed at achieving low-cost production of basic petrochemicals in order to raise the international competitiveness of chemical and related industries.

Although the First Plan contained substantial financial and other incentives for investment, MITI adopted strong administrative guidance in following the guidelines of the First Plan to screen, alter, and choose specific
proposals submitted by various enterprises. MITI's principle for implementing the First Plan and subsequent programs was instrumental, not
political, and the Ministry accepted all the schemes that met the plan's
objective criteria. This rule, sometimes called "Equal Opportunities for
Worthy Competitors," encouraged oligopolistic competition for facility
expansion among large firms, and, in spite of all the instruments of administrative guidance, MITI would later find that it did not have any means
to regulate the excess competition and capacity expansion.

With MITI's approval the first petrochemical complex was thus
launched in Kawasaki near Yokohama in 1957 when Nippon Petrochemicals, a subsidiary of Nippon Petroleum, constructed an ethylene
producing facility of twenty-five thousand tons a year. The complex
had other specialized downstream chemical producers such as Showa
Petrochemical, Furukawa Petrochemical, Asahi Electrochemical, Nippon
Soda, Asahi-Dow, and Nippon Zeon. The intergroup combination of constituent enterprises became a prototype of subsequent petrochemical
complexes.

MITI approved four more ethylene centers: Mitsui Petrochemical
in Iwakuni, which started operation in 1958; Sumitomo Chemical in
Niihama in the same year; and Mitsubishi Petrochemical in Yokkaichi in
1959. By this time, the three old *zaibatsu* groups, Mitsui, Mitsubishi, and
Sumitomo, had clearly caught up technically with new entrepreneurial
zaibatsu companies. They also financially recovered from postwar reorganization, although Mitsui and Mitsubishi had to establish those two
new companies as a joint venture within each group. These joint ventures would later become problematic when the restructuring of petrochemicals became necessary in the 1970s. Sumitomo Chemical avoided
this trouble by making its own investment, although the company had
to narrow down its product portfolio to polyethylene and ammonias.

Actually, polyethylene was the most profitable and dominant product
in the First Plan. Manufacturing technology of polyethylene was imported
from various sources and even within one complex two different

methods competed against each other. Polyethylene and other polymer materials became new industrial materials that found wide markets in housing construction, machinery, transportation equipment, and other miscellaneous uses.

By the end of 1960 when all facilities approved by the First Plan started operation, the petrochemical industry became a noticeable field where many companies aimed to enter. They included not only chemical companies but also petroleum, synthetic fiber, and even steel manufacturers. MITI thus adapted the Second Plan to meet the demand for investment in petrochemicals. The first phase of the Second Plan consisted of the expansion of four ethylene centers and complexes. Furthermore, MITI also approved five new ethylene centers, four of which were organized by petroleum companies. These were Tonen Petrochemical in Kawasaki, Daikyowa Petrochemical in Yokaichi, Maruzen Petrochemical in Chiba, and Idemitsu Petrochemical in Idemitsu.

The only nonpetroleum company whose ethylene production was approved was Mitsubishi Kasei. Because Mitsubishi Petrochemical, a joint venture of several Mitsubishi enterprises, including Mitsubishi Kasei itself, had already entered ethylene production in the First Plan, those two companies within the same group became a sort of competitors.

Many of the companies participating in these complexes were organic chemical enterprises who still possessed manufacturing technology, such as fermentation and carbide manufacturing, which became obsolete thanks to the emergence of petrochemicals. Some of these companies smoothly moved to new technological bases, whereas some, for example, Chisso, experienced difficulties. In general, compared with the almost uniform success of the enterprises of the First Plan, the outcome of the Second Plan was mixed. This was mostly because the composition of the participants of the Second Plan was much more diverse, smaller, and independent and therefore the coordination of their businesses was more difficult.

In addition to the mixed results of the Second Plan, MITI was seriously concerned with the international competitiveness of Japan's petrochemical industry, particularly because the liberalization of capital markets in 1967 potentially opened the Japanese market to foreign producers. The response of MITI was to raise the level of minimum production capacity of ethylene producing facilities to three hundred thousand tons a year. By setting the entry barrier relatively high, the Ministry aimed to discourage new proposals for investment and facilitate the concentration of production among a few efficient producers.

Given the oligopolistic rivalry and managerial inclination, MITI's new guideline ironically encouraged new investment. For existing producers, the participation in the projects at the three-hundred-thousand-ton level became a matter of survival, because the economies of scale of new facilities would wipe out conventional producers with high production costs. Existing producers were thus forced to add ethylene-producing capacities, which they did mainly through joint ventures among themselves regardless of group affiliations. For instance, Mizushima Ethylene was established as a joint venture of Mitsubishi Kasei, Asahi Chemical, a member of the DKB group, and Nippon Mining. In all, nine ethylene centers actually began their operations by 1972, whereas MITI originally expected that at most four such investments were possible.

Although the scale of three hundred thousand tons certainly lowered the production costs of ethylene as MITI expected, the sudden drastic rise of supply created an excessive capacity as early as 1971. Compared to the total domestic ethylene facility of around 2.3 million tons in 1968, the level was expected to rise up to more than 5.0 million tons in 1972. This imbalance of supply and demand was partially caused by the slower growth of demand from such industries as consumer electronics, automobiles, and synthetic fibers, whose diffusion reached a certain peak in the early 1970s. Another factor of the slower growth of ethylene demand was that petrochemicals had by then replaced conventional organic chemicals as a raw material so that replacement demand gradually declined.

Regardless of their different origins and entry strategies, however, profitability of the "all-around" petrochemical companies stayed high until the late 1960s. Although subsequent depressed periods overshadowed the pre-1971 period, this phase is crucial in understanding the development of the Japanese chemical industry. The key reason for high performance was the overall underestimation of basic petrochemical demand in the Japanese market and of technical competencies of Japan's chemical and engineering firms. As downstream chemical users such as synthetic fibers, plastics, and synthetic rubber developed rapidly in the domestic market, demand for upstream basic chemicals remained high. As supply was relatively tight thanks to the government's entry control, companies operating in upstream parts could enjoy high profitability.

Phase Three

The third phase, up to the present time, turned out to be the difficult period of financial and strategic struggles and industry reorganization.

Individual companies continued to search for new growth and technology sources, whereas MITI faced difficulties in guiding the industry.

The Oil Shock of 1973 drastically worsened the excess capacity of petrochemicals. Through the Second Oil Shock of 1979, the Japanese petrochemical industry, particularly its upstream part, became structurally depressed, and, in a sense, never recovered to be reasonably profitable. All the producers rushed to rationalize the operation of basic petrochemicals, whereas strategic focus shifted toward fine and specialty chemicals.

In addition to huge excess capacity, there are several economic reasons for this long recession in the industry, which is unusual when compared to other industries. First is the so-called naphtha problem, which even became a political issue. As the primary raw material, Japanese petrochemicals became dependent on naphtha whose prices rose drastically because of the oil shocks. Relative to American producers who could utilize natural gas as a raw material, the competitive position of the Japanese industry deteriorated. Furthermore, domestic naphtha prices were set higher than international prices, and Japanese chemical producers could not import naphtha freely. On the demand side, there were two issues. First, as the Japanese economy started experiencing business slumps in the 1970s and later in the 1990s, domestic demand did not grow rapidly and smoothly. Foreign markets also were uncooperative, because some developing nations built their own petrochemical facilities and took over markets that had been Japan's export outlets. The naphtha problem was more or less solved in the mid-1980s when MITI finally sided with the petrochemical interest. By the late 1970s, MITI knew of the cost problem of chemical companies, which partially came from artificially high prices of naphtha in Japanese market. For various policy and political reasons, this problematic situation continued.

Finally in 1982, however, MITI allowed the freer import of naphtha and the prices of domestic and import naphtha dropped. Even as the naphtha problem was almost over and the import of naphtha actually substantially went up in the early 1980s, the depression in the petrochemical industry worsened. Because the industry itself seemed incapable of finding an agreeable solution to excess capacity problems, MITI responded by organizing a group for the chemical industry within the Industry Structure Council in order to reduce the domestic capacity of a few major basic petrochemicals. Based on the recommendation of the council, for instance, a depression cartel was formed, through which the capacity

Table 11.5. Japanese Exports and Imports of Chemical Products: 1995

	Exports (Million Yen)	Imports (Million Yen)	Balance (Million Yen)
Taiwan	366,099	53,415	312,684
South Korea	391,074	90,838	300,236
Hong Kong	216,311	3,742	212,569
China	192,079	124,270	67,809
Great Britain	52,310	121,535	−69,225
Switzerland	30,935	109,473	−78,538
France	51,266	151,004	−99,738
Germany	103,180	286,342	−183,162
United States	452,928	663,247	210,319
World total	2,829,276	2,309,160	520,116

Note: For exports, five largest trading partners are included. Nine nations are listed because the United States is both the largest exporter and importer.

Source: Compiled and calculated from Somucho Tokeikyoku, *Nippon Tokei Nenkan, 1997* (Tokyo: *Nippon Tokei Kyoukan, 1996*), Tables 12-8 and 12-9.

of ethylene producing facilities was supposed to be cut by 36%, from 6.35 million tons a year to 4.06 million tons. In actuality the cartel achieved 88% of the targeted reduction. Other targeted products were polyolefin, vinyl chloride, ethylene oxide, and styrene monomer.

A common strategy of chemical companies since the early 1970s has been the restructuring and rationalization of petrochemical facilities, the expansion of overseas markets, and the diversification into fine and specialty chemicals and pharmaceuticals. In none of these measures was the process of restructuring easy.

The exit from products with excess facility and depressed prices did not proceed smoothly and rapidly. Or, in general, financially troubled companies could go out of business all together. This did not happen on the large scale for three reasons. First, Japanese companies, most of which were management-controlled, rarely considered this option, because, more than anything else, senior managers, wanted to keep their own employment opportunities. Second, discipline from financial markets was very weak. Given the huge stake that was a result of past investment in large-scale facilities, commercial banks did not want chemical companies to default on their debt obligations. On the capital market, by contrast, there were no active shareholders to force management to adapt certain policies. If any, those commercial banks were often instrumental in advancing more money for restructuring. Third, permanent employment naturally worked as a huge exit barrier.

Moving to other growing industry areas was impossible, because all major industries were well covered thanks to the group structure. Actually, except for Sumitomo Chemical, even moving within chemical fields was not easy, because many groups had some large chemical enterprises specializing in narrow niches.

Considering these institutional forces, the most trouble-free strategy for reorganization and growth can be product development based on technological capabilities. But, historically, this was not the strength of Japanese chemical companies. Japanese companies have been technology learners, not generators.

THE DEVELOPMENT OF ALL-AROUND CHEMICAL ENTERPRISES

The five Japanese companies discussed in this section, Mitsubishi Chemical, Sumitomo Chemical, Mitsui Chemicals, Showa Denko, and Ube Industries, are representative of the Japanese chemical industry in the sense that they are the largest firms that have integrated from basic ethylene cracking to various derivatives and intermediate products. Although in terms of profitability these enterprises are not necessarily impressive, they are the most full line and diversified of Japanese chemical producers in terms of their product portfolio. Whenever the term "all-around chemical enterprise" (*sogo kagaku gaisha*) is used in the Japanese context, it usually refers to these five firms. The chairperson of the petrochemical industry's association has always been chosen from among these five companies.

In contrast to the American, German, and British or other national large chemical enterprises, the five Japanese companies discussed in this section have evolved as technological borrowers and learners. Although giant diversified firms from the three advanced chemical economies mainly developed as technological generators and large-scale commercializers of new products and processes, the Japanese firms developed by importing already tested and commercialized technologies from those Western nations, often from the very enterprises mentioned in three previous sections. As has been the case in many other industries, the critical capabilities for the commercial success of Japan's chemical enterprises have been the quick learning and incremental improvment of imported technology.

In addition to their distinct nature of technology acquisition, the five leading Japanese enterprises differ from their Western counterparts in terms of their group affiliation. All five Japanese firms are parts of large

Table 11.6. Research and Development Activities of Representative Japanese Industries: 1995

Industry	Number of Companies	Number of Researchers	Intra-Firm Expenditures on R&D (Billion Yen)	R&D Expenditure as Percent of Sales (%)	R&D Expenditure per Researcher (Million Yen)
Manufacturing, total	12,019	362,360	8,365	3.4	23
Chemicals	1,559	61,257	1,549	5.3	25
Industrial chemicals and fibers	534	21,177	551	4.2	25
Oil and paints	270	9,004	156	4.4	17
Drugs and medicine	386	20,091	633	7.8	32
Electrical machinery	2,134	145,367	3,065	5.9	21
Transportation equipment	427	35,668	1,220	3.2	34
Machinery	2,070	34,127	697	3.2	21
Instruments	555	18,267	334	5.5	18

Note: Five industry groups listed are the five largest in terms of their R&G expenditures.

Source: Compiled and calculated from Somucho Tokeikyoku hen, *Nippon Tokei Nenkan, 1997* (Tokyo: Nippon Tokei Kyokai, 1996), p. 726.

diversified corporate or business groups (*kigyo shudan* or *keiretsu*). Mitsubishi Chemical belongs to the Mitsubishi Group, although the company, even after its merger with Mitsubishi Petrochemical in 1994, shares chemical markets with other important chemical-related enterprises within the Mitsubishi Group such as Mitsubishi Gas Chemical, Mitsubishi Plastics, Asahi Glass, and Mitsubishi Rayon. The characteristics of being technology borrowers and group affiliates have constituted two major factors in determining the growth strategy of these five Japanese firms. The following sections will discuss the strategic developments of the five, which can be coherently divided into three phases: up to World War II, when a group of aggressive entrepreneurs (including Mori of Showa Denko) from outside the established *zaibatsu* groups (Mitsui, Mitsubishi, and Sumitomo) led the Japanese chemical industry; from the 1950s to the early 1970s, when industrial policy instruments introduced by MITI induced massive investments in petrochemical facilities by the five and a few other enterprises, making the industry second only to that of the United States in terms of size; and from the early 1970s to the present time, when the five struggled to find an appropriate strategy to achieve profitability and further growth, and industrial policy has become less effective in reorganizing the industry. The following descriptions are centered on Mitsubishi Chemical, which not only is the largest chemical firm in Japan but also represents the common patterns of growth and struggle of other full-line enterprises.

TO 1945: DIVERSE ORIGINS AND DEVELOPMENTS
BEFORE PETROCHEMICALS

The five enterprises examined here can be divided into two categories in terms of the historical lineages of the groups to which they currently belong. Sumitomo Chemical, Mitsui Chemical, and Mitsubishi Chemical, respectively, originated in one of the three biggest *zaibatsu* combinations before World War II. By contrast, Showa Denko and Ube Industries did not belong to established *zaibatsu* groups but instead represented a new set of enterprises that expanded its chemical businesses since the 1920s and 1930s. The two companies would thus become members of postwar groups that were formed around large commercial banks, such as Fuji Bank and Sanwa Bank.

Japan's chemical industry before World War II thus contained these two basic types of large enterprises: firms within old diversified *zaibatsu* groups whose involvement in chemicals was long but somewhat hesitant and enterprises controlled by emerging and aggressive entrepreneurs. The caution of the old *zaibatsu* resulted mainly from the entrepreneurial conservatism shared by the owning family members and senior managers, who emphasized the maintenance of family fortune and income. It was a small group of aggressive entrepreneurs who took this opportunity to advance their interest in chemicals. Such industrial groups as Nissan, Nichitsu, Nisso, and Mori became organized around large electro-chemical firms. By successively importing the latest technology available, these enterprises became leaders of Japan's chemical industry.[1]

Among the three old *zaibatsu* groups, Sumitomo's involvement in chemicals started earliest and, by the end of World War II, Sumitomo Chemical had established itself as the most mature firm among the five enterprises discussed here. Sumitomo's chemical manufacture started in an ironical context, the utilization of a hazardous waste gas. Since the Edo period, the Sumitomo family had operated a huge copper smelting facility in Besshi-Niihama on the Shikoku Island. The smelting process perilously affected the local population because of its heavy emissions of sulfur dioxide, but only in the early twentieth century was research conducted to find a way to reduce the attendant health hazards. In 1913, Sumitomo Fertilizer Factory was established to introduce a new technology that could transform sulfur dioxide into sulfuric acid and calcium superphosphate, which were to be utilized as fertilizers.[2] The chemical operation of the Besshi-Niihama mine was reorganized in 1925 as an independent company, Sumitomo Fertilizer Manufacturing, whose size grew substantially through public offering of its stocks. New sources of finance were needed for the company to enter ammonia production, which was launched in 1928. As it added a new product line of nitric acid in 1934, the firm changed its name to Sumitomo Chemical Company.

Sumitomo Chemical became one of the biggest and most comprehensive chemical producers when, again as a part of wartime industrial reorganization, it acquired Japan Dyestuff Manufacturing in 1944. The company's product line then consisted of inorganic, organic, and agricultural products and pharmaceuticals.

[1] Masaharu Udagawa, *Shinko Zaibatsu* (Tokyo: Nippon Keizai, 1984); Hidemasa Morikawa, *Zaibatsu: The Rise and Fall of Family Enterprises* (Tokyo: Tokyo Daigaku Shuppankai, 1990).
[2] Takashi Iijima, *Nippon no Kagaku Gijyutsu: Kigyoshi ni miru sonoKozo* (Tokyo: Kogyo Chosakai, 1981); Masahiro Shimotani, *Nippon Kagaku Kogyoshi Ron* (Tokyo: Ochanomizu Shobo, 1992).

Mitsui's involvement in the chemical industry originated immediately before the outbreak of World War I when the Miike mine coking plant delivered the first domestically produced synthetic dyestuffs. Thanks to a shortage of imported dyes during the war, Mitsui's manufacturing operation expanded in spite of the poor quality of its products. The coking and dyestuff plant became Miike Dyestuff Works, a separate division of Mitsui Mining. After initial struggles with the processing of the by-products of the Koppers-type oven, the dyestuffs division became financially successful around 1926. By the time it became legally independent in 1941 as Mitsui Chemical Industries, Mitsui had become the largest dyestuffs manufacturer in Japan.

In contrast to Sumitomo and Mitsui, Mitsubishi's extensive involvement in chemicals waited until as late as 1934, when Koyata Iwasaki, the fourth head of the Mitsubishi *zaibatsu*, formed Nippon Tar Industries. Mitsubishi bought the Makiyama coking factory in northern Kyushu and reorganized and modernized it to become the main Kurosaki plant for dye making of the newly formed Nippon Tar Industries. Originally, the company was controlled by two Mitsubishi enterprises, Mitsubishi Mining and Asahi Glass. Nippon Tar Industries changed its name to Nippon Chemical Industries in 1936 to diversify into such coal chemicals as coke and related products, fertilizer, and ammonia and its derivatives.[3]

Nippon Chemical Industries grew rapidly in the wartime economy, when the company started explosive manufacturing for armed forces and followed Japan's military expansion by launching the production of cokes, magnesium, and agricultural chemicals in northern China. In 1942, the company took over Shinko Rayon, which had been controlled by Mitsubishi since 1934. In 1944, Nippon Chemical Industries absorbed Asahi Glass, another Mitsubishi-affiliated company, as a part of wartime economic rationalization, and became Mitsubishi Chemical Industries, a major diversified manufacturer of such organic and inorganic products as coal-based chemicals, agricultural chemicals, glass, and rayon.

Showa Denko, by contrast, represents the set of emerging aggressive enterprises known as "new *zaibatsu*," which became involved in chemical manufacturing in the 1920s and 1930s as changing technology in electrochemicals and the entrepreneurial conservatism of old *zaibatsu* groups created opportunities for innovating firms to enter the chemical industry. Showa Denko was a core firm of the Mori group. The company

---

[3] Takashi Yamaguchi and Ikue Nonaka, *Asahi Kasei and Mitsubishi Kasei: Sentan Gijutsu ni kakeru Kagaku* (Tokyo: Otsuki Shoten, 1991).

was different from the other four in that its technological background was electrochemical rather than coal chemical. It was formed in 1939 when two enterprises of the group, Japan Electrical Industries and Showa Fertilizer, merged as a part of wartime economic rationalization. Japan Electrical Industries was an electrochemical firm that had originated in 1908 as Sobo Marine Products to manufacture iodine and potassium chloride from kelp. In 1926, the company was reorganized as Japan Iodine and produced a wide range of electrochemicals such as calcium carbide, electrodes, abrasives, and ferro-alloys. It became an original producer of aluminum in Japan in 1934, and the following year the enterprise changed its name to Japan Electrical Industries. In the meanwhile, Showa Fertilizer was founded in 1929 within the Mori group for manufacturing fertilizers such as ammonium sulfate, which the company pioneered in Japan.[4]

Ube Industries belongs to yet another group of large firms that developed in local areas so that they are often called "local *zaibatsu*." Ube Industries was a war-related merger of four local companies of the Ube region in the Yamaguchi prefecture in Western Japan. The oldest and largest one of these was Okinokyma Coal Mine. Another company, Ube Cement, was originally founded in 1924. Coal mining and cement manufacturing occupied most of the new company's activities, although Ube Industries' future lay in two other small companies. Shinkawa Iron Works specialized in manufacturing coal-mining machinery, whereas Ube Nitrogen Industries was originally organized in 1933 and soon started producing ammonia and sulfuric acid from coal. This enterprise further enhanced its technological capabilities when in 1936 it succeeded in producing nitric acid and distilling gasoline from the low-temperature carbonization of coal.[5]

World War II had a leveling effect on the five firms discussed here in terms of their technological and organizational capabilities. For the three old *zaibatsu* groups, World War II played a critical role in their catch-up to other new and more technologically advanced electrochemical manufacturers of the new *zaibatsu*. Because the war eventually stopped the inflow of new technical information, old *zaibatsu* chemical firms could catch up with those leaders by nurturing their coal-based chemical technology.

---

[4] Udagawa, *Shinko Zaibatsu*; Barbara Molony, *Technology and Investment: The Prewar Japanese Chemical Industry* (Cambridge: Harvard University Press, 1990).

[5] Hidemasa Morikawa, *Chibo Zaibatsu* (Tokyo: Toyo Keizai Shuppansha, 1988).

World War II, by contrast, had a destructive impact on the newly emerged groups and enterprises in two ways. First, since the late 1920s, many of the groups and enterprises, seeking a cheaper source of energy, aggressively transferred their operations to Korea and northern China. As the war ended, the groups not only lost a substantial part of their production facilities but also became a political target for their colonial aggression. Second, because the organizational design created by these new groups was not coherent, the groups could not weather the turmoil that resulted from the war. As groups, therefore, they simply collapsed, although many constituent companies, such as Asahi Chemical, Hitachi, and Nissan Automobile, could survive because of their technical and organizational competencies.

1945–1970: THE EMERGENCE AND DEVELOPMENT OF PETROCHEMICALS

The five Japanese companies discussed here became instrumental in propelling the nation's chemical industry into the age of petrochemicals. They not only invested extensively and continuously in large petrochemical facilities since the mid-1950s, but also transformed the entire structure of Japan's chemical industry. Regardless of their different origins and entry strategies, the growth rate and profitability of these full-line petrochemical companies stayed high until the late 1960s. The critical reason for this performance was the overall underestimation of basic petrochemical demand in the Japanese market and the technical competencies of Japan's chemical and engineering firms. As downstream chemical users, such as producers of synthetic fibers, plastics, and synthetic rubber developed rapidly in the domestic market, demand for upstream basic chemicals remained high. As supply was relatively tight – thanks to the government's entry control – companies operating upstream could enjoy high profitability.

Boom years after the 1950s resulted in the convergence of basic investment strategies adopted by many petrochemical enterprises. While continuously trying to locate and import the latest product and process innovations in the United States and Europe, they invested heavily in their own incremental process improvement capabilities.[6] With their sharpened process learning and improving competencies, these companies

---

[6] Mark Mason, *American Multinationals and Japan: The Political Economy of Japanese Capital Controls, 1899-1980* (Cambridge: Harvard University Press, 1992), pp. 209-18.

competed against each other by building ever larger facilities for ethylene production in order to exploit the cost savings that resulted from scale economies. This strategy was made possible by the substantial operating profits and also the availability of low-cost loans from the government and the so-called main banks of groups.[7]

In terms of Japan's commercial entry in petrochemicals, Mitsui played a pioneering role when in April 1958 Mitsui Petrochemicals' Iwakuni ethylene plant started its naphtha-cracking facility constructed with the Stone & Webster technology. Mitsui's entry into ethylene production had originally been conceived in the context of coal-chemical technology, which had been Mitsui's strong background. Financial commitment and technical difficulties associated with a naphtha-cracking facility were initially thought to be too heavy a burden for Mitsui to get into petrochemicals.[8]

When Toyo Rayon as the major user of basic petrochemical products and also a member of the Mitsui group, however, raised some serious concern about the future of coal-based chemicals, the group reluctantly reached an agreement on the establishment of Mitsui Petrochemicals. Because there still was strong inclination toward coal chemicals within Mitsui Chemical and the initial cost was too huge for the company, a separate firm was organized by the investment of seven Mitsui members: Mitsui Chemical, Toyo Koatsu, Mitsui Mining, Toyo Rayon, Miike Synthetic, Mitsui Metal Mining, and Mitsui Bank. Koa Oil, a Caltex-associated oil refining company, also had a minority stake. In reality, the negotiating process among the constituent Mitsui companies, leading to the formation of Mitsui Petrochemicals worked for those oft-antagonists to reunite and reintegrate into the group as a unified business entity.

After World War II, Mitsubishi Chemical Industries was divided into three parts under the Enterprise Reconstruction and Reorganization Law of 1950, which intended to deconcentrate the economic power of old *zaibatsu* groups. Because Mitsubishi's name was prohibited, the basic chemical operation became Nippon Chemical Industries, whereas Shinko Rayon took over rayon manufacturing, and glassmaking was separated as Asahi Glass. Shares of the three companies became public and were listed on all the major stock exchanges in Japan. In spite of these

---

[7] Tsunetada Kawade and Mitsuisa Bono, *Sekiyu Kagaku Kogyo* (Tokyo: Toyo Keizai Shimposha, 1970, new ed.); Tokuji Watanabe (Tokyo: Iwanami Shoten, 1972, second ed.); Hisao Hamasato, *Ronshu Nippon no Kagaku Kogyo* (Tokyo: Nippon Hyoronsha, 1994).

[8] Iijima, *Nippon no Kagaku Gijyutsu.*

efforts at deconcentration and ownership dispersion, all three enterprises would be gradually reunited within the reorganized Mitsubishi group, particularly after 1952 when the American occupation was formally terminated. The reorganized Nippon Chemical Industries changed its name back to Mitsubishi Chemical Industries in 1952. In the same year, Shinko Rayon became Mitsubishi Rayon. Ever since, those three enterprises kept their independent identities, although they engaged in a certain degree of strategic coordination within the Mitsubishi Group.

Mitsubishi's involvement in petrochemicals had its origins in 1954 when Mitsubishi Chemical and Shell established a joint venture, Mitsubishi-Shell Petrochemical, with the narrowly focused scheme of producing isopropyl alcohol, aceton, and derivatives. The limited product line partly came from Shell's reluctance to commit to extensive petrochemical product portfolios in Japan. Shell's attitude was shared by other foreign oil companies operating in Japan, which were understandably uncertain about both the future growth of Japan's petrochemical markets and the technical capabilities of domestic chemical firms.[9]

Since the early 1950s, MITI, by contrast, aimed to nurture more comprehensive and integrated ethylene-based petrochemical combines. The Ministry thus declined the proposal of Mitsubishi-Shell Petrochemical and asked Mitsubishi to come up with an alternative plan.[10]

When a huge piece of coastal land in Yokkaichi near Nagoya became available in 1955, therefore, Mitsubishi Chemical as a central chemical company of the group suddenly had to draw a concrete plan to launch a large-scale petrochemical production. Because Mitsui and Sumitomo had already initiated their petrochemical program earlier in the same year, competitive pressure became strong. Shell, as a partner, remained dubious about the feasibility of the Mitsubishi's idea and originally withdrew from participating in the plan. Because of Shell's reluctance, the Mitsubishi group as a whole had to support the petrochemical project. Mitsubishi Chemical alone was simply not large and strong enough to bear the financial burden.[11]

In 1956, thus, Mitsubishi Petrochemical was independently formed with the equity holding by such core Mitsubishi enterprises as Mitsubishi Chemical, Mitsubishi Rayon, Asahi Glass, Mitsubishi Bank, Mitsubishi

[9] Akira Kudo, "Sekiyu Kagaku," in Shin'ichi Yonekawa et al., eds., *Sengo Nippon Keieishi*, Vol. II (Tokyo: Toyo Keizai, 1990), pp. 279–336.

[10] Ibid.

[11] Mitsubishi Yuka Kabushiki Kaisha, *Mitsubishi Yuka Sanjyunen Shi* (Tokyo: Mitsubishi Yuka, 1988).

Metal Mining, and Mitsubishi Trading. Shell, after observing the serious-ness of Mitsubishi, eventually offered technical assistance on ethylene and styrene monomer in exchange for the 15% equity stake in the new company.[12]

As was the case with Mitsubishi Chemical, Sumitomo Chemical became a target of the economic reconstruction policy after the war. Sumitomo's name was prohibited, and the company had to rename itself Nisshin Chemical Industries in 1946. In the case of Nisshin Chemical, however, the company expanded its product portfolio in 1949 when it acquired the facility of Sumitomo Aluminium Reduction, which had been dissolved. In 1952, as did Mitsubishi Chemical, Nisshin Chemical revived its old name, Sumitomo Chemical. The company strategically positioned itself as the only chemical company within the Sumitomo Group, which was unusual among group-affiliated chemical firms. Sumitomo Chemical as well as Sumitomo Bank and Sumitomo Metal became the core of the group.[13]

Sumitomo had been a pioneer in petrochemical-related technology in Japan. During World War II, Sumitomo had a strong tie to the engineering department of Kyoto University, which was experimenting with synthetic rubber and high-density polyethylene. Although Sumitomo, too, was hes-itant to move from coal chemicals to petrochemicals, competitive pres-sure from other groups forced Sumitomo to enter into petrochemicals in 1955. Sumitomo Chemical bought high-density polyethylene know-how from ICI, which was followed by the introduction of the Stone & Webster naphtha-cracking technology.[14]

Sumitomo's entry into petrochemicals was different from Mitsui and Mitsubishi in that Sumitomo Chemical alone, rather than a group, invested on the relatively small scale in the new technology. This was partly because of Sumitomo's tradition of "one company for one trade," and also probably because the group did not possess any rayon or syn-thetic fiber manufacturers to serve as the captured outlet for ethylene and derivatives, as did Mitsuibishi and Mitsui.

Showa Denko entered into organic chemical areas when it absorbed an associated company, Showa Synthetic Chemical Industries in 1957, and further into petrochemical fields by establishing Showa Petrochemical

[12] *Ibid.*; Kudo, "Sekiyu Kagaku."
[13] Eleanor M. Hadley, *Antitrust in Japan* (Princeton, NJ: Princeton University Press, 1970); Kudo, "Sekiyu Kagaku."
[14] Iijima, *Nippon no Kagaku Gijyutsu;* Shimotani, *Nippon Kagaku Kogyoshi.*

in the same year. When Nippon Petrochemical, a subsidiary of Nippon Oil, expanded a petrochemical complex in Kawasaki near Yokohama, to facilitate ethylene production in 1959, Showa Petrochemical participated in the complex by building a plant for high-density polyethylene production. In addition to the activities of Showa Petrochemical, Showa Denko itself constructed a petrochemical complex in Oita in 1969.[15]

Being a merger of local diverse enterprises of the Ube region in the 1930s and independent until the 1950s, therefore, Ube Industries became a member of the group organized by Sanwa Bank, which also was formed originally as a merger of three local commercial banks in the Osaka region.

Ube Industries converted itself from a coal-mining and cement company to a widely diversified chemical company in the 1960s, when coal's significance as an energy source and raw material drastically declined in the Japanese economy. As early as 1960, Ube Industries launched its fifteen-year plan of diversifying from coal-related businesses, when other coal mining companies were still trying to survive in that business. Although the diversity of original businesses of Ube Industries functioned positively for this refocusing strategy, the early timing of strategic adaptation certainly helped the company's subsequent growth. Furthermore, the company's new affiliation with the emerging Sanwa Bank group worked instrumentally, because the group's presence was then particularly weak in the upstream part of the chemical industry, whereas Sekisui Chemical in plastics and Teijin in textiles guaranteed a stable and captured chemical demand within the group.[16] Ube Industries entered into petrochemical production in 1964 when the company participated the development of the Chiba petrochemical complex where Maruzen Petrochemical, a subsidiary of Maruzen Oil, established an ethylene production facility of forty-four thousand tons per year.

Ube Industries then founded a secular supply of ethylene that was processed into low-density polyethylene. In contrast to the other four major diversified chemical companies, Ube Industries hereafter relied on outside suppliers for the company's ethylene needs. In 1967, Ube Industries built a new chemical plant in Sakai near Osaka that produced ammonia, urea, and caprolactam and added a new facility for polypropylene in

[15] Showa Denko Kabushiki Kaisha, *Showa Denko Sekiyu Kagaku Hattenshi: Showa Yuka no Setsuritsu kara Kappei made* (Tokyo: Showa Denko, 1981).

[16] Iijima, *Nippon no Kagaku Gijyutsu.*

the following year. Technical know-how of polyethylene and polypropylene was supplied by Dart and ICI. Texaco and Kellogg gave technical assistance for the ammonia-producing plant.

By 1970, when Ube Industries closed its last coal-mining facility, the Sanyo anthracite mine, the enterprise had invested more than two hundred billion yen in the restructuring and refocusing processes. A half of this investment went to petrochemical and related areas, whereas cement manufacturing took a third of that amount. As a result, in 1972 petrochemical products, particularly caprolactam and fertilizers, accounted for around 40% of the company's sales revenues, whereas cement's share also rose to close to 40%. Although the cement business itself became structurally depressed in the 1970s, Ube Industries by then had a strong machinery division whose technological capabilities were nurtured mainly through the company's construction and operation of chemical and cement facilities.

The investment competition for larger and newer production facilities among these five and other large enterprises could not last forever, however, simply because sooner or later the speed of capacity expansion exceeded the growth of demand. As early as 1970 or 1971, Japan's market could no longer absorb the vastly expanded supply of ethylene. Japan's basic chemical markets started experiencing overcapacity and structural depression.[17]

1970 TO THE PRESENT: CONTINUING STRUGGLES OF PETROCHEMICALS

If a common investment pattern among chemical companies during the boom years of the 1950s and 1960s was the construction of ever larger petrochemical plants, generic strategies since the early 1970s have been the restructuring and rationalization of petrochemical facilities, the expansion of overseas markets, and the diversification into fine and specialty chemicals and pharmaceuticals. None of the five enterprises, however, experienced an easy time of reorganizing their large-scale operations of upstream petrochemical fields and refocusing their strategy into promising areas.[18]

---

[17] Tokuji Watanabe and Yasuharu Saeki, *Tenki ni tatsu Sekiyu Kagaku Kogyo* (Tokyo: Iwanami Shoten, 1984).

[18] Watanabe and Saeki, *Tenki ni tatsu Sekiyu Kagaku Kogyo*; Yoshio Tokuhisa, *Kagaku Sangyo ni Miraiwa Aruka* (Tokyo: Nippon Keizai Shinbunsha, 1995).

Table 11.7. Ten Largest Chemical-Patenting Countries: 1985–1995

Country	1985	1990	1995	Annual Change, 1985–1995
United States	11,557	13,122	15,259	2.8
Japan	3,733	5,473	6,138	5.1
Germany	2,179	2,706	2,486	1.3
France	716	917	1,020	3.6
Great Britain	854	953	903	0.6
Canada	358	432	564	4.6
Switzerland	458	475	435	−5.0
Italy	296	409	414	3.4
Taiwan	18	52	305	33.0
South Korea	9	52	305	42.0
World total	21,257	26,017	29,433	3.3

Note: Countries are ranked by the number of U.S. patents issued in 1995. Origin of patents is based on address of inventor whose name is first on patent application.

Source: Compiled from "Facts and Figures for Chemical R&D," *Chemical and Engineering News*, August 26, 1997, p. 71.

Although restructuring of commodity petrochemicals turned out to be difficult because of continuing competitive investment behavior of those large enterprises, new product development and international growth were not their conventionally strong functions. For this reason, many new strategic moves were carried out through joint ventures and projects. A general pattern was that in Japan and other advanced economies the partners of those joint ventures were high-technology firms whose technology and products were sought by Japanese companies, whereas in emerging economies Japanese enterprises entered into cooperation with local nonchemical firms in conventional product lines.[19]

Ever since the economic necessity for the reorganization and consolidation of chemical businesses became evident in the early 1970s, it took more than twenty years to consolidate two of Mitsubishi's petrochemical-related companies, Mitsubishi Chemical and Mitsubishi Petrochemical. The two firms merged in October 1994 to form new Mitsubishi Chemical, the eighth largest chemical producer in the world. The delay of this merger symbolizes the difficulty of the reorganization of Japan's

[19] Mike Ward, *Japanese Chemicals: Past, Present and Future* (London: Economic Intelligence Unit, 1992).

petrochemical industry within the country's business and institutional contexts.

Between the two, Mitsubishi Chemical experienced a less stressful time since the 1970s, because the company historically had a diversified product portfolio in petrochemicals, carbon products and inorganic chemicals, and the functional product category such as electronics and pharmaceuticals. Although Mitsubishi Chemical, as was the case of all chemical companies, suffered seriously in petrochemical businesses, two other business lines of the enterprise stayed relatively healthy. Mitsubishi Chemical has been placing particular emphasis on pharmaceuticals, high-performance materials, and electronic products such as optical disks.[20]

Mitsubishi Petrochemical, in the meanwhile, was the largest petrochemical manufacturer in Japan. As such, it was hit hard by the oil shock and subsequent economic turmoil. Although the company continued to improve the efficiency of its petrochemical operation, Mitsubishi Petrochemical expanded the production of electrochemicals and related fine and engineering chemicals and other specialty items. The enterprise particularly emphasized the application of life-science in areas such as biochemicals, pharmaceuticals, and agricultural chemicals.[21] A Mitsui enterprise, Mitsui Toatsu, was the first company among the major petrochemical producers that recognized the seriousness of the business recession and thus suspended dividend payment in 1972, before the oil shock further worsened the situation in the following year. The company introduced a strategy of exiting from commodity petrochemicals and expanding its fine chemicals and specialty chemicals business lines. They also tried to expand the overseas sales, not only in Asia where the company had many joint ventures but also in the United States and Europe.[22] In order to cope with the trouble of commodity petrochemicals, another Mitsui enterprise, Mitsui Petrochemical adopted a strategy of expanding its specialty product lines such as containers, packaging materials, and industrial supplies. The critical issue for these two enterprises, however, still remains. If the Mitsubishi group finally succeeded in 1994 in combining its two petrochemical interests, the Mitsui group has not achieved a similar goal. In spite of repeated talks generally urged by the group and

---

[20] Ibid.

[21] Mitsubishi Yuka Kabushiki Kaisha, *Mitsubishi Yuka Sanjyunen Shi* (Tokyo: Mitsubishi Yuka, 1988); Takashi Yamaguchi and Ikue Nonaka, *Asahi Kasei and Mitsubishi Kasei: Sentan Gijutsu ni kakeru Kagaku* (Tokyo: Otsuki Shoten, 1991).

[22] Ward, *Japanese Chemicals.*

the government, the senior management of the two companies refused to merge to consolidate and rationalize their operations.

Mitsui Toatsu, therefore, independently advanced its strategy of rationalization, diversification, and internationalization. In 1978, for instance, the company organized the Fine Chemical Division, which, in addition to developing new fine chemicals, started marketing dyes, agrochemicals, and fine industrial chemicals. The division utilized the company's joint venture with BASF, Mitsui Badish Dyes, to capture southeast Asian as well as domestic markets. Furthermore, in 1981 Mitsui Toatsu formed a joint project with Genex of the United States to advance generic engineering products, which was sold through a subsidiary, Mitsui Pharmaceuticals. Recently, furthermore, Mitsui Toatsu succeeded in developing high value-added products such as drawing-type optical disk, CD-Rs, and superabsorbent resins, as the sales of the company's multiuse resins peaked.

Mitsui Toatsu became active in overseas expansion since the early 1970s. The company had been actively seeking opportunities to form joint ventures in foreign markets, particularly those of emerging economies. In 1966, Mitsui Toatsu, in cooperation with Mitsui & Co., a general trading company of the group, participated in the establishment of two joint ventures with local interests in Asia: Singapore Adhesives and Chemical and Thai Plastic and Chemical. Mitsui Toatsu, again with Mitsui & Co., founded a plastics company, Siam Resin and Chemical in 1974. In the same year, the enterprise also had a joint venture in Indonesia for the production of adhesives and plastics. The company then entered in the South Korean market in 1976 by investing in the petrochemical business of First Chemical Industries.

In the early 1970s, Sumitomo Chemical had a well-balanced product portfolio that covered basic chemicals, organic and inorganic, specialty chemicals, and aluminum production. But Sumitomo Chemical was relatively slow in responding to the deteriorating business environment, particularly that for fertilizers. The business suffered and Mitsubishi Chemical was surpassed as the country's largest chemical producer. In order to recover from the business slump, Sumitomo Chemical had to rationalize its depressed operations. In 1977, aluminum manufacturing was spun off, and basic fertilizer business was curtailed substantially. In 1983, the Ehime Plant had to scale down its ethylene and ethylene derivatives to concentrate production at the newer and more efficient Chiba Plant.

In terms of positive growth strategy, Sumitomo Chemical expanded its business lines in fine chemicals, particularly agricultural chemicals and pharmaceuticals, and new materials and fibers. In the early 1980s, for instance, Sumitomo Chemical established a joint venture with Hercules for the production of carbon fiber. In 1983, Sumitomo Chemical spun off pharmaceutical manufacturing to Inabata & Co., with which the company later established a joint marketing enterprise, Sumitomo Pharmaceuticals.

Sumitomo Chemical became active from the mid-1970s in developing overseas businesses. Through its agrochemical products, the company developed a significant presence in Asia and became a major player in a new petrochemical complex, PCS, in Singapore. In 1987, in a joint venture with Chevron Chemical, Sumitomo Chemical established Valant USA, which became instrumental in marketing Sumitomo Chemical's products in the U.S. markets. In the same year, the enterprise entered in an agreement with Rohm & Haas to manufacture new methacrylates. Sumitomo Chemical further strengthened its European marketing networks by founding two subsidiaries, Sumitomo Chemical (U.K.) and Sumitomo Chemical Netherlands, in 1988 and another, Sumitomo Chemical France, in 1990.

Even when compared to a difficult reorganization at Sumitomo Chemical, Showa Denko's restructuring was particularly troublesome since the two strategic areas the company had invested in up to the early 1970s, petrochemicals and aluminum and ferro-alloy refining, both became structurally depressed. As early as 1972, the company had to suspend its dividend as the financial standing deteriorated rapidly. Showa Denko spun off its aluminum operation in 1976, as Showa Light Metal, which was eventually closed in 1986. Showa Denko also had to absorb its petrochemical subsidiary, Showa Petrochemicals in 1979 as a part of rationalization program.[23]

Showa Denko sought to find a growth niche in such diverse fields as housing construction and sales, pollution control, new ceramics and materials, agricultural products, and biotechnology. The company, for instance, formed Showa Unox in 1972, which specialized in the disposal control of sewage for large cities and factories. Results are mixed at best. Pollution control remained a popular but unprofitable business, and building material business had to be spun off in 1995.

Compared to the other four enterprises, Showa Denko was slow in its internationalization drive in chemicals. Although the company acquired a minority interest in an aluminum-refining venture in New Zealand in

[23] Showa Denko, 1981.

1969, in Venezuela in 1973, and in Brazil in 1976, those ventures did not perform well due to internationally depressed aluminum prices. Showa Denko had only two small overseas manufacturing enterprises in plastic molding: one in Taiwan and another in Singapore.[24]

As was the case of Showa Denko, Ube Industries also suffered from low profitability in two major industries, petrochemicals and cement, from Kaiser Cement and Gypsum, a subsidiary of Kaiser Industries, as late as 1976 excess cement capacity was clearly visible. In order to recover from its miserable financial performance, Ube Industries adopted two basic strategies: diversification and internationalization.

In terms of diversification, Ube Industries had many joint ventures with enterprises whose technology was attractive to the company. For instance, in 1984 Ube Industries and Marubeni, a general trading company, entered into a joint venture with Wormser of the United States to manufacture fluidized-bed combusters. In 1989, Ube Industries and Kemira, a specialized chemical company from Finland, formed Kemira-Ube to produce hydrogen peroxide in Japan. In the following year, Ube Industries established two joint ventures with two technology-oriented chemical firms: UBE-EMS with Ems-Chemi of Switzerland and Ube Rexene with Rexene of the United States. Ube Industries also was active in working with other Japanese companies, particularly in the pharmaceutical business. Ube Industries established joint ventures with Takeda Pharmaceutical and Sankyo Pharmaceutical, both of which were the major manufacturers, in order to employ their marketing extensive networks for pharmaceutical products developed by Ube Industries.[25]

CONCLUSION

The basic characteristics of the Japanese petrochemical companies were formed during the boom years since the mid-1950s. While continuously trying to locate and import the latest product and process innovations in the United States and Europe, they invested heavily in their own incremental process improvement capabilities. With their sharpened process learning and improving competencies, those companies competed against each other by building ever larger facilities for ethylene production to exploit cost savings resulted from scale economies. This strategy was made possible by the substantial operating profits and also the

[24] Ward, *Japanese Chemicals.*
[25] Ibid.

Table 11.8. International Technology Transfer of Japanese Industries: 1975–1994

	1975	1980	1985	1990	1993	1994
Manufacturing industries, total						
Exports	58.9	133.3	205.6	320.7	394.2	452.6
Imports	1764.9	233.2	288.6	368.3	359.6	367.8
Balance	−106.9	−99.9	−83.0	−47.6	34.5	84.8
Chemical industry (including pharmaceuticals)						
Exports	21.5	31.9	38.2	58.2	59.3	64.1
Imports	26.9	39.3	37.4	54.0	61.4	59.0
Balance	−5.4	−7.4	0.8	4.2	−2.1	5.1
Electrical machinery industry (including computers)						
Exports	7.3	23.0	59.5	97.0	127.4	140.5
Imports	38.2	61.7	84.2	259.9	159.2	177.4
Balance	−30.9	−38.7	−24.7	−62.9	−31.8	−36.9
Transportation equipment industry						
Exports	6.3	21.8	32.4	92.0	127.4	164.2
Imports	35.7	40.3	59.7	52.3	40.4	35.6
Balance	29.4	−18.5	−27.3	39.7	87.3	128.6

Source: Compiled and calculated from Somucho Tokeikyokuhen, *Nippon Tokei Nenkan, 1996* (Tokyo: Nippon Tokei Kyokai, 1996), p. 728.

availability of low-cost loans from the government and the so-called main banks of groups.

Ever since the economic necessity for the reorganization and consolidation of chemical businesses became evident in the early 1970s, it took more than twenty years to consolidate two of Mitsubishi's petrochemical-related companies, Mitsubishi Chemical and Mitsubishi Petrochemical. Actually, the Mitsubishi situation is better than that of the Mitsui group, which has not solved a similar situation. The substantial delay of these reorganizations symbolizes the difficulty of the restructuring of Japan's petrochemical industry within the country's business and institutional contexts.

Ironically, all of the reasons for the difficulty of reorganizations worked positively when the Japanese petrochemical industry developed from the mid-1950s: growth-oriented aggressive management, committed employees, involved commercial banks, silent capital markets, and so on. The Japanese chemical industry is still a puzzle.

REFERENCES

Hamasato, Hisao, *Ronshu Nippon no Kagaku Kogyo* (Tokyo: Nippon Hyoronsha, 1994).

Iijima, Takashi, *Nippon no Kagaku Gijyutsu: Kigyoshi ni miru sono Kozo* (Tokyo: Kogyo Chosakai, 1981).

Kawade, Tsunetada, and Mitsuisa Bono, *Sekiyu Kagaku Kogyo*, new ed. (Tokyo: Toyo Keizai Shimposha, 1970).

Kudo, Akira, "Sekiyu Kagaku," in Shin'ichi Yonekawa et al, eds., *Sengo Nippon Keieishi*, volume II (Tokyo: Toyo Keizai, 1990), pp. 279–336.

Mason, Mark, *American Multinationals and Japan: The Political Economy of Japanese Capital Controls, 1899–1980* (Cambridge, MA: Harvard University Press, 1992).

Mitsubishi Yuka Kabushiki Kaisha, *Mitsubishi Yuka Sanjyunen Shi* (Tokyo: Mitsubishi Yuka, 1988).

Molony, Barbara, *Technology and Investment: The Prewar Japanese Chemical Industry* (Cambridge, MA: Harvard University Press, 1990).

Morikawa, Hidemasa, *Chiho Zaibatsu* (Tokyo: Toyo Keizai Shuppansha, 1988).

Morikawa, Hidemasa, *Zaibatsu: The Rise and Fall of Family Enterprises* (Tokyo: Tokyo Daigaku Shuppankai, 1990).

Shimotani, Masahiro, *Nippon Kagaku Kogyoshi Ron* (Tokyo: Ochanomizu Shobo, 1992).

Showa Denko Kabushiki Kaisha, *Showa Denko Sekiyu Kagaku Hattenshi: Showa Yuka no Setsuritsu kara Kappei made* (Tokyo: Showa Denko, 1981).

Tokuhisa, Yoshio, *Kagaku Sangyo ni Miraiwa Aruka* (Tokyo: Nippon Keizai Shinbunsha, 1995).

Tokuhisa, Yoshio, *Sekiyu Kagaku Kogyo shi Kankei Ronbunshu* (Fujisawa, Kanagawa: Author, 1995), mimeo.

Udagawa, Masaharu, *Shinko Zaibatsu* (Tokyo: Nippon Keizai, 1984).

Ward, Mike, *Japanese Chemicals: Past, Present and Future* (London: Economic Intelligence Unit, 1992).

Watanabe, Tokuji, *Sekiyu Kagaku Kogyo*, second ed. (Tokyo: Iwanami Shoten, 1972).

Watanabe, Tokuji, and Yasuharu Saeki, *Tenki ni tatsu Sekiyu Kagaku Kogyo* (Tokyo: Iwanami Shoten, 1984).

Yamaguchi, Takashi, and Ikue Nonaka, *Asahi Kasei and Mitsubishi Kasei: Sentan Gijutsu ni kakeru Kagaku* (Tokyo: Otsuki Shoten, 1991).

European Followers

12

The Rise and Fall of the Italian Chemical Industry, 1950s–1990s

VERA ZAMAGNI

In no branch of advanced industry today is Italy as badly placed as in the chemical industry.[1] A glance at Table 12.1 is enough to demonstrate the point. Until 2003,[2] Italy had a comfortable surplus balance of trade for manufacturing as a whole, with a few negative items, among which chemical products and pharmaceuticals remain consistently among the worst performers. In 2003, the import penetration of the Italian chemical market was 47%, whereas branches of foreign corporations producing in Italy employ 40% of the total labor force. In terms of employment, the Italian chemical industry moved from 6% of the manufacturing industry in 1951, to 7% in 1981, to fall back to 4.3% in 2003 (but value added is 7.5% of total VA of the manufacturing industry). As can be seen in Table 12.2, this performance is a remarkable exception among the advanced industrial countries, all of which enjoy a comfortable and often increasing export surplus (even discounting for mounting prices); only the United States, surprisingly, has shown a deficit, beginning in 2002.

This sad picture of the Italian chemical industry could have been the result of a conscious choice against the pollution and the unsafe

---

[1] In this chapter, I cover both the chemical and pharmaceutical sectors. When pharmaceuticals are excluded, it will be noted explicitly.

[2] After the appreciation of the Euro, the situation of the Italian competitiveness has worsened remarkably across the board.

Table 12.1. Trade Balances (Billion Euros)

	1985	1990	1998	2003
Textiles, clothing	5.9	7.4	13.5	12.0
Leather, shoes	4.3	4.9	6.5	6.2
Machinery	7.6	10.0	23.8	32.8
Metal products	2.7	3.5	9.3	−2.3
Wood and furniture	1.0	1.5	3.8	−2.0
Rubber and plastic products	0.9	1.3	3.9	4.1
Means of transport	−1.0	−3.3	−5.6	−9.9
Food and beverages	−4.1	−4.4	−4.2	−3.5
Chemical products and pharmaceuticals	−2.8	−5.5	−8.3	−9.4
Manufacturing industry	13.9	8.8	40.4	35.2

Sources: Istat, ICE, Bankitalia, Federchimica.

environment connected with that industry, especially in the past. After all, Italy owns such a large share of the world's beauties of nature and art that their preservation would be worth the sacrifice. Unfortunately, this is not the case. Italy tried hard to have a chemical industry and especially a petrochemical industry. If success has been meager, the reason is not a lack of investment in the sector. The Italian efforts to establish a chemical industry date from the period of World War I,[3] and continue through reconstruction and the "economic miracle" at the time of the rise of the petrochemical industry.

This chapter tries to unravel the reasons behind the Italian failure in the chemical industry, as seen from the viewpoint of domestic and international developments. There are basically two sequences of events responsible for this unsatisfactory performance. First, in the international context of rapid petrochemical growth, there was a rush in the 1950s and 1960s to build as many petrochemical plants as possible, without overall planning and coordination. When the oil crises hit in the 1970s and early 1980s, the industry went bankrupt. Second, although it proved possible to rescue something from this first phase, a wholly domestic problem in the 1990s squandered most of what had been left. Montedison had

---

[3] On the history of the Italian chemical industry, see V. Zamagni, "L'industria chimica in Italia dalle origini agli anni '50'," in Franco Amatori, and Bruno Bezza, eds., *Capitoli di storia di una grande impresa: la Montecatini 1888-1966* (Bologna: Il Mulino, 1991). On the interwar period, see R. Petri, "Technical Change in the Italian Chemical Industry: Markets, Firms and State Intervention," in Anthony S. Travis et al., eds., *Determinants in the Evolution of the European Chemical Industry, 1900-1939* (Netherlands: Kluwer, 1998), pp. 275-300.

Table 12.2. Trade Balances of Chemical Products* (Billion Euros)

	Germany	France	UK	Belgium	Netherlands	Italy	EU15	Japan	USA
1980	8.3	1.8	3.2	2.2	3.7	−2.2	26.4§	0.4	8.7
1998	21.8	7.3	6.1	7.3	8.7	−8.7	36.2	5.7	12.0
2003	28.7	10.1	6.5	14.4	14.1	−9.4	65.7	8.2	−8.5

Source: CEFIC, *National Statistics.*
* Excluding pharmaceuticals §1985.

tried to unify the petrochemical industry under its control but failed and exited the field. Belated and inappropriate state intervention will turn out to be an important part of the story, but the inadequacy of Italian entrepreneurship in the field also will be apparent.

THE RUSH TO BUILD PETROCHEMICAL PLANTS AND THE ENSUING CRISIS (1950s–1970s)

In a recent article on the Italian chemical industry during the "economic miracle," Francesca Fauri[4] outlined the beginning of the petrochemical scramble, during which the Italian chemical industry was allowed to grow quantitatively at a rather fantastic pace. A host of concurrent reasons gave several players sufficient incentives to move in that direction. The following section will look at these companies and their motivations.

Montecatini was the interwar quasi-monopolist of the Italian chemical industry that would have been expected to hold its grounds. Although this was indeed the case, the lack of an adequate successor to its great interwar CEO, Donegani, and the persistent allegations of monopolistic practices from leftist parties and the press, kept Montecatini from acting on an appropriate scale, and from lobbying effectively for a national policy that would have safeguarded its position.

In 1950, Montecatini opened in Ferrara[5] what was effectively the first petrochemical plant in Europe. However, despite the brilliant innovations that the Nobel Prize winner, Giulio Natta, developed precisely in that plant (moplen, or isotactic polypropylene), it wasn't until 1957 that this

[4] Francesca Fauri, "The 'Economic Miracle' and Italy's Chemical Industry, 1950-1965: A Missed Opportunity," *Enterprise and Society*, I (2000): 279-314.

[5] R. Petri, "Il polo chimico ferrarese," in Pier Paolo D'Attorre and Vera Zamagni, eds., *Distretti, imprese e classe operaia. L'industrializzazione dell'Emilia-Romagna* (Milan: Angeli, 1992).

plant came to produce on an acceptable scale. The decision to build a larger plant in Brindisi (Puglia) was taken belatedly and in several steps,[6] in the midst of financial difficulties produced by the loss of markets[7] and the mismanagement of the company. This worsened the company's financial difficulties and prompted its merger with Edison (see later) in 1966 to form Montedison. But the most myopic decision on its part was perhaps letting AGIP[8] achieve full control of ANIC, the oil cracking joint venture that Montecatini had founded with AGIP and AIPA in 1936.[9]

ANIC soon emerged as a strong competitor to Montecatini, first in fertilizers, and, subsequently, in petrochemical productions. Having become a division of the newly formed state-owned Italian oil company, ENI, ANIC did not suffer from the financial difficulties that made life difficult for Montecatini. It built a new, large, petrochemical plant in Ravenna, which opened in 1957, and that was superior to Montecatini's Ferrara plant (first built in 1936) because of the more recent and more coherent technological layout. Other plants were built in the South (Gela, Sicily, 1960; Pisticci, Basilicata, 1960).

Edison was the most important and prosperous private Italian electrical company. As a result of the campaign for nationalization of electricity that started soon after the end of the war, Edison diversified into chemicals with little knowledge of the field and unsatisfactory financial results, which were hidden by its large electrical profits. When nationalization of electricity was accomplished in 1963, Edison was left with huge amounts of capital and a few chemical plants. The obvious solution from a financial point of view appeared to be a merger with Montecatini, which did indeed take place in December 1965, forming Montedison.[10] On paper, Montedison seemed to recover a position of leadership in the Italian chemical sector (it supplied about 19% of the domestic market in 1968, a market that was already largely penetrated by foreign multinationals)[11].

---

[6] The first section of the plant was opened in 1962.

[7] Between 1953 and 1964, Montecatini's share of the Italian fertilizers market decreased from 80% to 30%.

[8] AGIP was the state oil company that Mussolini wanted to create in 1926, due to be liquidated by Mattei after the war. But Mattei saw a future for it, and with the aim of strengthening its structure, collected around it some other related companies, including ANIC, and formed in 1953 a new state-owned corporation, named ENI.

[9] This argument is developed by B. Bottiglieri, "Una grande impresa chimica tra stato e mercato: la Montecatini degli anni '50,'" in Amatori and Bezza, *Capitoli di storia di una grande impresa.*

[10] The company was first called Montecatini-Edison, and became Montedison in 1970.

[11] As much as 47% of the Italian domestic chemical market was in foreign hands.

Internationally, Montedison's turnover ranked fourth after DuPont, ICI, and Hoechst. But Montedison's production was achieved in a myriad of smallish plants scattered across the country, and the reorganization of this chaotic "chemical empire" proved exceedingly difficult. As a result of its weakness and to more political reasons, in 1968 ENI acquired a large share of Montedison stocks, making this company a half-public and half-private hybrid.[12]

Originally known as Società Italiana Bakelite, SIR was founded in Lombardy in 1922. It was restructured in 1931 into SIR (Società Italiana Resine) for the production of early types of plastic materials. In 1949, SIR was bought by the engineer, Nino Rovelli, the head of a small local engineering company. He started enlarging SIR by opening a plant for the production of intermediates in Sesto San Giovanni in 1956, and other plants in Solbiate Olona. In 1960, he decided to make the big jump into petrochemicals by starting the building of a large complex at Porto Torres (Sardenia).[13] For a small company such as SIR, it would have been utterly impossible to find enough resources to carry out a huge project if it were not for the combination of two factors: first, the state policy of industrialization of the South through the Cassa per il Mezzogiorno and the granting of substantial subsidies to private and state-owned companies to localize there;[14] and, second, the availability of state-owned long-term financial institutions ready to administer soft loans and other credit facilities in connection with government guidelines. In the case of SIR, the principal financial institution involved was IMI, but ICIPU and CIS also were quite active, as was Mediobanca with Montedison. The selection of Sardenia was the result of local political pressures to obtain a share of the pie, after ANIC-ENI and Edison had already selected Sicily, Calabria, and Basilicata, and Montecatini (and later ANIC), Apulia. The preferential treatment granted by IMI to SIR was probably due to IMI's chairman, Stefano Siglienti, who came from Sardenia.

Created in 1915, Rumianca specialized in the electrochemical production of soda and nitrates for fertilizers; in the postwar period, it

---

[12] On the presence of ENI in the chemical sector, see Vera Zamagni, "L'ENI e la chimica," *Energia*, XXIV (2003): 16–24.

[13] AsIMI, Libro verbale Comitato Esecutivo, note 38, p. 28.

[14] The chemical industry received a disproportionate share of the Cassa del Mezzogiorno subsidies: 23% of the total in the 1950s, 35% in the 1960s, and as much as 42% in the early 1970s. See L. Mattina and A. Tonarelli, "Lo sviluppo della chimica. Gruppi di interesse e partiti nell'intervento straordinario," in L. D'Antone, ed., *Radici storiche ed esperienza dell'intervento straordinario nel Mezzogiorno* (Rome: Bibliopolis, 1996).

produced DDT and other products for agriculture. The jump into the petrochemical industry was made one year after SIR (1961), following the same approach. The place selected was again Sardenia, at Assemini (near Cagliari). Because Rumianca's financial difficulties started sooner than SIR's, it requested help from SIR, effectively forming one group from a financial point of view, although not from a technical one, as was the case with Montecatini-Edison.[15] SIR-Rumianca accounted in 1968 for a very modest 2.7% share of the Italian chemical market (as against 19% for Montedison).

Liquichimica was a branch of Liquigas, founded in Milan in 1936 to produce liquified gas. In 1968, it became Liquichimica, on the basis of an oil refinery it had built in1952, and Isor, a petrochemical company with a small plant in Piedmont (Robassomero, Turin) that it had absorbed. It is on this quite modest basis that Liquichimica launched itself into the scramble to obtain subsidies for locating new plants in the South and built a plant in Augusta (Sicily). In 1972, its plans for further enlargement were still quite vague and took shape in Sardenia in the interior of the island, Ottana, paradoxically during the first oil crisis.

To the already long list of companies engaged in enlarging capacity in the Italian petrochemical sector offered earlier, one must add other Italian chemical companies mainly specialized in allied fields, for example, SNIA, the champion of artificial fibers, and some multinationals such as Solvay, which occasionally tried to build their own plants for the supply of basic products and intermediates.

There was no need for the oil crisis to hear the alarm bell ringing. It was already clear in 1968 that the Montecatini-Edison merger was very problematic because of a pronounced inadequacy in the management. The state, which already owned some shares of Montedison through IRI since the 1930s,[16] took the surprising decision of authorizing IRI, and above all ENI, to acquire other Montedison shares up to the point of exceeding the shares syndicated by the larger private owners.[17] This

---

[15] See R&S, *L'industria chimica* (Milan: Capriolo, 1970).

[16] Corresponding to 20% of Montecatini's capital before the merger. IRI was the huge industrial state holding created in 1933 as a result of the bailing out of the largest Italian universal banks. See Vera Zamagni, *An Economic History of Italy 1860-1990* (Oxford: Clarendon Press, 1993), chapter 9.

[17] It might be useful to recall that in Italy, as well in Germany and in Japan, there is a tradition of cross-ownership of companies by the part of other companies (and banks). The "public" is either absent (in most cases) or not influential. In other words, in Italy there has been no tradition of public companies.

was done with the aim of changing the management of the company,[18] but it was the beginning of a tortuous period in which the government tried without the slightest success to put the chemical industry on some reasonable track.

The change in Montedison's management, the coordination between the two souls of the new company, and the relationship between public and private owners made management of the company exceedingly complex, and the company continued to register losses. But what was worse was that Montedison's weakness, and the indecisiveness of ANIC-ENI, which had lost its big boss, Mattei, in an air crash in 1962, gave the new contestants, SIR-Rumianca and Liquichimica, the go-ahead to make massive investments. The complacency of members of the government who granted large amounts of subsidy authorizations for the enlargement of capacity, on the basis of which the state-owned financial institutions were lavish with soft and ordinary loans compounded the problem. Characteristically, funds were requested by a host of small, legally separate, but technically interconnected, companies and granted in successive waves so that each company received at any one time a small amount of money. This was done primarily because subsidies had to be granted by law to small companies.[19] But the financial institutions looked favorably on this expedient because it allowed separating the responsibility for guaranteeing each plant and ensured that construction of the plant proceeded effectively. The subsequent loan was, in fact, not granted if the previous one had not been used already. But this was an approach that favored carelessness by both the industrial companies and the financial institutions, lack of coordination, and a low level of transparency in the overall planning of plants.

Even the chemical plan that the government issued in 1971 did not interrupt this rush because there was no way to win the collaboration of the major companies in the effort to develop coordination and diversification. Each company remained busy securing through political lobbying the largest possible share of capacity authorizations and subsidies.[20] Only

---

[18] Alves Marchi and Roberto Marchionatti, *Montedison 1966-1989* (Milan: F. Angeli, 1992), p. 44.

[19] The soft loans and the capital subsidies admitted by the bills 634/1957 and 634/1959 for the industrialization of the South were aimed at small and medium-size companies, which were defined in legal terms without considering economic, technical, and financial integration. Only SIR-Rumianca created almost fifty companies to abide with these rules.

[20] R. Giannetti, "Imprese e politica industriale. La petrolchimica italiana negli anni '70,'" in in L. D'Antone, *Radici storiche ed esperienza.*

Table 12.3. Data on the Major Italian Chemical Companies in 1971

	Montedison	ANIC	SIR-Rumianca	SNIA	Italy
Capacity in basic chemicals (kton)	3,910	1,463	1,120	15	7,179
Capacity in inorganic products (kton)	5,005	1,395	716	125	7,362
Capacity in plastic products (kton)	1,210	265	400		2,245
Capacity in fibers (kton)	250	58	21	245	653
Turnover of the entire group (bln lire)	2,023	226	222	368	—

Source: Report of the chairman of Montedison, suppl. to *Mondo Economico*, 30 September 1972.

at the end of 1972 was there a parliamentary inquiry into the chemical industry. On that occasion, hearings were held with the chairmen of the most important companies from whom it was possible to glean the interesting data visible in Tables 12.3 and 12.4. Table 12.3 shows how large Montedison was with reference to the other companies: It held 54% of the capacity in basic chemicals, as much as 68% of that in inorganic chemicals (this as a result of its long past history); 54% of the capacity in plastic materials, and 38% of that in fibers (where the champion was SNIA, a company that was not, however, doing very well because of its attachment to the prewar technology of artificial fibers from cellulose). Although Liquichimica is not even cited in the tables because its size was far too small, its chairman was invited to the hearings. By its own admission, in 1971 it boasted a turnover of eighty billion lire,[21] most of which was in gas!

Table 12.4 shows that authorizations to new investments, most of which were subsidized, were subdivided in a much different proportion, with SIR-Rumianca on top, ANIC next, and Montedison placing only third. Montedison's chairman commented that in the case of SIR-Rumianca, this investment effort was 347% of the stock of capital already in existence at the end of 1969, and, in the case of ANIC, 127%. In the case of Liquichimica, the ratio could not even be produced, while in the case of Montedison, this was a much more reasonable 20%. The different risk of the new investment is apparent. Notwithstanding, the chairman of SIR was insisting that Montedison was in crisis, and that the State had

---

[21] Report of the chairman of Liquigas-Liquichimica, *Mondo Economico*, suppl. 11 Nov. 1972.

Table 12.4. Amount of Chemical Investments Approved by the
State between 1969 and 1972 (Billion Current Lire) Billion Lire %

SIR-Rumianca	716	32
ANIC	604	27
Montedison	501	22
Liquichimica	205	9
Other	225	10
Total	2,251	100

Source: As for Table 12.3.

a "duty" to guarantee "pluralism." A really bold passage in his report
demonstrates how strong his political ties must have been:

> SIR does not accept the "role of subordinate chemical company." SIR is ready,
> as its past witnesses, to live its future in a milieu of fair competition, on a
> foot of equality and freedom with all companies, as it is guaranteed by our
> Constitution.[22]

The hearings convinced the government to cut subsidies to the chemical industry as a whole but were absolutely useless in terms of proving
that there was little solidity in the investment plans of a number of the
companies involved. With the first oil crisis, the drama exploded. Montedison and ANIC-ENI reported enormous losses (see Table 12.5), but their
size and experience allowed them to hold their ground. SIR-Rumianca
and Liquichimica, instead, became helpless, caught in the middle of
their huge enlargement plans. Given the disproportionate involvement
of state-owned financial institutions in their funding, another perverse
effect was that these institutions, partly to conceal their mistakes in the
hope that the situation would soon improve, and partly because of political pressure to avoid disclosing bankruptcies, continued to keep the
failed companies afloat, using far too large a share of their resources in
the process.

Four of the long-term state-owned financial institutes that were
strongly involved in financing the chemical bubble from the middle of

[22] From the report of the chairman of SIR, *Mondo Economico*, suppl 18 Nov. 1972, p. 6. "La
SIR non è disponibile per entrare nel 'ruolo industrie chimiche subalterne': è disponibile,
come il suo passato testimonia, per vivere il proprio futuro in un clima di leale concorrenza,
su un piano di uguaglianza e libertà tra tutte le imprese, come è previsto dalla nostra
Costituzione."

Table 12.5. Net Gain or Loss as a % on Turnover

	1974	1975	1976	1977	1978	1979	1980	1981
Montedison	3.5	−3.8	−2.2	−16.6	−8.1	0.0	−4.9	−126.3
ANIC	1.1	−6.3	−13.3	−24.0	−19.8	−4.4	−7.7	−26.1
Bayer	4.1	3.7	3.9	3.2	3.1	3.3	2.9	3.0
Hoechst	3.8	2.8	3.8	2.6	2.8	3.2	3.3	2.9
ICI		6.9	6.7	5.4	7.0	8.2	0.2	3.3
Hercules		12.6	13.7	6.6	7.2	7.8	5.9	10.1

Source: C. Cazzola et al., *La crisi dell'industria chimica italiana: raffronti internazionali*, Milan, Angeli, 1984.

the 1960s to the second half of the 1970s were: ICIPU, CIS,[23] IMI, and Mediobanca. Asso and De Cecco[24] give a clear picture of what ICIPU did (see Table 12.6). ICIPU[25] gave large loans to all chemical companies. The amounts granted to SIR-Rumianca and Liquigas are truly amazing but were only part of what these firms were able to get.[26] Montedison's long-term debts went from 644 billion lire in 1971 to 1,598 billion in 1979,[27] whereas short-term debts also went from 928 billion lire to 2,126 billion versus 743 billion of own capital. In addition to ICIPU, Montedison was financed mainly by Mediobanca and also by IMI. In late 1977, IMI declared that it had a standing credit of 300 billion lire with Montedison, 230 billion with ANIC, and 1,069 billion with SIR-Rumianca,[28] whereas it had no stake in Liquichimica. On December 31, 1977, SIR-Rumianca reported 2,043[29] billion lire long-term debts, and 1,017 billion in short term debts, versus 245 billion of own capital. By

---

[23] Some of the activities of CIS in the financing of Sardenian chemical industry are illustrated in G. Piluso, "Il Banco di Sardegna 1953–1994," in G. Toniolo, ed., *Storia del Banco di Sardegna* (Bari: Laterza, 1995).

[24] P. F. Asso and M. De Cecco, *Storia del Crediop. Tra credito speciale e finanza pubblica* (Bari: Laterza, 1994).

[25] ICIPU was an institute founded in 1924 by Alberto Beneduce to finance public utilities. Since 1963, its orientation changed toward industrial finance and the chemical industry especially.

[26] It appears that Liquigas, which was mostly financed by ICIPU, received loans amounting to as much as 1,228 billion of current lire in a period of six years.

[27] They continued to rise later to 1,977 billion lire in 1981.

[28] AsIMI, Verbali del Consiglio di Amministrazione, v. 18, 5 Oct. 1977, p. 105, and 12 Jan. 1978, p. 145. IMI had devoted an increasing amount of its industrial credit, a field in which IMI was a champion among the state-owned, long-term financial institutes, to metallurgy and chemicals. The chemical industry accounted for 20% of total IMI credit already in 1958–63 and reached 28% in 1970–78. See IMI, *Misure e modalità di trent'anni di sviluppo industriale in Italia* (Rome: 1986), p. 94.

[29] AsIMI, Verbali, 16 May 1978. IMI had a share of little more than 50% in the indebtedness of SIR-Rumianca.

Table 12.6. Loans by ICIPU to the Chemical Industry 1963-1977
(Billion Current Lire)

	Soft Loans	Ordinary Loans	Total
SIR-Rumianca	390	131	521
Montedison	236	275	511
Liquigas	127	181	308
ENI	77	30	107
Total	831	617	1,448
% chemical loans on total ICIPU loans	20	15	35

Source: Asso-De Cecco, p. 581.

the end of 1978, the situation had become much worse, as can be seen in Table 12.7.

By 1979, ICIPU was bankrupt and IMI was in dire straits.[30] Not without major drama, they were rescued by the state. ENI was put in charge of absorbing those plants of SIR-Rumianca and Liquichimica that could be rescued.[31] This increased the difficulties of the state-owned company and enlarged its losses in the early 1980s, burdening it with a host of inadequate plants scattered all over the place,[32] precisely during the second oil crisis. In 1982, ENI formed a new petrochemical division called Enichimica (changed to EniChem in 1985) and financed the restructuring operation largely through the substantial profits of its main oil division.[33]

While ENI was busy digesting SIR-Rumianca and Liquichimica, the political milieu conceived the idea of reprivatizing Montedison (1981) and building up an agreement between ENI and Montedison that would strengthen their respective market position. This agreement, drafted by Montedison in April 1982 and substantially endorsed by ENI, was signed in March 1983.[34] The results of the reorganization of the two groups

[30] Mediobanca, too, denounced major losses: −3.4% of turnover in 1975, −3.1% in 1976, −11.7% in 1977, −7.7% in 1978, −0.5% in 1979, −3.5% in 1980, −8% in 1981.

[31] Many of these plants, especially those of Liquichimica, were never opened!

[32] ENI, *I problemi e le prospettive dell'ENI*, Libro bianco presentato dal presidente dell'ENI al Parlamento, Rome, 1983.

[33] It must be noted that the chemical division of ENI accounted only for 1% of the revenues on average, although in many years it accounted for a disproportionate amount of ENI's budget losses.

[34] Similar types of restructuring were taking place elsewhere in Europe. See Ashish Arora and Alfonso Gambardella, "Evolution of industry structure in the chemical industry," in Ashish Arora, Ralph Landau and Nathan Rosenberg, eds., *Chemical and Long-Term Economic Growth. Insight from the Chemical Industry* (New York: Wiley, 1998), pp. 379-414.

Table 12.7. Standing Credits to SIR-Rumianca Dec. 31, 1978 (Billion Current Lire)

IMI	1,234	54
CIS	269	12
ICIPU	228	10
ICCRI	323	14
7 other banks	224	10
Total	2,278	100

Source: AsIMI, *Verbali del Consiglio di amministrazione*, vol. 20, May 14, 1979, p. 3.

in the petrochemical field are summarized in Table 12.8. Montedison also received a large injection of liquidity. However, plants remained too numerous and dispersed, and the ex-post reorganization of product lines meant that the same plant could be producing for more than one company. For example, the Montedison Ferrara plant came to host also EniChem Polimeri, producing LD polyethilene and ABS.

Paradoxically, it was not the case that the growth of the Italian chemical industry had been excessive by world standards. In fact, the Italian share of the chemical production of the Western industrially advanced world, which had reached 5.6% in 1962, went down to 4% at the end of the 1970s.[35] The problems were the lack of profitability of plants built piecemeal, which prevented an adequate competitiveness in both the domestic, and the international markets, and the excessive number of players. Notwithstanding the costs and the pending problems, the 1983 agreement really seemed the end of the nightmare. The public-private duopoly seemed to satisfy everybody for the time being and both companies had recovered profitability by 1984.

ENI-MONTEDISON'S FAILED EFFORTS AT CONSOLIDATION, 1980s–1990s

Although EniChem lived its modest life within the big oil company, ENI, Montedison in the first half of the 1980s engaged in reorganizing itself as a truly international corporation under the chairmanship of Mario Schimberni. In 1980, Montedison was still the tenth chemical company in the world by size of turnover, after three German, one British, one

[35] L. Morandi and G. Pantini, *Dialogo sull'industria chimica. Vie e modelli di sviluppo* (Milan: Etas libri, 1982), p. 20.

Table 12.8. Market Situation of ENI and Montedison After the 1983 Agreement (% Share of the European Market)

	ENI	Montedison	Major European Competitor
LD polyethylene	13		Basf 11
HD polyethylene	10		Hoechst 25
Polyvinylchloride	16		Solvay 15
Polypropylene		19	Hoechst 15
Polystyrol		17	Basf 24
Abs	15		B.Warner 25

Source: Marchi-Marchionatti, p. 189.

Dutch-British, and four U.S. companies.[36] Between 1981 and 1983, new people with international experience took over 80% of the high-level management positions. A joint venture with Hercules was signed in 1983 to form Himont, a world leader in the production of third generation polypropylene.[37] The rest of the company was reorganized as shown in Table 12.9.

There was an effort also to move out of basic chemicals into specialties, but it could not be as rapid a process as was desired. Because of the very negative burden of liabilities, the investment effort in research and new plants was inadequate, both in Montedison as well as in ENI. Schimberni, however, was confident in the future and wanted to make Montedison one of the first Italian examples of a public company. He did not consider that times had not yet changed in the Italian financial milieu; indeed, while he was cultivating his dream, someone else was thinking differently.

Between October 1986 and March 1987, Montedison was taken over by Ferruzzi, an agroindustrial company led by Raul Gardini.[38] The operation increased the indebtedness of Ferruzzi, without relieving that of Montedison (in spite of the sale of the department store Standa). In

---

[36] ANIC was not included in the first twenty companies. Its importance grew after the absorption of SIR-Rumianca and Liquichimica and after the 1983 agreement with Montedison. In 1985, Montedison was eleventh in the world for R&D expenditure, and Enichimica fourteenth. See G. Bertini, P. Delmonte, and G.Rosa, "Ristrutturazione e cambiamento nell'industria chimica: l'esperienza degli anni Ottanta," *Rivista di Politica Economica* (1987): 905-32.

[37] Hercules sold its participation to Montedison in 1987.

[38] Raul Gardini was the son-in-law of Serafino Ferruzzi, the founder of the Ferruzzi company, who died in an airplane accident in 1979. See the entry "Gardini" written by Franco Amatori in the Italian Biographical Dictionary.

Table 12.9. Montedison in Mid-1980s

Sectors	% Turnover	Divisions or Affiliates
Basic chemicals	38	Montepolimeri (up to 1983, then Himont), Montedipe
Fibers	7	Montefibre
Fertilizers and other products for agriculture	7	Fertimont (up to 1986, then Agrimont), Farmoplant
Specialties, dyes	19	Ausimont, Montefluos, Acna
Pharmaceuticals	9	Farmitalia-Carlo Erba (up to 1986, then Erbamont), Antibiotics (acquired in 1987)
Energy	2	Selm
Services and sundries	18	Iniziativa Meta, Standa (sold in 1987), Fondiaria (acquired in 1986), Technimont

Source: My elaborations from Marchi-Marchionatti.

December 1987, Gardini took over from Schimberni as the chairman of Montedison and conceived the idea of a second agreement with ENI to further rationalize basic productions to reap all possible economies of scale. At the time, Montedison still ranked seventeenth among the world chemical companies by turnover.[39] In January 1989, the joint venture Enimont was born, with 40% ownership by each of the two partners, plus 20% left to the market, with the understanding that the two major partners would not acquire any further slice of the company, at least until 1992. The world positioning of Enimont in various productions is reported in Table 12.10. It can be seen that Enimont was the leader in six out of fifteen products, coleader in one, and second in three. This allowed Enimont to be included among the top ten world chemical groups, an acceptable position that later would be strengthened by the acquisition of other divisions of Montedison and ENI, by new investment, and by internationalization.

Not all problems, however, were solved. Dispersion of plants, indebtedness, complementarities with the divisions left in the two separate companies, even some omitted competitors (Himont, world leader in polypropylene) were all unsettled problems. Above all, public-private relations had not found an equilibrium. Gardini was not a man to let the

---

[39] See Fred Aftalion, *History of the International Chemical Industry* (Philadelphia: University of Pennsylvania Press, 1989).

Table 12.10. Shares of Capacity in Western Europe in 1989

	Leader	**Second Producer**	**Enimont**
Ethylene	Enimont	Dow 9	10
Benzene	Shell 13	Dow 11	9
Butadiene	Shell 17	BP 14	13
Phenol	Phenol Chemie 33	Enimont	23
Styrol	Dow 21	Basf 15	12
Caprolactame	Basf 34	Enimont, DSM 24	24
N-paraffins	Enimont	Exxon 15	48
Intermediates	Enimont	Petresa 20	30
Polyethylene	Enimont	BP 10	12
PVC	Enimont, EVC 20	Solvay 16	20
Polystyrol	Basf 22	Enimont, Dow 18	18
Elastomers	Enimont	Shell 15	27
Acrylic fibers	Enimont	Courtaulds 23	27
Polyester fibers	Hoechst 23	Enka 18	12
Fertilizers	Norsk Hydro 17	Basf 12	8

Source: My elaborations from Marchi-Marchionatti.

opportunity of reunifying the chemical industry under Montedison go untried. First, he persuaded some friends to buy most of the 20% shares of Enimont left on the market. In March 1990, he and his friends held the majority of Enimont capital, and he declared his readiness also to buy ENI's share, thus becoming the boss of the Italian chemical industry. The press quoted him as saying: "I am the Italian chemical industry."

The problem was how to amass the money to pay for this project. Gardini lobbied Parliament for a huge fiscal bonus, which, in the end, was not granted. Banks made clear that Montedison and Ferruzzi indebtedness was already too large. In November 1990, Gardini had to admit that he had lost the game, and there was nothing else to do but sell his share of Enimont to ENI, at a price that everybody considered too high. It was the high point in corruption of the Italian political parties, and the amount of money paid in bribery on that occasion was the largest in absolute terms for any corrupt act that was uncovered.[40] For the second time in ten years, ENI had to reorganize its chemical division, again renamed EniChem. The capital and running costs of this operation were

---

[40] A sad drama took place in that connection. Clean Hands - the judges offensive to put the Italian political corruption under control - arrested the head of ENI, Cagliari, who committed suicide after 103 days in prison. Gardini, instead, preferred to kill himself before being arrested on 23 July 1993.

Table 12.11. EniChem Performance 1992–1999 (Million Euros)

	1992	1993	1994	1995	1996	1997	1998	1999
Turnover	5,663	5,503	5,969	7,201	5,283	4,985	4,048	4,096
Net profits before taxes	−806	−1,378	−434	566	117	124	−229	−553
No. of employees	32,963	28,913	23,501	21,358	16,839	15,950	14,442	13,908

Source: ENI, *Yearly Reports.*

enormous and are shown in Table 12.11. Many plants were closed and employment was drastically cut, with dramatic social problems.[41] Financially, ENI did not have serious problems, because it could finance the operation with its oil profits. EniChem concentrated on the core business, selling the plants for fertilizers and other products for agriculture, fibers,[42] and detergents, mostly to foreign multinationals, and retaining a strong position in ethylene, polymers, and elastomers. The share of its chemical division in its total turnover was 10.5% in 1999 and remains pretty small by international comparison, whereas it is still today (2003, with the name Polimeri Europa) the largest chemical company in Italy.

It is highly instructive to note that ENI, commenting on the very negative results of 1998–99 that destroyed the illusion of having returned Enichem to profitability, after having quoted the negative international factors, wrote: "The negative international trend is worsened by the structural problems connected with a too large number of scattered plants, with less than optimal size and fixed costs that are higher than those of our competitors. It is inevitable, therefore, to act for the further rationalization of production, concentrating it in those plants that, thanks to the appropriate size and the favourable location, will allow to produce in conditions of competitiveness."[43] The original sin of the Italian petrochemical industry is not overcome yet! In later years, it proved impossible to reverse the situation, and today ENI is trying to sell its chemical division.

If Enichem reorganization proved difficult, that of Montedison ended in disaster. The company was left with Himont (polypropylene), the pharmaceutical divisions (Erbamont and Antibioticos), the specialties division (Ausimont), energy, and services (especially the insurance

---

[41] In some cases, such as in the Crotone plant, there were even serious riots against the closing of plants.

[42] Montefibre was sold in 1996 to the Italian group, Finlane, and is still in operation.

[43] ENI, *Yearly Report,* 1999.

company, Fondiaria). Both Montedison and its mother company Ferruzzi had enormous debts.[44] The banks, which bore most of the debt burden, with Mediobanca on the forefront, merged the core business of Ferruzzi (Eridania-Beghin-Say) into Montedison, and Montedison into a new company, Compart (including an insurance division with Fondiaria and a shipbuilding division). The new management aimed at decreasing debt by selling all the pieces that could be sold. Fondiaria was first; then the pharmaceutical sector was sold to Pharmacia between 1993 and 1996. The jewel of the company, Himont, was not considered for sale at first. In 1995, there was even a last effort to strengthen its line of production, through a joint venture with Shell, producing Montell, by far the largest world producer of polypropylene. But in September 1997, Montell was sold to Shell for 1,860 million Euros.[45] Ausimont, focused on the production of fluoropolymers, meforex, and perossides, was sold in 2002 to Solvay. Today, not only is Montedison no longer a chemical company, but its name has been discarded. The agroindustry business (formerly, Ferruzzi) also was sold and the only remaining sector was energy, renamed Edison, a company that is today a major player in the process of privatization of electricity in Italy. History has come full circle, with the "revenge" of the electrical component of Montedison, the only one that has remained in existence! This demise of chemicals by Montedison might be considered inevitable if seen from the angle of the years after Gardini, but it was certainly not inevitable before. The glorious Ferrara plant has been minced up among several corporations: Basell, EniChem, Norsk Hydro, Polimeri Europa (joint venture between EniChem and Union Carbide), Snam (ENI), Edison, and the minor Crion and P-Group and is always on the brink of having lines of production shut down or displaced.

What is left of the Italian chemical industry? After the exit of Montedison and the confinement of EniChem to the lines of production connected with its oil business, the Italian chemical industry is mostly made up of small and medium size firms, in the style of the country's industrial structure (see Graph 12.1). The total domestic turnover of the sector is comparable to the United Kingdom's, but, in the United Kingdom, there is at least ICI and Astra-Zeneca, with their foreign subsidiaries, whereas nothing similar exists in Italy. An interesting recent report on the chemical

---

[44] The press talked of 15,500 million Euros!

[45] Shell formed in 1997 a joint venture with BASF that finally took over Montell in 2001, changing its name to Basell.

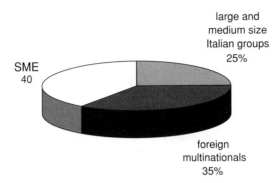

Graph 12.1. Share of production in the Italian chemical industry in 2003.
Source: 2004 Report by Federchimica (pharmaceuticals are not included). The large and medium-size Italian companies include thirty-four groups with >100 million Euro turnover. Only three of these have >1 billion Euros and five >500 million turnover.

industry by the research center Confindustria CEPS[46] underlines the fact that, although the chemical industry is still a sector of strong European leadership, American corporations are more profitable than the European ones and, having more funds for investment, are threatening that leadership. The reasons for this gap in profitability are higher labor costs and lower productivity, higher energy costs and, above all, lower capital intensity in Europe. This, in turn, has an explanation in "the peculiarity of the United States chemical industry, which has a more homogeneous and larger market, leading both the chemical industry and manufacturing as a whole to adopt more capital-intensive processes able to fully exploit economies of scale, and to create a larger quantity of standardized products than in Europe."[47] Now, this diagnosis applies completely to the Italian chemical industry, which is certainly the one in Europe showing the denounced shortcomings to the highest degree. The report suggests that the reaction in Europe to this has been "the widespread diversification towards chemicals with higher value added, i.e. specialty and fine chemicals able to meet customer/consumer need."[48] Can we say that this has been the reaction in Italy as well?

Table 12.12 gives a detailed picture of the present structure of the Italian chemical industry, where it can be seen that basic products

---

[46] CEPS, *Chemical Industry and Regulation*, by V. Maglia and C. Rapisarda (Rome, Sept 1999), mimeo.
[47] Ibid., p.35.
[48] Ibid.

Table 12.12. The Structure of the Italian Chemical Industry in the
Early Twenty-First Century

	Production in 2003 (Billion Euro)	Value Added in 2003 (Billion Euro)	No. of Operating Units 2001	Employment 2001 (000)	Trade Balance 2003 (Billion Euro)
Basic chemicals	19.0	3.1	461	52	−7.4
Fibers	0.8	0.1	51	9	−0.5
Paints, vaernishes	4.3	1.0	351	19	+0.4
Detergents, cosmetics	5.9	1.3	374	25	+0.9
Chemicals for agriculture	0.7	0.2	31	2	−0.1
Other chemicals	15.2	3.7	469	26	−1.7
Pharmaceuticals	19.8	6.9	357	71	−1.0
Total	65.8	16.3	2094	204	−9.4

Source: My elaborations from Istat and Federchimica.

perform worse than secondary products, but that only two out of the
seven categories of products listed – namely, paints, detergents and cosmetics – have a surplus export over import. With the exception of detergents, where large foreign multinationals produce on the Italian domestic
market because costs of long-distance transport are too high, the other
two sectors are typically branches where SME can prosper, but SME with
not too advanced technology.[49] On the whole, Italy seems to concentrate not on specialties but in niches not requiring advanced research
and high technology. This is hardly a surprising finding, given the general situation of the country! Further study of the matter is not the subject
of the present chapter and is left to future research.

In conclusion, the sad story of the Italian petrochemical industry
is the result of a host of deficiencies in the Italian manufacturing system. Although these problems can crop up anywhere, they usually do
not occur all at the same time. For an unfortunate destiny, the Italian
chemical industry combined them all. They included the problem of the
large corporation: Italy does not have a talent for large corporations. The

[49] There is an interesting report on these emerging medium-size chemical companies produced by Andrea Colli. See Federchimica, *Il volto nuovo della chimica italiana. Gli
approfondimenti, part II Dinamiche di sviluppo delle medie imprese chimiche*, March
2004. Also, on the same topic, A. Colli and V. Maglia, "Medie e piccole imprese nella
chimica italiana", *L'Industria*, XXVI (2005), no. 2: 321–360.

millennia of economic activity performed on an artisanal level, and local civic traditions have worked against the build up of large corporations (Italians prefer to work on a self-employment basis) and large plants (each city wants to have local economic activity). Managerial talents for governing large corporations have, accordingly, been very slow in developing, with a few exceptions (Fiat, metallurgy, oil). Now, it is recognized by everybody that the chemical industry cannot be run effectively outside the large corporation.[50] The Italian story is an example of this, demonstrating that the lack of success in building a viable Italian large corporation doomed the sector to failure. In contrast, in all fields where the large corporation is not indispensable, Italy could organize itself in a reasonable, and sometimes outstanding, way. State intervention presented an additional problem. Again, in the style of the country, state intervention has always been spread out "like rain" (to translate literally an Italian expression), that is, offering small subsidies to a large number of agents. In the case of the chemical industry, in spite of the repeated advice that the state should behave differently, this approach has proved devastating. Subsequently, the public-private relations that in other cases were played cooperatively, in this case were played very conflictually in the duopoly ENI-Montedison, up to the final tragic denouement. The conclusion to the book by Arora, Landau, and Rosenberg, states, with reference to Japan:

It appears that cozy relations between politicians, the financial institutions . . . and private companies can lead to misallocation of capital and possibly corruption, which the more impersonal markets are likely to avoid. One consequence of this misallocation seems to have been the creation of overcapacities and excessive real estate building.[51]

This statement can be applied very well to the Italian chemical industry. Characteristically, the Italian state has been responsible for excessive fragmentation of the Italian chemical industry and the excessive build up of capacity. All the later setbacks are connected to this earlier mistake, including the fact that the drive by Italian companies to be international players has been far too modest. The problem is that the strategy of international mergers and the formation of international groups in the

---

[50] Among the many writings on this issue, see the essay by Landau, in Arora-Landau-Rosenberg, *Chemicals and Long-Term Economic Growth*. "Large companies are needed not only for marketing and manufacturing skills, but also to provide finances for research and development and for plant construction," p. 177.

[51] Ibid., p. 517.

chemical industry today is among the most pressing tendencies and appears to be crucial for the survival and profitability of companies. I think there is only one way out of this disgraceful legacy. The international chemical industry is today on the brink of another major technological revolution, with biochemistry and DNA manipulation in the forefront. Connections with previous technological complexes are loose, and a new start is not impossible. Italy should start up massively on the new lines. But the disheartenment as a result of its legacy, the persisting handicap when it comes to managing large corporations, and the lack of a government capable of backing innovative research might prevent success even for the future.

13

The Global Accommodation of a Latecomer

The Spanish Chemical Industry since the Petrochemical Revolution

NÚRIA PUIG

This paper analyzes the development of the Spanish chemical industry during the twentieth century.[1] One of the ten largest producers in the world for the past forty years, Spain has remained, however, a country without major, innovative companies, and therefore an eager importer of chemical technology. Using a business history perspective, this chapter examines how Spain's late economic modernization, changing industrial and trade policies, the lack of appropriate scientific institutions, and the strategies of both foreign and local companies hindered the formation of technological and entrepreneurial capabilities in the Spanish chemical industry during the second and third Industrial Revolutions. In order to state whether Spain can play some relevant role in the context of growing

[1] This chapter is based primarily on empirical research conducted in various public and private archives during the last six years. The author would like to thank especially those who helped her at Bayer Hispania Industrial (Barcelona), Bayer AG (Leverkusen), Ministerio de Asuntos Exteriores (Madrid), Schering AG (Berlin), Schering España (Madrid), Fundación Juan March (Madrid), Perfumería Gal (Alcalá de Henares), Antonio Puig SA (Barcelona), La Seda de Barcelona (Barcelona), Ajuntament de El Prat de Llobregat (Barcelona), Archivo General de la Administración (Alcalá de Henares), CEPSA (Madrid), BASF (Ludwigshafen), and Sociedad Española de Participaciones Industriales (Madrid). For the sake of brevity, archival references have been omitted, as have been references to fundamental books and articles on the history of the world chemical industry already discussed in other chapters. Financial support from two publicly funded Spanish research projects (DGES PB96-0301 and SEC2000-1084) is gratefully acknowledged.

European integration and globalization, some recent, although modest, success stories are briefly outlined.

The modern chemical industry emerged during the second Industrial Revolution. A typically scale-oriented, resource-based, and capital-intensive industry, it soon was represented by a few large international firms with headquarters in the most industrialized countries of the world. Since the late 1930s, oil rapidly substituted for coal as the industry feedstock. Massive investments in both the chemical and oil refining industries gave rise to an unprecedented amount of basic upstream petrochemicals, and encouraged the international transfer of technology. New firms, within or outside the original countries, emerged and joined those founded in the previous hundred years. This scenario was dramatically changed by the oil crisis of 1973. Excess capacity and falling profits led most firms to implement alternative growth strategies in the following years. As a consequence, there was a shift from upstream production to downstream production, best represented by fine and specialty chemicals. Again, most of the veteran companies were the main actors in this change, transforming themselves from producers of basic chemical commodities into high-tech manufacturers, and undergoing a series of concentration processes.

The world chemical industry can be described as a dynamic oligopoly because both its geography and corporate identity have remained strikingly stable during the past 150 years. How firms create their corporate capabilities while continuing to innovate and adapt to new challenges constitutes one of the most interesting issues in the history of the chemical industry. Although the two postwar periods weakened the once monopolistic position of German companies, providing opportunities for already advanced nations as well as for newcomers such as Japan and Italy, the world chemical production continues to be highly concentrated in a few countries and companies. The ongoing transition toward a knowledge-based, research-intensive industry obviously will accentuate this feature of the chemical industry.

Spain does not belong to the founding nations of the modern chemical industry. During the last forty years, however, around 2% of the world chemical output has been produced in Spain, which ranks eighth in the world. Its relative position also has gradually improved, particularly when compared to Italy. Yet, when compared with Italy, whose industrial dismantling is analyzed in another chapter in this volume, what is most striking about the Spanish industry is that not a single Spanish company has ever made a relevant contribution to the major innovation cycles of the chemical industry. Not a single Spanish company, therefore, is to be found among the largest firms of the world. The main goal of this chapter is to provide a convincing explanation for this particular path.

In trying to understand why Spain did not succeed in becoming a first comer to the European industrial revolutions, most scholars have pointed to the insufficient size of the domestic market (and thus the slow development of a poor, agrarian country), as well as to the controversial industrial and trade policies implemented by successive governments (and, thus, the attitudes of bureaucrats and entrepreneurs). Natural conditions (resources and communications) and the size and quality of the educational and scientific institutions have been considered (but not quantitatively evaluated), too. It is only recently that scholars have started to look at the role played by firms, both local and foreign, in this general process of frustration or retardation of the overall economic modernization. This chapter focuses on the evolution of the largest chemical firms during the past century. Their development offers sharp contrasts with that of the global first movers, and constitutes an excellent basis for understanding the peripheral, noninnovative role played by the Spanish chemical industry until now. Explicit comparisons with other latecomers that were able to join the world leaders during the twentieth century should help us to evaluate the effects of the previously mentioned factors on the Spanish chemical industry. It is the accumulation of technological and entrepreneurial capabilities that we will mainly address.

SPAIN AND THE INTERNATIONAL CHEMICAL INDUSTRY, 1880–1939

Until the 1930s, the Spanish chemical industry was represented by a few modern firms and a multitude of tiny, preindustrial establishments. We can identify the former in Table 13.1, in which the largest companies in 1959 have been ranked by the only indicator available, capital stock. Some well-known international firms, such as Nobel, Solvay, and Griesheim, already had been established there by the turn of the century.[2] The table shows that seventeen firms had been founded before the Spanish Civil War. The Great War, during which Spain remained neutral,

---

[2] On Nobel, see Gabriel Tortella, "La primera gran empresa química española: la Sociedad Española de la Dinamita (1872-1896)," in *Historia económica y pensamiento social. Homenaje a Diego Mateo del Peral* (Madrid, 1983), 431-53. Solvay has been recently analyzed by Ángel Toca, *La industria química de los álcalis en España. La empresa Solvay y su planta de Torrelavega (1904-1935)*, Ph.D. thesis, Universidad Española de Educación a Distancia (Madrid, 2001). Griesheim has been explored by Javier Loscertales, *Deutsche Investitionen in Spanien, 1880-1920* (Stuttgart, 2002).

was decisive in accelerating the industrialization process in this Mediterranean country. The aim of foreign firms had been to exploit – with or without Spanish partners – local natural resources, and eventually to serve a small, although comfortable, market. Only two major Spanish enterprises stood along with those subsidiaries: Cros and La Unión Resinera Española (LURE). The former, founded by a French immigrant to the Barcelona area, produced sulfuric acid and other chemicals demanded by the growing Catalan textile factories.[3] LURE, in the Basque Country, was a major forest owner and a resin and turpentine producer.[4] The soap and perfume industry made a brilliant start with Perfumería Gal, founded in 1901 in Madrid.[5]

World War I gave breath to new initiatives in more modern fields such as synthetic dyes, nitrogen fertilizers, and pharmaceuticals.[6] The mighty I.G. Farben was present in the Electro-Química de Flix EQF (with Cros), and Fabricación Nacional de Colorantes y Explosivos FNCE, and had its own pharmaceutical firm (Química Comercial Farmacéutica QCF Bayer) and sales company (Unión Química Lluch UQLL, later Unicolor).[7] The

[3] The only available indirect studies on Cros are: Jordi Nadal, "La debilidad de la industria química española durante el siglo XIX. Un problema de demanda," *Moneda y Crédito* 186 (1986), 33–70; and "La consolidació pel biaix dels adobs. 1914-1939," in *Història econòmica de la Catalunya contemporània* 6 (Barcelona, 1993), pp. 149–66.

[4] Rafael Uriarte, *La Unión Resinera Española (1898-1936)*, Fundación Empresa Pública, Working Paper 9602 (Madrid, 1996).

[5] Núria Puig, "The Search for Identity: Spanish Perfume in the International Market, 1901–2001," *Business History* 45, 3 (2003), pp. 90–118.

[6] The synthetic dyes industry has been analyzed by Núria Puig, *Los orígenes de una multinacional alemana en España: Fabricación Nacional de Colorantes y Explosivos, 1881-1965*, Fundación Empresa Pública, Working Paper 9904 (Madrid, 1999); "El crecimiento asistido de la industria química en España: Fabricación Nacional de Colorantes y Explosivos, 1922-1965," *Revista de Historia Industrial* 15, 105–36 (1999); and "Business and Government in the Rise of the Synthetic Dyes Industry: the Case of FNCE, 1922-1965," *Proceedings of the 3rd European Business History Association, September 1999*, eds. Anne-Marie Kuijlaars, Kim Prudon, and Joop Visser (Rotterdam, 2000), 137–58. The origins of the nitrogen fertilizers industry can be traced in: Francisco Bustelo, "Notas y comentarios sobre los orígenes de la industria española del nitrógeno," *Moneda y Crédito* 63 (1957), 23–40. As for the pharmaceutical industry, it has been scarcely explored yet. Núria Puig, *La nacionalización de la industria farmacéutica en España: el caso de las empresas alemanas, 1914-1975*, Fundación Empresa Pública, Working Paper 2001/2 (Madrid, 2001).

[7] There are some commemorative histories of IG's subsidiaries: Sociedad Electro-Química de Flix, *Historia de Electro-Química de Flix SA, 1897-1965* (Barcelona, 1966); Pere Muñoz, ed., *Centenario de "la fábrica." De la Sociedad Electro-Química de Flix a Erkimia, 1897-1997* (Barcelona, 1997); Química Farmacéutica Bayer, *Apuntes para una historia de Química Farmacéutica Bayer SA* (Barcelona, 1992); Unicolor, *Historia de Unicolor* (Barcelona, 1967). See also Núria Puig and Javier Loscertales, "Las estrategias de crecimiento de la industria química alemana en España 1880-1936: exportación e inversión directa," *Revista de Historia Económica* XIX 2 (2001): 345–87.

Table 13.1. The Fifty Largest Spanish Chemical Firms in 1959

Firm	Founded	Location	Capital Stock (Million Pesetas)	Main Shareholders	Main Product
1. UNIÓN ESPAÑOLA DE EXPLOSIVOS (UEE)	1896	Madrid	1,000	Urquijo group	Explosives fertilizers
2. CROS	1904	Barcelona	780	Cros group	Inorganic chemicals fertilizers
3. ENERGÍA E INDUSTRIAS ARAGONESAS	1918	Madrid	561	Urquijo and Mateu groups	Fertilizers
4. SNIACE	1939	Madrid	524	Snia Viscosa Altos Hornos group	Chemical fibers
5. SOLVAY	1904	Barcelona	500	Solvay	Inorganic chemicals
6. UNIÓN QUÍMICA DEL NORTE DE ESPAÑA (UNQUINESA)	1939	Biscay	449.3	Lipperheide group	Organic chemicals
7. FABRICACIÓN ESPAÑOLA DE FIBRAS SINTÉTICAS (FEFASA)	1940	Madrid	434	National Institute of Industry	Chemical fibers
8. SEFANITRO	1941	Biscay	312	Altos Hornos and Lipperheide groups	Fertilizers
9. HIDRO NITRO	1940	Madrid	350	Altos Hornos group	Fertilizers
10. SOCIEDAD IBÉRICA DEL NITRÓGENO (SIN)	1923	Madrid	350	Urquijo group	Fertilizers
11. COMPAÑÍA INSULAR DEL NITRÓGENO	1958	Madrid	300	CEPSA and Cros groups	Fertilizers
12. CARBUROS METÁLICOS	1897	Barcelona	225	Urquijo group	Industrial gas

	Year	Location	Amount	Group/Owner	Sector
13. NITRATOS DE CASTILLA	1940	Biscay	222.2	Altos Hornos group	Fertilizers
14. COMPAÑÍA ANÓNIMA ESPAÑOLA DEL AZOE	1941	Madrid	200	Spanish group	Fertilizers
15. COMPAÑÍA ESPAÑOLA DE INDUSTRIAS ELECTRO-QUÍMICAS	1942	Madrid	150	Spanish group	Inorganic chemicals
16. PRODUCTOS QUÍMICOS IBÉRICOS	1942	Madrid	150	Cros and Urquijo groups	Organic chemicals
17. INDUSTRIAS DEL ACETATO DE CELULOSA (INACSA)	1948	Madrid	165	Mateu group	Chemical fibers
18. FOSFORERA ESPAÑOLA	1956	Madrid	125	Cros and Spanish groups	Chemicals
19. LA UNIÓN RESINERA ESPAÑOLA (LURE)	1898	Biscay	125	Spanish group	Wood derivatives
20. CEPLÁSTICA	1953	Biscay	122	Lipperheide group	Plastics
21. ANTIBIÓTICOS	1949	Madrid	120	Spanish laboratories (Abelló, Ibys, Uquifa, Leti and Zeltia)	Pharmaceuticals
22. DERIVADOS DEL COK	1951	Madrid	120	Urquijo group	Organic chemicals
23. PRODUCTOS QUÍMICOS SINTÉTICOS	1944	Madrid	112.5	Urquijo group	Organic chemicals
24. FORET	1927	Barcelona	110	Foret family	Chemicals
25. LA SEDA DE BARCELONA	1925	Barcelona	100	AKU	Chemical fibers
26. PERLOFIL	1951	Madrid	100	Urquijo group	Chemical fibers
27. SOCIEDAD ELECTRO-QUÍMICA DE FLIX	1897	Barcelona	100	Cros group	Inorganic Chemicals

(continued)

Table 13.1 (*continued*)

Firm	Founded	Location	Capital Stock (Million Pesetas)	Main Shareholders	Main Product
28. SOCIEDAD ANÓNIMA AUXILIAR DE LA INDUSTRIA QUÍMICA	1941	Madrid	100	Cros group	Inorganic chemicals
29. COMPAÑÍA ESPAÑOLA DE INDUSTRIAS ELECTROQUÍMICAS	1942	Madrid	100	Spanish group	Inorganic chemicals
30. SANDOZ	1924	Barcelona	100	Sandoz	Pharmaceuticals
31. ELECTROMETALÚRGICA DEL EBRO	1904	Barcelona	100	Cros group	Inorganic chemicals
32. GENERAL QUÍMICA	1948	Biscay	90	Lipperheide group	Chemicals
33. SOCIEDAD ESPAÑOLA DE LA SEDA ARTIFICIAL	1930	Burgos	81	Alday family	Chemical fibers
34. INDUSTRIAS QUÍMICAS TEXTILES	1952	Madrid	80	Lipperheide group	Chemical fibers
35. COMPAÑÍA ESPAÑOLA DE PENICILINA Y ANTIBIÓTICOS (CEPA)	1949	Madrid	80	Urquijo group	Pharmaceuticals
36. MINERALES Y PRODUCTOS DERIVADOS	1942	Biscay	75	Lipperheide group	Inorganic chemicals
37. PENIBÉRICA	1956	Navarre	75	Lipperheide group	Pharmaceuticals
38. INDUSTRIAS QUÍMICAS CANARIAS	1940	Madrid	67.5	CEPSA group	Petrochemicals
39. LA ALQUIMIA	1940	Barcelona	66	Spanish group	Chemicals
40. COMPAÑÍA IBÉRICA DE DETERGENTES	1957	Madrid	60	Fierro group Lever	Detergents

				Spanish group	Inorganic chemicals
41. COMPAÑÍA ARAGONESA DE INDUSTRIAS QUÍMICAS	1942	Barcelona	60		
42. AGRA	1945	Biscay	60	Lipperheide group	Agrochemicals
43. ZELTIA	1939	Pontevedra	60	ICI Fernández family	Pharmaceuticals
44. FABRICACIÓN NACIONAL DE COLORANTES Y EXPLOSIVOS (FNCE)	1922	Barcelona	56	FNCE families BASF and Bayer	Dyes
45. SA ABONOS MEDEM	1921	Madrid	55	Cros group	Fertilizers
47. PRODUCTOS QUÍMICOS SCHERING	1924	Madrid	52.5	Urquijo group	Pharmaceuticals
48. UNIÓN ESPAÑOLA DEL ÁCIDO ACÉTICO	1952	Madrid	50	Urquijo group	Chemicals
49. PÓLVORAS Y ARTIFICIOS	1942	Madrid	50	Urquijo group	Explosives
50. ALTER	1939	Madrid	50	Alonso family	Pharmaceuticals
51. INSTITUTO DE BIOLOGÍA Y SUEROTERAPIA	1919	Madrid	50	Urgoiti family	Pharmaceuticals
52. FABRICACIÓN ESPAÑOLA DE ESPECIALIDADES FARMACÉUTICAS (FAES)	1933	Biscay	50	Urquijo group	Pharmaceuticals
53. DOCTOR ANDREU	1935	Barcelona	50	Andreu family	Pharmaceuticals

Notes: Only the most relevant Spanish groups mentioned in the text have been named in the table. The Basque heavy industry (Altos Hornos de Vizcaya) and the banks supporting stood behind many of the Spanish firms, particularly the large fertilizers manufacturers, had strong links with the Lipperheide group and some common investments with the Urquijo Bank.

Source: Anuario Financiero y de Sociedades Anónimas 1959 and author's elaboration.

Unión Española de Explosivos (UEE), a syndicate of Spanish explosive makers controlled by Nobel, went through a process of diversification (mainly in mining) and gradual nationalization.[8] LURE, which accounted for most of the (modest) Spanish chemical exports, also tried to diversify, entering into the chemical business through a short-lived partnership with the German Ruth. The major Swiss dye makers opened sales offices in the 1920s. The most impressive and successful development was that of Cros, which became a successful and vertically integrated manufacturer of phosphate fertilizers, with factories scattered all around the Peninsula. Like UEE, Cros went into the mining and explosives business, founding Cloratita, later sold to I.G. Farben. Energía e Industrias Aragonesas EIA and Sociedad Ibérica del Nitrógeno SIN, were the first producers of nitrogenous fertilizers, both with the financial support of local banks and foreign technology. They faced several difficulties, particularly international dumping. Carburos Metálicos was born in 1897 as a Swiss-Spanish joint venture and, like UEE, moved into Spanish hands in 1927. Foret, founded by a French immigrant, became a dynamic firm in the field of inorganic chemicals. IBYS, a creation of industry-minded scientists close to Ramón y Cajal, introduced in Spain the manufacturing of vaccines and serum.[9] Finally, artificial fibers were started by the Sociedad Anónima de Fibras Artificiales (SAFA), a Catalan joint-venture with the French Gillet group, and La Seda de Barcelona, a subsidiary of the Dutch concern AKU.[10]

Not only was the takeoff of the Spanish chemical industry dominated by foreign capital and technology, but it was very limited. By 1935, Spain was still a net importer of chemicals, the most relevant of which were nitrogenous fertilizers (about 50% of total imports, with natural nitrates from Chile and synthetic fertilizers from the Netherlands and the United Kingdom); pharmaceuticals (15%, mostly from Germany and, to a lesser degree, France and Switzerland); and artificial fibers (14%, with Italy as the largest provider, followed by the Netherlands and France).[11] Synthetic

---

[8] Gabriel Tortella, "La integración vertical de una gran empresa española durante la dictadura de Primo de Rivera. La Unión Española de Explosivos, 1917-1929," in García Delgado, ed., *Economía española, cultura y sociedad. Homenaje a Juan Velarde Fuertes* (Madrid, 1992), 359-93.

[9] Instituto de Biología y Sueroterapia IBYS, *Memorias 1919-1944* (Madrid, 1944).

[10] Sociedad Anónima de Fibras Artificiales SAFA, *Cincuenta años de la Sociedad Anónima de Fibras Artificiales SAFA* (Barcelona, 1973); Seda de Barcelona, *Història de la Seda de Barcelona. 75 anys* (Barcelona, 2000); Núria Puig, "Una multinacional holandesa en España: La Seda de Barcelona," *Revista de Historia Industrial* 21 (2002), pp. 123-58.

[11] According to official statistics, the ratio was 12:1. *Estadística del Comercio Exterior de España* (Madrid, 1935). See also Emilio de Diego, *Historia de la industria en España. La química* (Madrid, 1996).

dyes, one of the largest groups from the beginning of the century, had fallen to 6% of total imports, as a consequence of the I.G. participation in FNCE since 1927. The reverse of this import substitution process – common to all European countries in the interwar years – was a massive increase in the import of intermediates, particularly nitrated and chlorate derivatives. This would remain a characteristic of the Spanish chemical industrialization. The absence of quality coal, along with a narrow market, kept both local and foreign entrepreneurs away from the heavy organic industry.[12]

The rise of the modern Spanish chemical industry is usually associated with a law passed in 1917 to promote the national industry.[13] In Spain, as elsewhere, World War I marked the start of a long-lasting nationalistic industrial drive, successively approved by dictatorial and democratic governments until the outbreak of the civil war. Although its impact was extremely modest, the law, in fact, provided those "companies of national interest" (and most of the chemical business fell into this category) with the least expensive support available: tariff protection (more or less monopolistic), control over the domestic market, and limitation of foreign investment (up to 25% of capital stock). Thus, in Spain, state intervention was very weak when compared with other nations. There was no public investment (in firms or in research institutions), there were neither subsidies nor specific growth strategies, and the establishment of tariffs required tiring bargaining that, although trying to satisfy everybody, left everybody unsatisfied. Finally, foreign investment flourished (more at a commercial than at an industrial level) in spite of all efforts to boost national investments. Behind the protectionist drive stood the military establishment, the designer of Spanish industrial policy from 1917 through 1959.[14] Before 1939, however, its actual impact on the growth of the Spanish modern chemical industry did not go beyond its control of the bureaucracy created to implement the law, the elaboration of industrial statistics, and, last but not least, its role as consumer of war

---

[12] For a quantitative, comparative survey by branches, see Gary Goertz, *The World Chemical Industry around 1910: A Comparative Analysis by Branch and Country* (Geneva, 1990). The U.S. Department of Commerce seemed surprisingly hopeful about Spain, *The European Chemical Industry in 1932* (Washington, 1933), 49-52; and, later on, Alfons Metzner, *Die Chemische Industrie der Welt. Teil I: Europa* (Düsseldorf, 1955), 624-42. Ungewitter's very famous report instead regarded the coal deficit and the small market as serious handicaps: Claus Ungewitter, *The Chemical Industry* (Geneva, 1927).

[13] A good example is provided by the nationalistic treatise by Félix Suárez Inclán, *Industrias esenciales. Organización económica del Estado* (Madrid, 1922).

[14] Elena San Román, *Ejército e industria. El nacimiento del INI* (Barcelona, 1999).

chemicals. Foreign capital, finally, found its way into the chemical sector in spite of all legal restrictions.

The chemical business community, largely concentrated in Barcelona, made some organizational efforts and tried to influence the tariff policy. The most visible results were the first company directory and specialized journal, as well as a promising school of directors of chemical industries closed down in 1925 by the antiregionalist general Primo de Rivera, in power since 1923.[15] Up until the 1960s, when the sleeping university system went through a modernizing process, the only institution devoted to the education of industrial chemists and the implementation of (foreign) chemical technology was the modest Institut Químic de Sarrià, founded in 1916 by the Jesuits.[16]

The main stimulus for the nascent Spanish chemical industry of the first third of the century was therefore the sustained economic growth of the country following the Great War. One of the few outsiders to the gold standard, Spain grew inward-looking, and the business community identified itself with the dominant doctrine of economic nationalism, the granting of private monopolies over the domestic market becoming thus a major goal of entrepreneurial action. The chemical industry shared this view but had necessarily to look to the international chemical industry for technological, and eventually, financial partners. Indeed, the evolution of the world chemical industry – whose many cartels excluded Spain – had a stronger influence on the Spanish chemical business than the outlined policies.[17] The story of FNCE constitutes an excellent example of the many limitations faced by Spanish firms in the interwar period.

Founded in 1922 by five Catalan dye manufacturers, FNCE had three clear goals: to gain control over the Spanish market, to fight German takeover, and to diversify into further innovative fields.[18] After a rather disappointing start, its partners looked for strong international partners. In exchange for technological support, the Spaniards offered up the

---

[15] *Anuario de Industrias Químicas* (Barcelona, 1932); *Química e Industria* (Barcelona, 1924–37). Most of the initiatives of the time were due to Josep Agell, professor of organic chemistry at various technical schools and a highly respected leader of the Spanish chemical industry through the 1960s. After the civil war, this indefatigable organizer became managing director of the chemical textile firm SAFA.

[16] Núria Puig and Santiago López, "Chemists, engineers, and entrepreneurs. The Chemical Institute of Sarrià's impact on Spanish industry (1916–1992)," *History and Technology* 11, 345–59.

[17] League of Nations, *International Cartels. A League of Nations Memorandum* (New York, 1947).

[18] For FNCE's story, see references in note 6.

Spanish (and eventually Portuguese) market. The best deal went to I.G. Farben, which felt confident enough to require a share of 50% and full control of the technical department of the joint venture. By the end of 1926, and thanks to the diplomatic skills of FNCE's chairman, a former supplier of the Army, the Spanish partners had been able to get from the government the status (and the privileges) of a "national" enterprise. The deal with the I.G. caused enthusiasm among the Spanish partners, who saw more the promises of technical support and diversification than the limitations of an agreement with the world's first chemical firm. And there were many limitations. First, UQLL, the former Spanish sales offices of the I.G. and a 100% German company, was in charge of the commercialization of imported and locally manufactured products. Second, FNCE was to pay a fee to EQ de Flix, a Spanish electrochemical business owned by Cros and I.G. Farben. Flix, in turn, was to close down its own dyes plant (built during World War I) and supply several intermediates at fair prices. And, third, an additional fee was to be paid to the Unión Española de Explosivos after the German company and Kuhlmann had reached a "noninterference agreement," according to which UEE and FNCE gave up their respective plans of going into the business of dyes and explosives. In retrospect, such conditions reveal the diverging interests of both parties, as well as Spain's absence from the international cartels. In the following years, the Spanish partners worked hard to get the necessary approvals from Germany and the Spanish government to make their plans for vertical integration (to produce intermediates) and diversification (to build a plant of synthetic fuel) come true. The I.G. was never really interested in the former, for it constituted the backbone of its exports, now favored by a tariff reduction. As for the synthetic fuel project, although praised by the army engineers who, along with the Spanish directors, paid a visit to the Leuna works, it was never approved by the Spanish government.[19] In these and in other cases, FNCE's effort was strictly managerial and eventually political, not technological. This fact supported the German view, apparent in any examination of German files, that the Spaniards were not technically capable partners. FNCE, however, proved a great business for both sides. The firm had a share of 80% of the growing Spanish market, and most of the profits, in the absence of R&D expenditures, were distributed among its shareholders. A comparison with the

---

[19] César Serrano and Antonio Mayorga, *Síntesis de los compuestos nitrogenados. Hidrogenación de los carbones. Informe sobre la obtención del nitrógeno y sus derivados* (Madrid, 1929).

Italian synthetic dyes story reminds us at least of three things. First, that perhaps Cros (a mining and chemical business, like Montecatini, but not a modest firm like FNCE), had been a suitable interlocutor for the I.G. Second, that without minimal technological capabilities (far below those of the unique Fauser), FNCE was condemned to remain outside the international cartels. And, third, that without a consistent support from the state at various levels, FNCE could in no way become the "national" enterprise their founders and most of the military establishment had dreamed of.[20]

The outbreak of the Spanish civil war in 1936 changed the minds and plans of many of the Spanish chemical industry players. In spite of the initial violence and confusion, it was a sweet time for the chemical business, and especially for the Germans, who moved down from Barcelona and Madrid to Seville, where they kept on importing and manufacturing chemicals, explosives, dyes, and drugs for both the rebel and the legitimate armies.

THE "NEW STATE" AND THE CHEMICAL BUSINESS, 1939–1959

The advent of Franco's regime after a devastating civil war marked the end of liberal capitalism and the start of a new era of state intervention, fast industrialization under autarkical premises, and widely spread corruption.[21] Yet, the ruling civil and military authorities' desire for self-sufficiency did not affect the chemical sector as much as it did other strategic industries. From the very beginning, the rise of public enterprise came along with the survival of private firms and the persistent cooperation with foreign capital and technology. High officials and experts in the regime recognized the backwardness of the country and made many exceptions to the general rule that forbade foreign participation in any Spanish firm beyond 25% of the capital stock and human resources.

In Table 13.1, we can see some of the effects produced by the new situation in the entrepreneurial landscape. Most of the new companies were related to the manufacture of cellulose and artificial fibers (SNIACE, FEFASA, INACSA, and INQUITESA), fertilizers (SEFANITRO, Hidro-Nitro, CIN, NICAS, and CAE Azoe), and pharmaceuticals (Antibióticos, CEPA, Penibérica, Zeltia, and Alter). By declaring the three

[20] Franco Amatori and Bruno Bezza, *Montecatini 1888-1966. Capitoli di storia di una grande impresa* (Bologna, Il Mulino 1990). The Japanese experience points in the same direction: Akira Kudo, "I.G. Farben in Japan: The Transfer of Technology and Managerial Skills," *Business History* 36 (1994): 159-83.
[21] Jordi Catalán, *La economía española y la segunda guerra mundial* (Barcelona, 1995).

fields "of national interest," the government became a partner, or kept the right to intervene in SNIACE, FEFASA, and Hidro-Nitro, and organized public contracts to regulate the manufacture of penicillin. Two Spanish companies, Antibióticos and CEPA (working under Schenley and Merck licenses, respectively), were granted the exclusive right to import and manufacture antibiotics in the Spanish market. The state intervened further through the Ministry of Industry or the mighty Instituto Nacional de Industria (INI), a public holding created in 1941 to foster industrialization under autarkical premises and headed by the military engineer, Suanzes, until 1963.[22] Whereas INI marched impulsively through the power, mining, steel, and motor industries, paying little attention to public finances or the interests of private companies, it stayed apart from most of the chemical sectors. The only fully state-owned firm was the Empresa Nacional Calvo Sotelo (ENCASO), a pharaonic work aimed at producing synthetic fuel out of bituminous coal, located in Puertollano, around two hundred kilometers south of Madrid. From its foundation in 1942 with the initial assistance of German technology, particularly BASF and Lurgi, through the 1960s, ENCASO consumed almost one third of INI's budget, and did not produce a single drop of synthetic fuel. However, with the assistance of the French Oil Institute since 1953, the state-owned firm was able to manufacture hydrogenation lubricants at its new research institute. Although the private distribution monopoly, CAMPSA, refused to sell them, those lubricants were for a very long time regarded and bombastically praised as a national innovation. They constitute, as a matter of fact, the link between the erratic start of the public petrochemical industry and the recent accomplishments of Repsol, ENCASO's heir and a private firm since 1985.

The relation with the first Spanish oil company, CEPSA (Compañía Española de Petróleos), was not easy either. Created in 1929, immediately after CAMPSA, CEPSA had to build its first refinery out of the monopoly's reach in the Canary Islands.[23] Yet, thanks to its international connections,

---

[22] Even though it remains controversial whether INI pushed industrialization or interfered with private firms: Pablo Martín Aceña and Francisco Comín, *INI. 50 años de industrialización en España*, (Madrid, 1991); Antonio Gómez Mendoza, *De mitos y milagros. El Instituto Nacional de Autarquía (1941–1963)* (Barcelona, 2000). From the story of INI's carbo- and petrochemical adventure, there is little doubt about the monopolistic ambitions and obstructions of the public holding.

[23] On Spanish oil policy, see CEPSA's surprisingly informative commemorative history: CEPSA, *Biografía de una realidad 1929–1954* (Madrid, 1954); Stanford Research Institute, *Análisis económico de la industria petrolera española* (Menlo Park, Ca., 1965); José María Marín Quemada, *Política petrolífera española* (Madrid, 1978); and Gabriel Tortella,

particularly American, and its knowledge of the oil industry, CEPSA would
be invited to be an advisor and partner to most of INI's carbo- and petrochemical enterprises. The relationship between the private company and
the state-owned holding perfectly mirrors the industrial dynamics of the
time, since, in spite of its privileged position, CEPSA had to wait well
into the 1960s before it was allowed to build its own refinery and petrochemical complex in Spain's mainland,in Algeciras, nearby Gibraltar.

Following a very expensive, intricate, and controversial path, indeed,
the autarkical dream would end with REPESA, a profitable oil refinery in
the Mediterranean (founded as early as 1949 as a separate firm in partnership with CEPSA and Caltex); a highly controversial inland oil refinery
in Puertollano, designed in 1958 but not finished until 1966, parallel to
the shutdown of the coal mines; and an ambitious petrochemical complex erected around it. In contrast to the refinery, and following the
rules established by the more liberal-minded bureaucrats in office since
the late 1950s, the public, as well as various Spanish firms (Cros and
Foret), and some multinational concerns (Montecatini, ICI, Phillips, and
Atlantic Richfield and Halcon), El Paular (1961), Alcudia (1963), Calatrava
(1963), and Montoro (1968), all held shares in the new petrochemical
firms. Puertollano also became one of the landmarks of Franco's ideolog-
ical industrial policy, which focused on the industrialization of Spain's
backward areas at the expense of more advanced regions like Catalonia
and the Basque Country. Faraway from the seaside and Spain's industrial
districts, but close to Madrid, Puertollano subsists today as one of the
plants of the first Spanish oil and petrochemical group, Repsol.

More important than the direct intervention of the state was the
conflicting responsibilities among several ministries over prices, import
licenses, and distribution of raw materials for the chemical industry. Such
practices intensified the bureaucratic tasks of private management, and
gave rise to a new class of political managers, many of them holding
top positions in the administration and sitting on the boards of several
companies. The paradigm of the ubiquitous manager in the chemical
sector was Antonio Robert, an industrial engineer, top officer, and businessman. The author of an autarky-praising book, he became director
general of industry in the early 1940s; sat on the boards of SNIACE,
Hidro-Nitro, and SIN (joined by INI between 1942 and 1951), among
other state-supervised firms. He also was the managing director of

Alfonso Ballestero, and José Luis Fernández Díaz, *Del Monopolio a Repsol YPF. La historia de la industria petrolera española* (Madrid, 2003).

Productos Químicos Sintéticos and Consorcio Químico Español, two firms formed by Cros, UEE, and the industrial banks, Urquijo, Herrero, and Hispano-Americano (known as the Urquijo group) to acquire the German chemical assets after World War II.[24] It was Robert who, after 1945, explicitly defended the idea that buying foreign technology was a cheaper and a more efficient way to grow than was research and innovation, thus describing what had been and has remained Spain's technological policy.[25]

The story of the Urquijo group also provides a fine example of the dynamics of the Spanish postwar chemical industry.[26] After a five-year-long process in which Robert was instrumental, Proquisa and the Consorcio bought Bayer and Schering, the core of the German pharmaceutical business in Spain. FNCE, however, went back into the hands of its founders, as did EQ de Flix, Cloratita, and other joint ventures of Cros. The I.G. Bayer became the basis of an ambitious coal-based chemical concern in La Felguera, in northern Spain, aimed at producing nitric, salicylic, and acetilsalicilic acids, and methanol. The plant was built with the technical cooperation of Montecatini and Kuhlmann, and the financial support of QCF Bayer, a healthy business working in close relationship with Leverkusen, and a captive customer of La Felguera production. Schering became the basis for another ambitious pharmachemical enterprise, the Compañía Española de Penicilina y Antibióticos (CEPA), directed by Antonio Gallego, a brilliant scientist from the Rockefeller Foundation. He had been called back by his brother, José Luis, the scientific director of Bayer in Spain from 1936 to 1943, and from 1950 onward. Before purchasing Productos Químicos Schering, the new Spanish owners made sure they could manufacture under Berlin licensing. As was the case with Bayer, the excellent sales network of the German firm was used to commercialize the first antibiotics manufactured by CEPA under Merck license, whereas Schering's rising profits kept CEPA going during its difficult start. Moreover, Bayer and Schering financed one of

---

[24] Antonio Robert, *Un problema nacional: la industrialización necesaria* (Madrid, 1943).

[25] On Spanish technological policy during the autarkic period, see Santiago López García, "El Patronato Juan de la Cierva (1939-1960)," *Arbor* 619, 625, 637 (1997, 1998, 1999): 201-38, 1-44, 1-32.

[26] María Jesús Santesmases, *Antibióticos en la autarquía: banca privada, industria farmacéutica, investigación científica y cultura liberal en España, 1940-1960*, Fundación Empresa Pública, Working Paper 9906 (Madrid, 1999); Núria Puig, *La nacionalización de la industria farmacéutica en España: el caso de las empresas alemanas, 1914-1975*, Fundación Empresa Pública, Working Paper 2001/2 (Madrid, 2001).

the few private scientific institutions of the time, the Instituto Español de Farmacología (IFE), located at the University of Madrid, where Gallego was professor of Physiology. IFE originally was an instrument for the transfer of technology, inspired and applauded by the new scientific authorities of the Consejo Superior de Investigaciones Científicas, as well as the only institution that trained industrial minded scientists. In the long run, however, it became more and more attached to the true academic interests of its director. Furthermore, Gallego persuaded CEPA's technological partner, Merck, to establish in 1954 in Madrid, a branch of the newly launched Screening Program, aimed at identifying natural active principles that were then synthetized in the United States. CEPA's Screening Program later became Merck's own subsidiary's research department, as CEPA's new ownership refused to continue supporting its scientific staff. This story sheds light on the roots of Spanish scientific and technical backwardness: the reticence – if not unwillingness – of entrepreneurs to invest in research and development and the tendency to regard it as a purely intellectual, business unrelated activity. Intelligent as it was, the Urquijo pharmachemical complex could not satisfy the expectations of its creators. Falling international prices of raw materials and intermediates, prospects of liberalization, the inexorable advent of petrochemicals, a scandalous smuggling of penicillin, and the proximate expiration of the license contracts with Germany combined to bring down the whole enterprise in the late 1950s and early 1960s. As with most of the international chemical arrangements in Spain, it was, in the end, a good business for both parties. And, no doubt, it was extremely useful in creating some entrepreneurial capabilities – and establishing further international contacts – among the liberal elite where the Urquijo team belonged. However, the scientific capabilities – the technological effort – required by the modern chemical industry were still absent.

Another interesting project in the heavy organic industry was UNQUINESA, the flagship of the chemical holding founded by Federico Lipperheide in the Basque Country after the war. An energetic businessman, Lipperheide had arrived from Germany in the 1930s, and soon seduced the industrial and financial elite of Bilbao (born out of the local heavy industry) that was to finance his plans. Politically adaptive, he courted the prominent members of the German bureaucracy and industry that toured Spain during World War II and established the necessary contacts to set up various carbochemical plants with German technical assistance. After the German defeat, he had no great difficulties in finding new partners and making the transition to oil, but the adverse natural and institutional conditions of the time led him (along with most of the Spanish big industrial firms) to integrate vertically at the price of huge investments and uneconomic production.

This brief review of the major Spanish chemical companies is consistent with the picture offered by the first industrial statistics available.[27] In the late 1950s, the Spanish industry was still dominated by the heavy inorganic industry and classical chemicals; yet, it was widely diversified, and new sectors such as pharmaceuticals and fertilizers had made remarkable advances. The self-sufficiency goals of the "new state" were only partially achieved, as imports outgrew exports fivefold in 1965.[28] Spain had become an exporter of inorganic chemicals, but it relied on imports for heavy organic chemicals (27%), fertilizers (15%), plastics (14%), and rubber (13%). Thus, the most advanced industry worked under expensive license agreements since research and development was practically nonexistent.[29]

CHEMICALS AND THE MODERNIZATION OF SPAIN, 1960–1974

In the 1960s and early 1970s, Spain experienced the most impressive economic and social development of its history – the "Spanish miracle." The industrialization of the country took place under the guidance of a new, liberal-minded group of administrators, the close surveillance of the United States (an active supporter of Spain after 1953), and international institutions that Spain had joined in the late 1950s, such as the World Bank and the OEEC. As the dictatorship relaxed, a technocratic government implemented a "stabilization plan" (1959) and a series of "development plans" (launched after 1964) aimed at integrating Spain into the capitalist world. Led by the industrial and tourism sectors, the Spanish economy registered the fastest growth in Europe. The prospects of becoming a member of the European Economic Community encouraged many businessmen to go abroad in search of partners and markets.

The chemical industry was, along with automobile manufacturing, the best-performing sector. The explanation can be found in at least three substantial changes: the advent of petrochemicals; massive foreign

---

[27] Those collected by the fascist corporation Sindicato Nacional de Industrias Químicas: *Anuario de Industrias Químicas* (Madrid, 1949–1975), *La Industria Química en España* (Madrid, 1961–1975); and the Organization for European Economic Co-operation: *The Chemical Industry in Europe* (Paris, 1954–1975), joined by Spain in 1958.

[28] *Estadística del Comercio Exterior de España* (Madrid, 1965).

[29] OCDE, *Examen des politiques scientifiques nationales: Espagne 1971* (Paris, 1971).

investment; and an unprecedented domestic demand for industrial and consumer chemicals. For one thing, the transition from coal to oil as a feedstock was relatively fast and appealing to new investors, the absence of appropriate coal having been an obstacle to the large-scale chemical industry since the nineteenth century. Yet, those who had gone into big coal-based projects in the years of autarky were of course in trouble. This was the case with the Urquijo group, Lipperheide, and INI. In spite of the Institute's continuous attempts to keep this field for itself, public and private refineries proliferated, and various petrochemical complexes arose, all of them with the technical and financial assistance of international firms. Engineering consulting firms played an important role, too. Most of them were formally constituted in the 1960s, even if some of them had been working in Spain since the early 1950s. This was the case with Foster Wheeler, the technical partner of INI, CEPSA, and Caltex in the construction of REPESA, an oil refinery in Escombreras. It also was true of A.D. Little, one of the technical partners of the Spanish consulting firm, SENER, since 1956. At least ten consulting companies specialized in the chemical industry, all of them relying on foreign technical assistance. American partners made up the overwhelming majority, but some German, French, and Dutch firms had their share, too. In terms of employed engineers and technical complexity, these firms represented a new institution. Both their history and actual impact on the economic modernization of Spain deserve an in-depth analysis.

Foreign investment became a powerful instrument for technological transfer. Most of the new joint ventures and entrepreneurial groups, were based, as in previous periods, on useful connections that gave them access to the required knowledge, capital, and authorizations, while sheltering them from financial disasters.[30] The favorable atmosphere created by the new economic bureaucracy, along with a relative exhaustion of opportunities in mature markets, brought international capital into Spain. Investing in Spain was especially easy for Swiss and German firms. The former had remained here since the 1920s, and the latter, in spite of postwar expropriations, had managed to keep control over the Spanish market through licensing and other comfortable arrangements. At least in these two cases, the nationalist industrial drive, begun in 1917 and

---

[30] In the sense that they made sure losses would be somehow socialized. Although the role of business groups in the Spanish chemical industry still deserves an in-depth analysis, it is very likely that both their nature and dynamics help explain the technological weakness of the Spanish industry.

strengthened after 1939, seems to have encouraged more or less veiled forms of direct investment, and thus an early familiarization with the Spanish market. Bayer is a case in point. In Spain since 1899, the German concern was able to use its Spanish subsidiaries and joint ventures as a platform for its international (European and Latin American) investments after World War II. The Anglo-Saxon firms, instead, went through a sometimes disappointing learning process. The position of French investors stood somewhere between the former and the latter. A regular way of investing in Spain was participating in or buying local "autarkical" firms, that is, now uneconomic companies. Those who already knew the Spanish market or had Spanish partners had an advantage, of course.

Rising living standards constitute the third major shaping force of the "chemicalization" of the Spanish economy in the 1960s and early 1970s. This explains the outstanding performance of plastics, synthetic fibers, detergents, and cosmetics. As for fertilizers and agrochemicals in general, their strong position is explained by the modernization of the agrarian sector, a much belated process underlying the dramatic urbanization of the Spanish population.

On the eve of the international oil crisis, our ranking looked somewhat different. Table 13.2 shows how the top positions were still held by the domestic pioneers, with UEE (part of the British mining concern, Río Tinto, and soon after, the owner of most of the Urquijo chemical business) and Cros taking first and second place. However, Solvay, the three big heirs of I.G. Farben, and the Swiss firms Ciba-Geigy and Sandoz were not far behind. Another change was the practical dismantling of the once mighty Lipperheide group. The German-Basque businessman made his best deal in 1960 by selling 50% of Unquinesa to Dow.[31] The American firm transformed this rusting and bureaucratized factory into a modern plastic-manufacturing plant and made important and pioneering investments in Tarragona. Today, the most important petrochemical district in Spain, Tarragona had become very much in demand because of its port, its proximity to Barcelona, and the many advantages offered by the local and national governments to private investors.[32] ENCASO did, in fact, fight hard to establish its own refinery (ENTASA) there. ENTASA soon came to feed a rising petrochemical complex where the major

---

[31] E. N. Brandt, *Growth Company: Dow Chemical's First Century* (East Lansing: Michigan State University Press, 1997).

[32] Núria Puig, *Bayer, Cepsa, Puig, Repsol, Schering y La Seda: Constructores de la química española* (Madrid, Lid, 2003), chapter 7.

Table 13.2. The Fifty Largest Spanish Chemical Firms in 1973

Firm	Founded	Location	Turnover (Million Pesetas)	Main Shareholders	Main Product
1. UNIÓN EXPLOSIVOS RIO TINTO	1896	Madrid	40,604	Urquijo group Río Tinto Zinc	Explosives
2. CROS	1904	Barcelona	15,403	Cros group	Inorganic chemicals
3. LA SEDA DE BARCELONA	1925	Barcelona	7,662	AKU and Urquijo group	Chemical fibers
4. BAYER HISPANIA COMERCIAL	1899/1925/1972	Barcelona	7,153	Bayer	Organic chemicals
5. HOECHST	1946	Barcelona	7,000	Hoechst	Organic chemicals
6. BASF ESPAÑOLA	1966	Barcelona	6,200	BASF	Organic chemicals
7. DOW UNQUINESA	1939/1960	Biscay	5,276	Dow Unquinesa	Organic chemicals
8. SNIACE	1939	Madrid	4,903	Spanish group Snia Viscosa INI	Fibers
9. CIBA-GEIGY	1920/1939	Barcelona	4,468	Ciba Geigy	Pharmaceuticals
10. SOCIEDAD ANÓNIMA DE FIBRAS ARTIFICIALES	1923	Barcelona	4,141	Vilà family Rhône Poulenc	Fibers
11. AISCONDEL	1943	Barcelona	3,924	Monsanto	Plastics
12. SOLVAY	1904	Barcelona	3,640	Solvay	Chemicals
13. LABORATORIOS E INDUSTRIAS IVEN	1961	Madrid	3,300	Basque group	Pharmaceuticals
14. LABORATORIOS DR. ANDREU	1935	Barcelona	3,081	Cros group	Pharmaceuticals
15. ENERGÍA E INDUSTRIAS ARAGONESAS	1918	Madrid	3,030	Urquijo group	Fertilizers
16. CAMP		Barcelona	2,900	Camp family	Detergents
17. FOSFÓRICO ESPAÑOL		Madrid	2,816	Urquijo and Cros groups	Chemicals
18. DISTRIBUIDORA INDUSTRIAL	1933	Tenerife	2,800	CEPSA group	Chemicals
19. SANDOZ	1924	Barcelona	2,774	Sandoz	Pharmaceuticals
20. LEVER IBÉRICA	1957	Madrid	2,735	Lever	Chemicals
21. EMPRESA NACIONAL DE FERTILIZANTES (ENFERSA)	1973	Madrid	2,731	National Institute of Industry	Fertilizers
22. FORET	1927	Barcelona	2,700	FMC and Foret family	Chemicals
23. INDUSTRIAS QUÍMICAS ASOCIADAS	1961	Madrid	2,500	Hoechst Shell Urquijo group	Chemicals
24. HISPAVIC INDUSTRIAL	1959	Barcelona	2,453	ICI	Plastics

	Year	City	Parent company/group	Revenue	Industry
25. HENKEL IBÉRICA	1954	Barcelona	Henkel	2,394	Detergents
26. ALCUDIA	1963	Madrid	ENCASO/INI ICI Foret	2,384	Plastics
27. RESINAS POLIÉSTERES	1957	Madrid	Rhône-Poulenc Dow-Unquinesa Urquijo group	2,000	Plastics
28. UGIMICA	1964	Madrid	Pechiney-Ugine-Kuhlmann	1,973	Chemicals
29. CARBUROS METÁLICOS	1897	Barcelona	Spanish group	1,897	Chemicals
30. SE OXÍGENO	1919	Madrid	L'Air Liquide	1,862	Chemicals
31. PRODUCTOS QUÍMICOS ESSO	1967	Madrid	Standard Oil N.J.	1,656	Petrochemicals
32. AMONÍACO ESPAÑOL		Madrid	Spanish group	1,628	Chemicals
33. IQ LUCHANA	1949	Biscay	CEPSA group	1,623	Plastics
34. ANTIBIÓTICOS	1949	Madrid	Spanish laboratories (Abelló, Ibys, Leti-Uquifa, Zeltia)	1,549	Pharmaceuticals
35. FERRER INTERNACIONAL	1947	Barcelona	Ferrer family	1,515	Pharmaceuticals
36. HISPANO QUÍMICA HOUGHTON	1970	Madrid	Houghton Puteaux	1,509	Chemicals
37. AGFA GAEVERT	1925	Barcelona	Agfa-Gaevert	1,500	Photochemicals
38. BILORE	1947	Guipúzcoa	Spanish group	1,500	Detergents
39. HIDRO NITRO	1940	Madrid	Oechiney-Ugine-Kuhlmann	1,469	Fertilizers
40. AVON COSMETICS	1965	Madrid	Avon and Gal	1,380	Cosmetics
41. FEDERICO BONET	1935	Madrid	Urquijo group	1,373	Pharmaceuticals
42. LILLY INDIANA	1953	Madrid	Lilly	1,368	Pharmaceuticals
43. PROCTER & GAMBLE ESPAÑA	1968	Madrid	Procter & Gamble	1,320	Cosmetics
44. LA CELLOPHANE		Burgos	UCB	1,305	Plastics
45. MANUFACTURA DE HULES		Barcelona	Cros group Solvay	1,255	Plastics
46. 3M ESPAÑA	1957	Madrid	3M	1,250	Plastics
47. PAULAR	1961	Madrid	ENCASO/INI Montecatini	1,250	Petrochemicals
48. TITÁN	1917	Barcelona	Folch family	1,237	Paints
49. FORMICA ESPAÑOLA	1962	Biscay	Urquijo group Formica International	1,211	Plastics
50. CALATRAVA	1963	Madrid	ENCASO/INI Phillips Petroleum	1,200	Petrochemicals
51. PETROQUIMICA ESPAÑOLA	1967	Madrid	CEPSA Continental Oil	1,200	Petrochemicals

Note: Explosivos, Río Tinto, and the Urquijo group merged from 1972 through 1978.
Sources: Ministerio de Industria y Comercio (1974) and Fomento de la Producción 1973.

international chemical firms, particularly the German, were present. The landing of multinational corporations was indeed the most remarkable event of that time. Along with Dow, Monsanto, FMC, and Avon were early explorers of the Spanish market, with local partners as well. European multinational firms such as ICI, Lever, Kuhlmann, Rhône Poulenc, and Houghton Putaux became more visible through their own subsidiaries. To understand how modest the Spanish landscape was by 1973, one must know that the fifty largest firms roughly amounted to one of the big German ones.[33]

The least active side of the otherwise impressive development of the Spanish chemical industry remained research and development. Certainly, the dreams of self-sufficiency of the military establishment advising Franco had nourished disparate creativity efforts in the public as well as in private sectors, with their main objective consisting of making the best out of Spanish mining and agricultural resources. In an era of liberalization and international technology transfer, and with no consistent scientific policy supporting them, however, most of those projects were necessarily sterile, and cannot be compared with the autarkical policies of Italy and Japan under fascist rule. Yet, this desolate panorama presents at least two exceptions: the petrochemical and pharmaceutical industries.

Ironic as it may seem, the persistent and uneconomic attempts to produce synthetic fuel from Spanish coal were responsible for the first research department ever established in the Spanish chemical industry, that of ENCASO.[34] Founded as early as 1945, it spent a long time on the training of technicians and the implementation of German war technology to reevaluate national agricultural resources, as the official doctrine put it. Things started to change when the French Oil Institute agreed to cooperate with ENCASO's research center in the manufacture of hydrogenation lubricants. By that time, Spain had become an ally of the United States. INI was desperately trying to get hold of American economic and technical assistance while keeping its former links with German firms such as BASF, which was able to continuously assist (and charge) ENCASO's many hydrogenation-related projects throughout the 1960s. The scope of technology available to INI thus widened considerably, as did ENCASO's research program. It was only in the 1970s that ENCASO's erratic path started to bear fruit, even if very

---

[33] Ministerio de Industria y Comercio, *Las 500 grandes empresas industriales españolas en 1973* (Madrid, 1974).

[34] José Luis Martínez Cordón, *Medio Siglo de I+D* (Madrid, 2002). A former director of Repsol's research department, this author interprets its history in a very positive light. Unfortunately, he offers little quantitative data.

modestly, by commercializing some lubricants and catalysts, and even licensing one of them to CEPSA's petrochemical firm, Petroquímica Española (PETRESA).[35] More cautious and financially bound than ENCASO, the private firm did not establish its research department until 1975. Assisted by Universal Oil, it did not have relevant results before 1990, when the French group, Elf, started to take over CEPSA. A careful analysis of both research departments should be undertaken in order to evaluate Spanish technological capabilities in a field where the entrepreneurial capabilities of Repsol seem to be sound.[36] In this company, it was not until the 1980s that the massive transfer of petrochemical technology that occurred twenty years earlier started to bear fruit. The main reason seems to have been the almost simultaneous departure of the Spain's foreign partners in the public petrochemical group (Montecatini/ Montedison, ICI, Phillips Petroleum, and Halcon), which favored both a change of focus in the firm's research activities (from oil to petrochemicals) and an extraordinary research effort in the areas of polyetylen, synthetic rubber, and poliolefines. Today, as in the past, Repsol's research center remains the largest industrial research institution in Spain, with over two hundred researchers.

The pharmaceutical industry followed a less tortuous and expensive path. In the hands of private firms, it experienced a learning process that, in the long run, has proved modestly rewarding in spite of its persistent technological dependence.[37] The duopoly officially created in 1948 to produce penicillin in Spain gave rise to two interesting companies, Antibióticos and CEPA. Although Montedison acquired the former in the 1980s, and the latter was slowly dismantled beginning in the 1970s when the Explosivos group took control of it, both were, for a time, effective instruments for technological transfer and learning. Why did they give up? The CEPA-Screening Program story, already chronicled, suggests that the absence of a nationwide scientific policy, plus the lack of true research interests by the best-placed entrepreneurs, might have aborted a promising process. By contrast, the establishment of a modern

---

[35] Repsol owns its name actually to a successful lubricant brand launched by ENPETROL, ENCASO's successor, in the 1970s.

[36] Given the comparatively low research budgets, it is possible that both departments are highly productive by international standards, as it happens in the academic area nowadays.

[37] The Spanish case is extensively analyzed in: Organisation for Economic Co-operation and Development, *Gaps in Technology. Pharmaceuticals* (Paris, 1969); Ayhan Çilingiroglu, *Transfer of Technology for Pharmaceutical Chemicals. Synthesis Report on the Experience of Five Industrialising Countries* (Paris: Organization for Economic Cooperation and Development, 1975).

national health system, new industrial regulation, and increasing liberalization created a favorable environment for the pharmaceutical industry
in the 1960s. Equally important was the skillful lobbying undertaken by
the largest Spanish laboratories since 1958.[38] The main outcome of such
development was the rise of fine chemicals and the first international
expeditions by local firms. Indeed, today's leading Spanish laboratories
built their technological and entrepreneurial capabilities at that time.
By following different strategies, all of them integrated backward and
became international, even if at a very modest scale.[39] The pioneering
and most aggressive firm was Ferrer Internacional, a classic postwar laboratory that had started manufacturing under foreign licenses but was
soon buying out European companies. As for research, measured by the
development of molecules, Almirall is the most dynamic nowadays. When
compared with the major international companies, of course, the Spanish firms fade. The fact that they are family firms also raises many doubts
about their future in a competitive European Union, even if they would
be able to merge.

THE GLOBAL ACCOMMODATION OF A LATECOMER, 1975–2000

The panorama analyzed in the previous sections has undergone, once
again, remarkable changes in the last quarter of the century. The end
of dictatorship in 1975, and the integration of Spain into the European
Union since 1986 – a major event in Spanish economic and political
history – contributed to the creation of an unprecedented climate of
international confidence and national optimism. To foreign investors,
Spain became more attractive than ever, thanks to relatively low labor and
energy costs, environmental permissiveness, rising demand, and proximity to major European markets. As the ongoing industrialization process
was fueled by international investment, the dismantling of state-owned
firms and large business groups sped up. The largest international companies have steadily increased their control over the Spanish chemical
industry through their own subsidiaries as well as through the massive
acquisition of Spanish firms. At the same time, the government made the

---

[38] The origins of Farmaindustria can be traced in the industry's journal, *Industria Farmaceútica* (1958-1965). J. Pedro López Novo, "La organización de los intereses empresariales en la industria farmacéutica española," *Papeles de Economía Española* 22 (1985),
144-60.

[39] Félix Lobo, "El crecimiento de la industria farmacéutica en España durante los felices años
del estado de bienestar," *Economía Industrial* 223 (1983), 121-33.

first enduring efforts to foster innovation, although it is too early to evaluate the overall impact of the innovation drive, which remains one of the weakest within the European Union. It is true, however, that the chemical industry is by far the most innovative, and the fastest-growing sector of the Spanish industry.[40] It is, also, too early to forecast how the few local firms that have consolidated or even strengthened their positions while following a common strategy of internationalization, will evolve.

The postpetrochemical adjustment has led to the productive structure shown in Tables 13.4 and 13.5. Basic chemicals, pharmachemicals, and transformed chemicals account each for almost one-third of total production value, whereas agrochemicals do not represent more than 5% of the total value. Self-sufficiency has been achieved up to 60–70%, and exports (most of them to the headquarters or subsidiaries of multinationals) have risen to 10–30% of the total output.[41] That the most remarkable features of the Spanish chemical industry remains technological and financial dependence on multinational concerns is best shown in Table 13.3, in which the fifty largest chemical firms in 1997 are ranked. The most visible change is the progress experienced by the chemical subsidiaries of two Spanish oil companies, Repsol and CEPSA. Privatized between 1989 and 1997, Repsol has become a large oil and petrochemical group, its latest international move being the acquisition of the Argentinian firm YPF. Its long and troubled research path already has been outlined. The oldest Spanish oil firm, CEPSA's petrochemical business includes Petroquímica Española and Intercontinental Química, both born as joint ventures with American partners and highly competitive from the beginning. American firms also assisted its research department until the whole group began to be acquired by the French concern Elf. As a whole, the petrochemical sector still appears dynamic and fast growing in Spain, with multinational firms as Dow, the big three Germans, Elf Atochem, and DuPont very consolidated in Tarragona and Southern Spain.

Meanwhile, the Spanish historical leaders (or their successors), Ercros and Erkimia, are still recovering from a stormy period of merger, diversification, foreign control (by the Arab group KIO), restructuring, downsizing, mismanagement, bankruptcy, and public intervention in the 1980s and early 1990s. Because they compete in very mature markets, their

---

[40] Accounting for some 25% of the total expenditure in industrial research and development. But note that such data very often include patents and technological assistance, remarkably high in the chemical sector.

[41] *La industria química en España 1997* (Madrid, 1998).

Table 13.3. The Fifty Largest Spanish Chemical Firms in 1997

Firm	Founded	Location	Turnover (Million Pesetas)	Main Shareholders	Main Production
1. REPSOL QUÍMICA	1944/1985	Madrid	196,500	Repsol	Petrochemicals
2. BAYER HISPANIA	1899/1925/1972	Barcelona	183,866	Bayer	Organic chemicals
3. DOW CHEMICAL IBÉRICA	1939/1960	Madrid	141,394	Dow Chemical	Organic chemicals
4. SOLVAY ESPAÑA	1904	Barcelona	108,848	Solvay	Inorganic chemicals
5. NOVARTIS HISPANIA	1920/1924/1939	Barcelona	102,199	Novartis	Pharmaceuticals
6. HENKEL IBÉRICA	1954	Barcelona	98,574	Henkel	Detergents
7. BASF ESPAÑOLA	1911/1966	Barcelona	89,447	BASF	Organic chemicals
8. ARBORA & AUSONIA	1978/1991	Barcelona	77,879	Ausonia Carulla family	Cosmetics
9. FERTIBERIA	1961	Madrid	68,109	Spanish group	Fertilizers
10. PRODUCTOS CAPILARES	1950	Madrid	62,900	L'Oréal	Perfumery
11. GLAXO WELLCOME	1970	Madrid	62,015	Glaxo Wellcome	Pharmaceuticals
12. PROCTER & GAMBLE ESPAÑA	1968	Madrid	60,707	Procter & Gamble	Detergents
13. ELF ATOCHEM ESPAÑA	1964	Madrid	59,796	Elf	Petrochemicals
14. ICI ESPAÑA	1923	Barcelona	57,600	ICI	Chemicals
15. FMC FORET	1927	Barcelona	53,879	FMC	Chemicals
16. ALMIRALL-PRODESFARMA	1944/1996	Barcelona	53,000	Gallardo family	Pharmaceuticals
17. QUÍMICA FARMACÉUTICA BAYER	1899/1935/1981	Barcelona	46,815	Bayer	Pharmaceuticals
18. ENERGÍA E INDUSTRIAS ARAGONESAS	1918	Madrid	45,883	Spanish group	Fertilizers
19. SMITHKLINE BEECHAM	1967	Madrid	44,644	Smithkline Beecham	Pharmaceuticals
20. PRODUCTOS ROCHE	1930	Madrid	44,000	Roche	Pharmaceuticals
21. TARGOR IBÉRICA	1997	Barcelona	44,000	BASF Hoechst	Plastics
22. BENCKISER ESPAÑA	1968	Barcelona	42,133	Beiersdorf	Cosmetics
23. ERCROS	1899/1904	Barcelona	39,084	Spanish group	Chemicals
24. CARBUROS METÁLICOS	1897	Barcelona	36,394	Air Products	Oxygen
25. LA SEDA DE BARCELONA	1925	Barcelona	35,643	Spanish group	Fibers, plastics
26. ANTONIO PUIG	1922	Barcelona	35,500	Puig family	Perfumery

27. INTERCONTINENTAL QUÍMICA	1972	Madrid	35,249	Elf Cepsa	Petrochemicals
28. ELANCO VALQUÍMICA	1974	Madrid	34,650	Lilly	Pharmaceuticals
29. RHODIA IBERIA	1997	Madrid	33,969	Rhodia	Chemicals
30. FERRER INTERNACIONAL	1947	Barcelona	33,685	Ferrer family	Pharmaceuticals
31. DU PONT IBÉRICA	1981	Barcelona	33,264	Du Pont	Chemicals
32. PETROQUÍMICA ESPAÑOLA	1967	Madrid	33,220	Elf Cepsa	Petrochemicals
33. 3M ESPAÑA	1957	Madrid	32,943	3M	Plastics
34. MYRURGIA	1916	Barcelona	32,940	Monegal family	Perfumery
35. UNIÓN ESPAÑOLA DE EXPLOSIVOS	1899	Madrid	31,800	Pallas Invest NL	Explosives
36. LILLY	1953	Madrid	31,700	Lilly	Pharmaceuticals
37. PRAXAIR ESPAÑA	1967	Madrid	31,600	Praxair	Oxygen
38. PHARMACIA & UPJOHN		Barcelona	31,551	Pharmacia & Upjohn	Pharma
39. MERCK SHARPE & DOHME DE ESPAÑA		Madrid	30,144	Merck & Co.	Pharmaceuticals
40. L'AIR LIQUIDE ESPAÑA	1927	Madrid	29,539	L'Air Liquide	Oxygen
41. BOEHRINGER INGELHEIM ESPAÑA	1921/1952	Barcelona	28,998	Boehringer Ingelheim	Pharmaceuticals
42. B. BRAUN MEDICAL	1955	Barcelona	28,750	B. Braun	Pharmaceuticals
43. CIBA ESPECIALIDADES QUÍMICAS	1920	Barcelona	28,444	Ciba-Novartis	Pharma-chemicals
44. BOEHRINGER MANNHEIM	1933	Barcelona	28,252	Boehringer Mannheim-Roche	Pharmaceuticals
45. LEVER ESPAÑA	1957	Madrid	27,985	Unilever	Detergents
46. LABORATORIOS DEL DR. ESTEVE	1929/1936	Barcelona	27,720	Esteve family	Pharmaceuticals
47. PFIZER	1962	Madrid	26,702	Pfizer	Pharmaceuticals
48. ANTIBIÓTICOS	1949	Madrid	25,100	Farmitalia/Montedison	Pharmaceuticals
49. ERKIMIA	1896/	Madrid	24,654	Ercros	Chemicals
50. ABBOTT LABORATORIES	1964	Madrid	24,400	Abbott	Pharmaceuticals

Note: Repsol Química is the outcome of the privatization (1987–95) of the petrochemical division of the state-owned firm ENCASO. ERT/Urquijo and Cros merged in 1988 into Ercros and Erkimia. In 1990, Elf started to participate in CEPSA. AKZO left its historical subsidiary, La Seda de Barcelona, in 1991. Puig acquired Myrurgia in 2000.

Source: Fomento de la Producción 1997, Ministerio de Industria y Minería (1998), Directorio de Consejeros y Directivos 1999.

Table 13.4. Structure of the Spanish Chemical Industry, 1958–1998 (%)

	1958	1974	1986	1998
Basic chemicals	26.7	27.5	28.4	21.5
Inorganic	12.6	11.1	9.3	4.3
Organic	14.1	9.5	10.0	9.6
Plastics	*	6.3	8.6	7.4
Rubber	*	0.6	0.3	0.2
Agrochemicals	10.2	9.3	9.0	3.7
Fertilizers	7.9	8.0	6.9	1.8
Plaguicides	2.2	1.3	2.0	1.9
Pharmaceuticals	11.6	15.4	13.1	20.5
Fine chemicals	*	1.5	3.1	3.7
Specialties	*	13.9	10.0	16.8
Industrial and consumer chemicals	43.5	47.6	45.5	52.8
Dyes	2.9	1.2	1.3	1.3
Paints	4.9	4.0	2.9	3.5
Adhesives	1.5	1.5	1.7	1.6
Perfumes	4.2	4.9	2.9	3.6
Detergents	4.9	3.1	5.0	3.5
Photochemicals	1.0	0.5	0.5	*
Explosives	2.7	1.1	*	0.5
Other	*	4.7	*	8.1
Rubber	10.3	11.0	8.9	9.0
Plastics	10.8	15.6	17.1	21.7
Fibers	7.8	*	3.8	1.5
Artificial	*	*	0.2	0.2
Synthetic	*	*	3.6	1.3
Total	100	100	100	100

Note: Output value.

Source: Sindicato Nacional de Industrias Químicas, Ministerio de Industria y Energía and author's elaboration.

future remains uncertain. In contrast, the explosives firm, UEE, was controlled by a Dutch-based firm, Fertiberia, and after having got hold of the major Spanish fertilizer firms, both public and private, encountered significant difficulties. It is now mainly in Spanish hands.

Energía e Industrias Aragonesas also remains a Spanish chemical group. The future is also unclear for La Seda de Barcelona, which was once the leading chemical fibers manufacturer and a historical partner of Akzo and, after World War II, the Urquijo group. La Seda underwent serious difficulties in the 1980s related to the international

Table 13.5. Output, Trade Balance, and Self-Sufficiency, 1974–98

	O 1974	TB 1974	SS (%) 1974	O 1986	TB 1986	SS (%) 1986	O 1998	TB 1998	SS (%) 1998
Basic chemicals	127	−51	71.3	838	−218	79.2	1,410	−640	68.6
Inorganic	51	−4	92.7	276	−46	85.7	284	−118	70.6
Organic	43	−23	65.1	297	−100	74.6	630	−361	63.6
Plastics	29	−15	65.9	255	−40	86.4	485	−93	84.0
Rubber	2	−9	25.0	10	−32	23.8	11	−71	13.4
Agrochemicals	43	1	102.3	266	−16	94.6	239	−72	76.6
Fertilizers	37	3	108.8	206	−7	97.1	118	−38	75.6
Plaguicides	6	−1	85.7	60	−9	86.9	121	−35	78.0
Pharmaceuticals	71	−8	89.8	389	−5	98.4	1,344	−230	85.3
Fine chemicals	7	−9	43.7	94	−14	87.0	242	−45	84.6
Specialties	64	0	100.0	295	9	102.7	1,102	−185	85.6
Industrial and consumer chemicals	220	2	100.9	2,112	65	103.0	3,456	−119	96.6
Dyes	6	−4	60.0	40	−10	81.6	86	16	122.8
Paints	19	−1	95.0	86	−4	95.5	228	−28	89.0
Adhesives	5	0	100.0	53	−5	91.3	105	−30	77.7
Perfumes	18	0	100.0	88	−5	95.6	233	−36	86.6
Detergents	11	−1	94.4	150	3	101.3	233	35	118.2
Photo chemicals	2	−2	50.0	15	−13	55.5	30	1	107.1
Explosives	5	0	100.0	*	*	*	530	−53	91.0
Other	55	−1	95.6	142	−1	99.3	*	*	*
Rubber	39	9	121.4	264	40	117.8	591	51	109.2
Plastics	60	2	102.8	505	10	102.0	1,420	−76	94.9
Fibers	*	*	*	114	−13	89.0	94	−40	69.6
Artificial	*	*	*	6	−6	50.0	12	1	109.0
Synthetic	*	*	*	108	−7	93.1	82	−41	66.6
Total	462	−56	89.0	3,719	−187	95.0	6,543	−1,103	85.4

Note: Milliard pesetas. Current prices.

Source: Sindicato Nacional de Industrias Químicas, Ministerio de Industria y Energía and author's elaboration.

crisis that stemmed from the emergence of new industrializing countries. The Dutch partners gave up in 1991, and now, after a costly salvation operation from various administrations and under new Spanish ownership, the firm is investing heavily in high-tech plastics and fibers.

The postpetrochemical industry is best represented by the pharmaceutical sector. It looks, at first glance, very much the same as it did
thirty years ago, with the Swiss (Novartis and Roche), German (Bayer,
Hoechst, Boehringer), and Anglo-American (Glaxo Wellcome, Smithkline
Beecham, Lilly, Merck Sharp and Dohme, Pfizer, and Abbott) laboratories
dominating the scene. However, and as already mentioned, the combined
effects of liberalization, pharmaceutical regulation, and vertical integration, have also favored four Spanish firms, Almirall, Ferrer, Esteve, and
Uriach (which would rank fifty-third). Coming out of different origins, all
of them share several characteristics: family ownership and management;
commitment to research and internationalization; and vertical integration. Moreover, the four laboratories are the result of interesting learning
processes, as they used their international contacts (as holders of foreign representations or licenses) to explore the international market at
an early stage. Because they are too small to compete successfully in the
world market, they have recently followed cooperative research strategies, at the national and international level, and gone into (national and
international) partnerships with biotechnological and cosmetics firms,
among others. Their future, as noted in the previous section, does not
exclude mergers or acquisitions.

Modest as it might seem, the other promising, downstream industry
is perfumery. Their two representatives are Antonio Puig and Myrurgia,
two family-owned firms that have undergone a remarkable internationalization process in the last decades, particularly Puig, which has owned
Myrurgia. Two different strategies underlay this process. Whereas Puig
started in the first postwar period selling foreign products in the domestic
market, Myrurgia was born as a large and creative manufacturer of toilet
soap and eau de cologne for the Spanish and Latin American markets.
Unlike some of their rivals, both firms successfully overcame the successive challenges of autarky, liberalization, and foreign competition, held
strong positions in the Spanish market, integrated backward, diversified,
researched, explored foreign markets in Europe and America, acquired
foreign firms and trademarks, and became multinational. This success
story is interesting because commercial skills – a key capability in perfumery – were extremely weak in Spain, as the Germans soon realized.
Even if arriving late, therefore, some Spanish firms have been able to
learn to compete in the world markets. Yet, family control and relative
size (Puig ranks tenth in Europe, thirty-fifth in the world), competing
with giants such as Procter and Gamble puts some question marks on its
future development.

SUMMARY AND CONCLUSIONS

The rise of the modern Spanish chemical industry roughly coincides with the petrochemical revolution. This revolution took place between the 1930s and 1970s in a reduced number of countries, led by a reduced number of companies whose accumulated corporate capabilities allowed them to keep on innovating and successfully competing in the world market until date. Not a single Spanish firm belongs to this historical club, and not a single major innovation of the last century is associated with a Spanish institution. Thus, the remarkable growth of the Spanish chemical industry that occurred in the last decades has taken place with the assistance of the dynamic oligopoly ruling the international chemical market. The subsidiaries of the largest multinational concerns, and, eventually, their local partners have indeed played a major role, transforming Spain into an important place for international chemical investment and manufacturing, but not for research, basic or applied. The history of the Spanish chemical industry is, therefore, substantially different from that of the other countries analyzed in this volume. A medium-sized chemical power, Spain has become a remarkable consumer and exporter thanks to its own late industrialization and the comparative advantages identified by most multinational companies during the past century. It is true that a few Spanish firms, founded in the first half of the twentieth century, have recently undertaken remarkable efforts to become international in such different areas as petrochemicals, pharmaceuticals, or perfumery, but their relative size and capital structure raise some doubts about their future. I have called this particular and strongly dependent development of the Spanish chemical industry the global accommodation of a latecomer.

It has been the aim of this chapter to provide a comprehensive answer to an obvious question: Why have there been no innovative, international chemical firms in Spain? Because the history of the Spanish chemical industry is an almost unexplored field, I have approached the topic by identifying the largest (foreign or national) firms operating in Spain during the twentieth century. Then I have analyzed their growth strategies in the light of the industrial policies implemented by the Spanish Administration, on the one hand, and the development of the international chemical industry and the world markets, on the other hand, during four successive periods. Like many other industrial latecomers, Spain faced severe difficulties in establishing a modern chemical industry, some of which were inherent in economic backwardness, others more specific. Strong foreign firms looking for some raw materials or a complementary market stood, therefore, behind the first relevant Spanish chemical companies. State assistance – in other nations an effective instrument to stimulate growth – initially took the shape of a highly rhetorical, inexpensive industrial nationalism. Then, under Franco it gave rise to some large state-owned firms that only interfered with private ones in the fields of carbo- and petrochemicals, fertilizers, and artificial fibers. As the interventionist drive and the economic principles of the dictatorship eased, the Ministry of Industry since the 1960s has tried to encourage private, national as well as foreign investment. A strikingly weak scientific system, however, along with chronic political uncertainty during most of the twentieth century, had rather hindering effects on the accumulation of innovative capabilities required to compete internationally.

Spanish entrepreneurs, therefore, had to look outside for ideas and technical and financial assistance. To potential partners, they were often able to offer a more or less officially granted monopoly of the domestic market. Even if modest, it was growing fast, and could serve as platform for Portugal, northern Africa, and Latin America. Moreover, Spaniards frequently behaved as loyal, complacent partners. Where they had little to offer was at the technological level. This helps explain why Spain was left out of the international cartels in the interwar period, and why not a single firm was seen as an interlocutor by any of the big concerns. The weak position of Spanish firms was made clear again in the two further periods of intense technological diffusion: the second postwar, and the establishment of the European petrochemical industry in the late 1950s and early 1960s. The obstinate defense of an outdated autarkical project, the self-protecting strategies of the largest Spanish groups, and the persistent dependence on foreign assistance contributed to frustrate the long learning processes before innovation. But, if there was a true interest in learning and becoming innovative, it was to be found only from the 1960s onward, and in very specific branches. Even the huge and most erratic investments undertaken by both public and private firms in the long Spanish postwar period, chronicled in this chapter, show how little confidence their managers had in their own capabilities (and probably in the framework they were playing).

In spite of this overall adverse trajectory, a few Spanish chemical firms have recently gone through a rewarding learning process and have become international. We have pointed out the cases of Repsol, four pharmaceutical laboratories, and a perfume manufacturer.[42] The modest

---

[42] The topic of survival and competitiveness is specifically addressed in Puig, *Bayer, Cepsa, Puig, Repsol, Schering y La Seda*.

Spanish chemical industry at the commencement of the twenty-first century is thus represented by an upstream big business, and five downstream medium-sized, family-owned firms. Otherwise, Spain remains a favorite place for foreign investment and one of the least innovative countries of the advanced world.[43]

---

[43] For an optimistic account of the present and future of the Spanish chemical industry, see Federación Española de la Industria Química FEIQUE, *La industria química en el siglo XXI: desarrollo sostenible y compromiso de progreso* (Madrid, 1999).

14

Some Final Observations

VERA ZAMAGNI AND LOUIS GALAMBOS

Are there lessons in history? Can we learn anything that will help us today, and maybe even tomorrow, by studying the chemical industry in the age of the petrochemical revolution? The authors of this volume were not given that charge, and none of them, with perhaps one exception, have seen prediction as an important part of what we were doing collectively. In this sense, they have been true to the nature of history as a humanity, rather than a social science.

But with their work in hand, we can perhaps nudge the boundaries of history as a humanity, looking first to the past and then to the present, with a glance to the future of this industry and the global economy it has helped to shape. As the essays in this book establish very clearly, the chemical industry has passed through two distinct eras and embarked in the recent past on a third. First came the transition in the second half of the nineteenth century from inorganic to organic chemicals; with this transformation, leadership in the industry shifted toward firms and nations capable of transforming scientific knowledge into new products and then achieving economies of scale and scope in production and distribution on a national and international scale. Firms were very large in order to benefit from economies of scale, and they were diversified in order to realize economies of scope.

Starting in the 1930s, a second era began to take shape as the coal base of the industry shifted to oil and launched the petrochemical revolution of the postwar era – the centerpiece of this book. This era was characterized by significant expansion and a long phase of dynamic innovation facilitated by elaborate networks of professionals who managed the links between the firms and their sources of scientific and engineering knowledge. In this industry, a particularly important accommodation to global opportunities for technology transfer were the specialized engineering firms (SEFs) that carried basic knowledge and the details of acquired techniques throughout the world. The SEFs promoted development and hastened the industry's move through a normal product life cycle that set the stage for the third era of transformation in the global chemical industry.

The outlines of this third era are less clear, but there is little question that major changes are taking place throughout this global industry. One aspect of this restructuring is taking place in petrochemicals, with firms attempting to achieve a global scale of operations in commodity markets or focus on specialty markets. Meanwhile, there has been a major shift in pharmaceuticals toward biochemicals and molecular genetics. Biotech has, in this case, given an advantage to the United States, where abundant sources of venture capital and enterprising scientists fostered the development of the new industry and a new pattern of cooperative innovation between large pharmaceutical firms and small companies in biotechnology. Germany slipped behind as this latter change took place and has yet to recover the strong position it held during the organic chemistry era of the industry's evolution.

Looking back over these historical experiences, our first and strongest conclusion is about the cyclical nature of innovation in this science-based industry. In this case, the cycles were very long and had unusually widespread effects, many of which could not have been anticipated in the early years of expansion in, for instance, petrochemicals. The petrochemical cycle returns us to Joseph Schumpeter and his analysis of the impact of major innovations on capitalist systems. In this case, there was an abundant history of creative destruction. Older forms gave way, and eventually even the "national champions" had to yield to a fundamental series of interrelated innovations.

There also was substantial creativity that was not destructive. The development of the elaborate networks that Gambardella, Cesaroni, and Mariani describe appears to have had positive effects, promoting innovation and the full exploitation of new scientific knowledge. So, too, with the engineering organizations that speeded technology transfer, opening the way for nations that had been only raw material suppliers to become producers of finished products. When chemical engineering was coming of age in America, no one could have anticipated that it would become a powerful force that would have the effect of evening out global economic relationships via advanced technology.

Amid all these changes during the petrochemical revolution, there remained certain constants that fit comfortably in the Chandlerian paradigm of scale and scope. The leaders in this industry, all up to global scale, were able to defend their positions by embracing process innovations and largely ignoring national boundaries, as did the successful German companies. In the recent past, the leaders have survived by radical restructuring in all branches of the industry, in pharmaceuticals as well as basic chemicals. In Japan, by contrast, business structures had the same impact as national governmental policy, keeping the chemical firms from joining the world leaders in large-scale production and ongoing process innovation. So, when national industries survived, they did so by changing, and "national champions" gave way when the political favors they received from their national governments could no longer fit the reality of a new technological setting and a rapidly globalizing economy. There may be a "lesson" there: In the long term, political barriers are unlikely to defend firms and industries that are unwilling or unable to transform themselves technologically, strategically, and organizationally as their environment shifts.

Other lessons involve a broadened definition of "efficiency." In chemicals, efficiency was promoted through creative finance as well as technology. The dangers of using debt financing emerged clearly in the national studies and the chapter by Da Rin. Efficiency in finance could thus be just as important as it was in operations when margins became narrow and competition fierce. Financial innovation and organizational innovation, à la Chandler, have been and will probably continue to be important aspects of successful competition in a global chemical industry that continues to be roiled by change.

The efficiency concept could be applied as well to the industry's responses to environmental and safety concerns. The world's leading chemical firms have responded creatively to the drastic changes that have taken place in our knowledge of the dangers of certain chemicals and in the regulatory systems for protecting workers and populations from those dangers. New regulations are coming, and new business policies will be needed. If Wyn Grant is right, there will be an entirely new structure of global regulations in the future. But, to date, the dire consequences predicted when regulation was being considered have never slowed the industry down appreciably; there have never been problems that the industry's engineers, scientists, and business executives could not solve – and remain in business.

Would we, then, invest in chemicals? The outlook is mixed and our authors are divided. Some are pessimistic, some are optimistic. We, too, are divided on this issue. One of us believes that production processes of the petrochemical type have become highly standardized and will follow a standard product life cycle and be progressively displaced out of the advanced economies. If they remain, in this perspective, they will not be among the most progressive and profitable corporations. Even in pharmaceuticals – the most profitable sector of the global industry – the pace of innovation has slowed recently, as have the increases in profits. Although none of our authors have as bleak a view of the future of the large bureaucratic organization as did Schumpeter, the story they provide us is clearly one that would give a venture capitalist pause.

But despite the abundant problems, one of us is optimistic. After all, the industry is still earning profits above the level of all manufacturing and still has considerable potential for innovation. As usual, the optimist sees the glass as half full. Although not favoring short-term investments in chemicals and counseling cautious selectivity about the middle term, the optimist thinks that the long-term prospects for the industry to enter another cycle of innovation and growth are excellent: The advice from this quarter is buy and hold – but first read the chapters in this book with care so that you can make an informed decision.

Appendix: The Chemical Industry
after World War II

A Quantitative Assessment

RENATO GIANNETTI AND VALENTINA ROMEI

INTRODUCTION

This appendix provides a quantitative picture of official data for the global sector and six countries: the United States, the United Kingdom, France, Germany, Japan, and Italy. In the first part, we examine production; in the second part, trade; in the third, productivity; and in the last, plant size and the main enterprises working in the field.

THE PRODUCTION OF CHEMICALS

The production of chemicals in the most industrialized countries grew steadily after World War II. Chemical production accounts for 15% of the total gross output in 1970, 19% in 1980, and 17% in 1990 (Table A.1). The peak year is 1982 with 22%.

After World War II, the production of petrochemicals increased rapidly, thanks to new supply opportunities (oil) and increasing demand. The organic chemicals industry was converted from coal-based into oil-based olefins such as propylene, an essential base material for the modern chemicals, ethylene, and butadiene. After the 1973 and 1979 increases in oil prices, firms producing chemicals searched for products that could provide higher added value, especially in fine chemicals such as

Table A.1. Chemical Production by Branch: France, Germany, Italy, Japan, UK and U.S. (Billion U.S. Dollars 1990)

	Industrial Chemicals	Basic Chemicals	Synthetic Resins	Other Chemicals	Drugs and Medicines	Petroleum Refineries	Petroleum Coal	Rubber	Plastic
1963	129.95	46.28	22.52	68.80	15.15	74.26	6.14	32.68	16.51
1970	229.17	78.85	50.59	120.76	34.71	132.93	11.27	47.74	48.46
1980	373.86	208.28	105.88	170.80	91.75	481.59	126.43	61.59	105.03
1990	445.50	164.62	95.35	286.43	121.13	337.24	21.06	76.50	232.99

Sources: Processed from *Statistical Yearbook* and *Industrial Statistical Yearbook.* Various years.

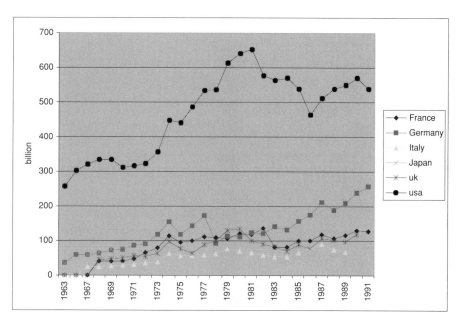

Graph A.1. The production of chemicals by country (1963-1991, billion U.S. dollars 1990).
Sources: Processed from *Industrial Statistical Yearbook.* Various years.

surface-active agents, flavors, perfumes, beauty products, and additives. Although petrochemicals were the leading products over the entire period, its lead was particularly notable before the 1980s. During the last decade of the period under study, the production of fine chemicals such as dyes, explosives, photographic chemicals, and plastic products grew at a faster rate (Graph A.1).

The increases in oil prices caused a diminished demand by user industries, the reduction of product substitution with chemical derivates, the reduced use of petroleum products and the increase in the imports of chemical products from developing countries, resulting in an "overcapacity" of petrochemical firms and the shrinking of the sector.[1]

---

[1] G. Mussati and A. Soru, "International Markets and Competitive Systems," in Alberto Martinelli, ed., *International Markets and Global Firms: A Comparative Study of Organized Business in the Chemical Industry* (London and Newbury Park: Sage Publications, 1991); Wyn Grant, "The Overcapacity Crisis in the West European Petrochemicals Industry," in Alberto Martinelli, ed., *International Markets and Global Firms: A Comparative Study of Organized Business in the Chemical Industry* (London and Newbury Park: Sage Publications, 1991); OECD, The Pharmaceutical Industry, Trade Related Issues (Paris, 1985). The OECD (1985) report puts the accent on the policies of investments made

Table A.2. Average Annual Growth Rates in the Production of Chemicals and Total Manufactures by Country (1963–1991, U.S. Dollars 1990)

	France	Germany	Italy	Japan	UK	U.S.
Total manufacture						
1963–1970	2.95	8.31	6.88	18.50	4.40	4.03
1971–1980	6.00	3.13	9.08	11.23	8.72	2.15
1981–1990	8.73	10.83	3.22	3.87	−1.27	0.16
1963–1991	**6.69**	**7.10**	**6.38**	**9.98**	**4.02**	**1.76**
Total chemicals						
1963–1970	0.06	17.74	5.38	19.61	3.30	4.27
1971–1980	12.87	7.17	10.94	12.48	12.74	7.73
1981–1990	2.46	8.37	0.54	4.65	0.46	−0.91
1963–1991	**6.57**	**9.68**	**5.93**	**10.73**	**6.30**	**3.24**

Source: Processed from *Statistical Yearbook* and *Industrial Statistical Yearbook*. Various years.

If we look at the different national experiences we can observe that the chemical industry, and manufacturing production as a whole, grew faster in Japan and Germany than in the other countries, whereas the United States was the worst performer both in manufactures and in the production of chemicals.

If we look at the different decades, we can also see that Germany was the only country to absorb relatively well, both in manufactures and in chemicals, the effects of the oil shocks in the 1970s, whereas the United States, Italy, and the United Kingdom were the worst performers (Table A.2).

Despite the difficulties of a direct comparison between countries,[2] several patterns of specialization can be observed. If we consider the specializations of the different branches, we can see that French production showed the best performance in petrochemicals, while Germany and Italy led in industrial chemicals. In Japan, synthetic resin

by petrochemicals firms that did not take into consideration the slowdown of demand: "the investment decided prior to the first oil crisis had produced surplus capacity by 1973/1974. Subsequently, though production collapsed in 1975 and took three years to recover its 1974 level, production capacity steadily grew. So, instead of being reduced, the 1974 surplus got worse," p. 22.

[2] France did not report the following subcategories: "basic chemicals," "synthetic resins," and "drugs and medicines." However, a direct comparison by categories is possible for France. Germany does not report the subcategories listed above. Moreover, the category "other chemicals" is included under "industrial chemicals," and the category "petroleum and coal" is included under "petroleum refineries." A direct comparison is possible only for "rubber" and "plastics." Italy had the same classification until 1988. A direct comparison is possible only for "rubber" and "plastics" also for Italy.

Table A.3. Specialization Index of the Gross Output of Chemicals[a] (France, Germany, Italy, Japan, UK, and U.S.)

	Industrial Chemicals	Basic Chemicals	Synthetic Resins	Other Chemicals	Drugs and Medicines	Petroleum Refineries	Petroleum Coal	Rubber	Plastic
FRANCE									
1970	1.41					2.00			
1980	0.80			1.36		1.26	0.02	1.18	0.99
1990	0.72			1.40	1.40	1.20		1.00	0.86
GERMANY									
1970	1.53					1.00		1.04	1.21
1980	1.70					1.01		1.05	1.24
1990	1.68					1.07		0.95	0.94
ITALY									
1970	1.72					0.81		1.14	0.68
1980	1.97					0.75		1.67	1.15
1989	1.03			0.79		0.79	2.61	1.31	1.21
JAPAN									
1970	1.01	1.27	2.21	1.07	1.74	0.59	1.66	0.91	1.87
1980	0.79	2.34	2.83	0.88	3.08	0.26	4.73	0.90	0.91
1990	0.82	1.14	1.67	1.22	1.30	0.68	1.13	1.34	1.40
UK									
1970	1.07	1.74	1.59	1.07	1.38	0.69	1.90	1.25	0.89
1980	1.09	1.30	0.77	1.37	0.96	0.91	0.30	1.27	1.20
1990	1.04	1.90	0.99	1.23	1.29	0.78	1.12	0.96	0.96
U.S.									
1970	0.74	1.28	1.05	1.43	1.21	1.05	1.14	1.09	0.89
1980	0.87	0.91	0.84	1.18	0.70	1.27	0.22	0.86	0.94
1990	0.88	1.38	1.25	1.15	1.09	1.16	1.54	0.82	0.83

[a] (The production of chemicals by branch and by country)/ (The total production of chemicals by country)/ (The production of chemicals by branch)/(The total production of chemicals).

Source: Processed from *Statistical Yearbook* and *Industrial Statistical Yearbook*. Various years.

411

Table A.4. Share in the Production of Chemicals by Country (1968–1970 1980–1989, Billion U.S. Dollars 1990, Percentage of the Production of the 6 Leader Countries)

	France	Germany	Italy	Japan	UK	U.S.
1968	6.93	11.07	4.34	11.49	8.05	58.12
1970	6.77	12.88	4.87	14.18	8.38	52.91
1980	9.13	8.49	5.35	18.19	10.23	48.60
1989	8.65	15.44	5.11	22.84	7.25	40.71
Average (1963–1991)	**8.79**	**12.96**	**5.34**	**18.28**	**7.98**	**49.60**

Source: Processed from *Industrial Statistical Yearbook*. Various years.

production was more important, whereas the United States and the United Kingdom had a more diversified production of chemicals.

Changes in competitiveness are evident in the share of the production of chemicals. The countries that depended more on the production of basic chemicals, such as France, the United Kingdom, and the United States, lost a share in the overall production of chemicals because of the market share increases of the countries producing specialties, such as Germany and Japan (Table A.3). Indeed, petrochemicals, fertilizers, and inorganic chemicals rapidly decreased their share in the overall production of chemicals after 1980. In 1969, petrochemicals products represented 23% of all chemicals. They reached a peak of 38% of total chemical production value in 1981, after which their share decreased to 21% in 1988. During that period, the production share of plastics increased from 8% in 1969 to 17% in 1990, whereas the production of drugs and medicines increased by about 2% (Tabel A.4).

Every country showed a similar growth trend in the different branches of chemicals: in the 1960s, the growth in chemicals was largely dependent on plastics; in the 1970s, on oil; and in the 1980s, on pharmaceuticals. But, differences among countries in long-range specializations changed their relative positions. In the following presentation, we will look at the different paths of development of the chemicals industry in the various countries.

UNITED STATES

Although the United States has been the leading producer of chemicals in terms of overall gross output, it has not been the largest per capita producer during the last fifty years. Basic chemicals, for the most part

Table A.5. U.S. Chemical Production by Branch (Billion U.S. Dollars 1990)

	Industrial Chemicals	Basic Chemicals	Synthetic Resins	Other Chemicals	Drugs Medicines	Petroleum	Petroleum and Coal	Rubber	Plastic
1963	76.99	46.28	22.52	68.80	15.15	67.57	6.14	24.57	13.10
1970	89.66	53.20	28.02	91.60	22.14	73.85	6.82	27.54	22.88
1980	157.97	91.89	43.07	98.11	31.25	298.21	13.37	25.81	47.73
1990	159.10	92.40	48.40	133.90	53.70	159.40	13.20	25.60	78.20

Source: Processed from *Industrial Statistical Yearbook*. Various years.

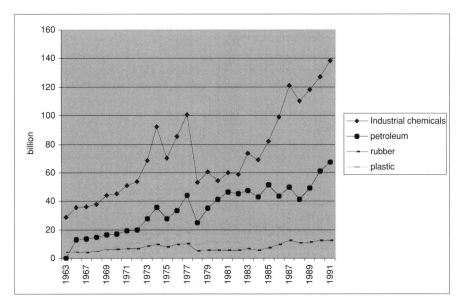

Graph A.2. German chemicals production by branch (1963–1991, billion U.S. dollars 1990). *Sources:* Processed from *Industrial Statistical Yearbook*. Various years.

fertilizers and inorganic chemicals, characterized the production of chemicals in the United States until the Great War. American leadership was based on the exploitation for a large market of abundant endowment of mineral and natural resources. The most important American chemicals firms, such as American Cyanamid, Allied Chemicals, and Union Carbide Chemicals, started as producers of basic chemicals. During the interwar period, the U.S. chemicals industry grew rapidly. The steady increase in economic performance, the introduction of new technologies from Europe, and the innovations derived by means of investments in research were the main factors of this growth. After World War II, the U.S. chemicals industry was based on large reserves of oil and natural gas, whereas Germany and the United Kingdom still had coke-based chemicals industries. Petrochemicals rapidly increased their applications in different industries. But, at the end of the 1970s, petrochemicals decreased their share in the production of chemicals in most industrialized countries. U.S. chemicals firms set up European petrochemicals branches, and acquired specialty firms that produced chemicals. The chemicals industry in the United States is a very diversified sector (Table A.5), but plastics, synthetic resins, drugs, and medicines are the most dynamic branches.

Table A.6. German Chemical Production by Branch (Billion U.S. Dollars 1990)

	Industrial Chemicals	Petroleum	Rubber	Plastic
1970	45.07	17.09	6.38	7.53
1980	54.01	41.45	5.50	11.08
1990	127.35	61.34	12.42	37.14

Source: Processed from *Industrial Statistical Yearbook*. Various years.

GERMANY

The German chemicals industry is, historically, a paradigm of a successful science-based sector (Graph A.2). It was established during the last quarter of the nineteenth century, and developed rapidly before World War I. The two world wars strongly affected the position of the German chemicals industry. After the wars, German know-how, trademarks and patents were confiscated. Nonetheless, after World War II, Germany was able to introduce the production of petrochemicals without converting coal-based chemicals firms at considerable cost. Germany showed the best performance in the international chemicals industry after the oil crisis of the 1970s.[3]

Over the years, Germany has kept a share of about 13% of the production of chemicals in most industrialized countries. Indeed, the chemicals sector was a dynamic one during the whole period, with its share in total production rising from 12% in 1963 to 15% in 1989. The early 1980s were the peak years for the production of German chemicals, with plastics and industrial chemicals as the most dynamic sectors (Table A.6).

GREAT BRITAIN

The British chemical industry was characterized by the production of specialties (Table A.7). During the past forty years, leading British goods have been plastics, but the production of petrochemicals has also played an important role. The British chemicals industry has not kept pace with its rivals. Its share in the production of chemicals of the most industrialized countries slowly decreased. Industrial chemicals production, like

---

[3] Fred Aftalion, *History of the International Chemical Industry* (Philadelphia: University of Pennsylvania Press, 1989); Ashish Arora, Ralph Landau, and Nathan Rosenberg, eds. *Chemicals and Long-Term Economic Growth: Insights from the Chemical Industry* (New York: John Wiley & Sons, 1998).

Table A.7. UK Chemical Production by Branch (Billion U.S. Dollars 1990)

Years	Industrial Chemicals	Basic Chemicals	Synthetic Resins	Other Chemicals	Drugs Medicines	Petroleum	Petroleum and Coal	Rubber	Plastic
1970	20.54	11.47	6.74	10.87	4.02	7.67	1.79	4.99	3.61
1980	41.54	27.62	8.33	23.95	9.03	44.76	3.89	8.03	12.85
1990	39.12	26.35	7.96	29.67	13.15	22.23	2.00	6.23	18.94

Source: Processed from *Industrial Statistical Yearbook.* Various years.

Table A.8. French Chemical Production by Branch (Billion U.S. Dollars 1990)

Years	Industrial Chemicals	Other Chemicals	Petroleum	Petroleum and Coal	Rubber	Plastic
1970	21.96		18.03			
1980	27.32	21.24	55.43	0.28	6.65	9.52
1990	29.82	53.10	37.70		7.13	18.62

Source: Processed from *Industrial Statistical Yearbook.* Various years.

other products, fell from about $41 billion during the decade of the 1970s to about $39 billion in the 1980s. Nevertheless, chemicals production decreased less than manufactures since the share of chemicals in the total gross output increased.

FRANCE

The chemical industry in France accounted for progressively larger shares of total gross output, reaching its peak years in the early 1980s, with 22% of the total gross output coming from chemicals. But after the 1980s, the growth in manufactures was about double that of chemicals and, in 1990 the chemical production accounted for 18% of total production (Table A.8). The most dynamic sectors during this period were drugs and medicines, which included France's long-standing specialization in perfumes.

JAPAN

The Japanese chemical industry remained largely dependent on foreign technologies for many years after World War II. The Japanese government played an important role in determining the structure and performance of the chemicals industry. Government policy was aimed at reducing imports of chemicals, in particular, of feedstock. The Ministry of International Trade and Industry encouraged competition among business groups and tried to rescue the petrochemicals producers after the 1970s. Japan's production of chemicals decreased rapidly after 1980, whereas plastics, paints, and photographic materials took the lead in production.[4] Chemicals improved their position in the output of Japanese manufactures, but less than in the United States, Germany, and the United Kingdom. Their share in the gross output of manufactures rose from 13% in 1963 to 14% in 1991, reaching a peak of 18% in 1983 (Table A.9).

[4] Aftalion, *History of the International Chemical Industry.*

Table A.9. Japanese Chemical Productions by Branch (Billion U.S. Dollars 1990)

Years	Industrial Chemicals	Basic Chemicals	Synthetic Resins	Other Chemicals	Drugs Medicines	Petroleum	Petroleum and Coal	Rubber	Plastic
1970	32.68	14.17	15.83	18.29	8.55	11.04	2.66	6.19	12.83
1984	58.50	33.38	20.93	41.22	18.05	61.58	5.25	12.76	38.37
1990	90.10	45.87	38.99	85.54	38.49	56.56	5.86	25.12	80.12

Source: Processed from *Industrial Statistical Yearbook*. Various years.

Table A.10. Italian Chemical Productions by Branch (Billion U.S. Dollars 1990).

	Industrial Chemicals	Petroleum	Rubber	Plastics
1970	19.29	5.25	2.65	1.6
1984	39.37	19.25	5.51	6.48
1990	34.92	15.61	5.10	14.12

Source: Processed from *Industrial Statistical Yearbook*. Various Years.

ITALY

The Italian chemical industry improved rapidly after World War II, because its coal-based chemicals industry was underdeveloped. Moreover, except for Natta's discovery of polypropylene, Italy largely depended on American and German technologies. In addition, in Italy plastics played the major role over the entire period. Italy was the poorest performer of the countries reviewed here. Its share in the production of chemicals in the most industrialized countries was almost stable and the share in total manufacture production was about 17%, with peaks of 20% during the 1980s (Table A.10).

TRADE

The chemicals balance of trade of the countries under consideration was characterized by a deficit in petroleum products, in contrast to a positive trade balance for all the other branches. Over the whole period, chemicals exports grew faster than overall exports, except for the 1960s and the 1980s. During the 1972–1982 period, exports of chemicals increased

Table A.11. Shares in the Exports of Chemicals by Country (France, Germany, Italy, Japan, UK, U.S. 1952–1992)

	France	Germany	Italy	Japan	UK	U.S.
1952	16.84	10.30	2.82[a]	1.20	20.35	48.48
1962	12.56	21.59	9.26	4.23	18.23	34.14
1972	14.32	28.35	10.32	9.47	14.15	23.39
1982	13.90	21.84	9.05	6.80	25.32	23.10
1992	16.02	26.12	7.85	9.83	16.96	23.22
1952–1992	**14.51**	**24.12**	**8.19**	**6.82**	**18.55**	**27.81**

[a] Index number partially distorted by the lack of petroleum data until 1958.

Source: Processed from *International Trade Statistics Yearbook*, Vol. I. Various years.

Table A.12. Export of Chemicals by Branch (France, Germany, Italy, Japan, UK, and U.S. Billion U.S. Dollars 1990)

	Petroleum and Products	Organic Chemicals	Inorganic Chemicals	Dyes, Tanning, Color Product	Medicinal Pharmaceutical Product	Perfume, Cleaning Product	Fertilizers Manufactured	Artificial Resins and Plastic Materials	Chemical Materials and Products, n.e.s	Rubber Manufactures n.e.s
1952	5,688	1,499	1,447	1,185	1,929	999	953	-	1,673	1,551
1962	5,866	3,970	2,286	1,702	3,107	1,365	1,758	3,906	3,774	1,047
1972	8,283	10,310	5,332	4,473	5,853	3,022	1,532	10,076	5,988	4,951
1982	50,401	28,452	13,565	5,708	12,024	6,312	4,600	23,089	14,835	9,730
1992	29,506	42,709	14,319	13,489	26,829	15,463	2,390	38,592	23,043	16,805

Source: Processed from *International Trade Statistics Yearbook*, Vol. I. Various years.

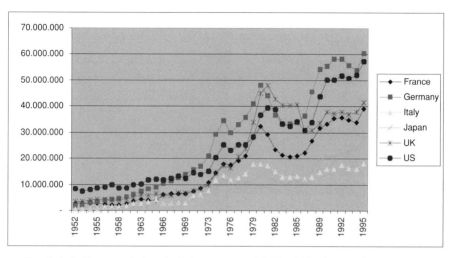

Graph A.3. Export of chemicals by country (1952–1995, thousand U.S. dollars 1990). *Sources:* Processed from *International Trade Statistics Yearbook*, Vol. I. Various years.

about twice as much as total exports (Table A.11). Organic chemicals and artificial resins were the main chemicals exports. During the 1980s, pharmaceuticals considerably increased their role. Fertilizers and inorganic chemicals showed the lowest growth rates. Growth in exports confirmed the facts observed in the production of chemicals. The countries more involved in basic chemicals also lost market share in the world exports of chemicals. The United States experienced the biggest loss in market share, France was relatively stable, whereas Germany showed the best performance after the 1980s, as did Japan (Table A.12; Graph A.3).

Specialties showed the fastest and most stable growth, whereas export of oil products was very variable throughout the entire period. Organic chemicals showed the fastest growth, followed by medicinal and pharmaceutical goods, artificial resins and plastics. The decrease in trade of petrochemical products was greater than the reduction of production after the "overproduction capacity" crisis because of the effects of new producers in Eastern Europe, in Saudi Arabia, and in the Middle East.[5] Even though Western European countries were less dependent on

[5] Greenwell Montagu (1995) estimated that in the mid-1980s "the rest of the world" accounted for only 13% of petrochemical production, while in 1995 the share rose to 23% of world production with Western Europe falling back from a third to a quarter.

Table A.13. U.S. Export of Chemicals by Branch (Million U.S. Dollars 1990 and Percentage of the U.S. Export of Chemicals)

	Petroleum and Products	Organic Chemicals	Inorganic Chemicals	Dyes, Tanning, Color Product	Medicinal Pharmaceutical Product	Perfume, Cleaning Product	Fertilizers Manufactured	Artificial Resins and Plastics	Chemical Materials and Products, n.e.s	Rubber Manufactures n.e.s
1952	3,704	790	370	466	1,048	288	141		772	626
	45.14	*9.63*	*4.51*	*5.67*	*12.78*	*3.51*	*1.72*		*9.41*	*7.63*
1962	1,686	1,060	692	322	1,152	529	435	1,266	2,033	649
	17.16	*10.79*	*7.04*	*3.28*	*11.73*	*5.38*	*4.43*	*12.88*	*20.70*	*6.61*
1972	1,381	3,068	2,041	489	1,248	614	594	2,035	1,797	724
	9.87	*21.93*	*14.59*	*3.50*	*8.92*	*4.39*	*4.24*	*14.55*	*12.84*	*5.18*
1982	8,532	7,439	4,528		3,208	1,274	2,448	5,315	4,858	1,368
	21.89	*19.09*	*11.62*		*8.23*	*3.27*	*6.28*	*13.64*	*12.47*	*3.51*
1992	6,261	11,053	4,136	1,850	5,446	2,710	2,390	9,581	5,817	2,561
	12.09	*21.34*	*7.98*	*3.57*	*10.51*	*5.23*	*4.61*	*18.49*	*11.23*	*4.94*

Source: Processed from *International Trade Statistics Yearbook*, Vol. I. Various years.

Table A.14. U.S. Specializations in the Chemicals Trade[a] 1952–1992

	Petroleum and Products	Organic Chemicals	Inorganic Chemicals	Dyes, Tanning, Color Product	Medicinal Pharmaceutical Products	Perfume Cleaning Product	Fertilizers, Manufactured	Artificial Resins and Plastics	Chemical Materials and Products, n.e.s	Rubber Manufactures n.e.s
1952	1.34	1.09	0.53	0.81	1.12	0.60	0.30		0.95	0.83
1962	0.84	0.78	0.89	0.56	1.09	1.13	0.73	0.95	1.58	1.82
1972	0.71	1.27	1.64	0.47	0.91	0.87	1.66	0.86	1.28	0.63
1982	0.73	1.13	1.45	–	1.16	0.87	2.30	1.00	1.42	0.61
1992	0.91	1.11	1.24	0.59	0.87	0.75	4.31	1.07	1.09	0.66

[a] (The export of chemicals by branch and by country)/ (The total export of chemicals by branch)/(The export of chemicals by country)/ (The total export of chemicals). Index used for tables n° 17, n°19, n°21, n°23, and n°25.

Source: Processed from *International Trade Statistics Yearbook*, Vol. I. Various years.

Table A.15. German Export of Chemicals by Branch (Million U.S. Dollars 1990 and Percentage of the German Export of Chemicals)

	Petroleum and Products	Organic Chemicals	Inorganic Chemicals	Dyes, Tanning, Color Product	Medicinal Pharmaceutical Product	Perfume, Cleaning Product	Fertilizers Manufactured	Artificial Resins and Plastics	Chemical Materials and Products, n.e.s	Rubber Manufactures n.e.s
1952	–	420	297	235	139	31	297	–	224	100
	–	24.07	17.05	13.50	7.99	1.79	17.02	–	12.85	5.75
1962	447	1,268	652	722	614	–	562	1,193	755	–
	7.20	20.41	10.49	11.63	9.88	–	9.04	19.20	12.16	–
1972	1,238	2,862	1,229	2,376	1,667	749	407	3,401	1,820	1,209
	7.30	16.88	7.25	14.01	9.83	4.42	2.40	20.06	10.73	7.13
1982	4,047	8,117	2,780	3,012	2,982	1,470	819	7,581	3,995	2,042
	10.98	22.03	7.54	8.17	8.09	3.99	2.22	20.58	10.84	5.54
1992	3,651	10,908	3,626	5,630	7,461	3,509	–	12,798	6,854	3,853
	6.26	18.71	6.22	9.66	12.80	6.02	–	21.96	11.76	6.61

Source: Processed from *International Trade Statistics Yearbook*, Vol. I. Various years.

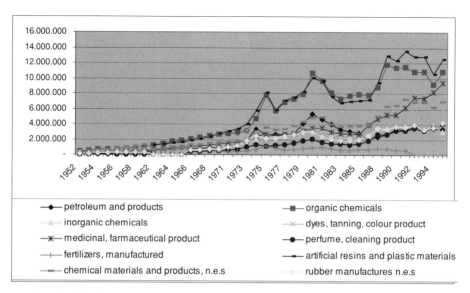

Graph A.4. German export of chemicals by branch (1952–1992, thousand U.S. dollars 1990.)

petrochemical production and had a dynamic production of specialties, they underwent a more severe crisis in the volume of export, especially in the United Kingdom and in France, because they were more dependent on international trade than their U.S. counterpart.[6]

On the contrary, the demand for medicinal and pharmaceutical products has been rising during the entire period, and countries that were net exporters in the 1970s maintained and enlarged their trade surpluses. Specialty productions competition largely relied on product innovation, marketing, and unusual standards of purity that increased the advantages of producers in industrialized countries.[7] These advantages can be better observed in the production figures rather than the trade figures, as, because of the important role played by foreign direct investments, "the pharmaceutical industry sends a surprisingly small proportion of its output across national frontiers despite the low volume and high unit value of its products and their low transport costs."[8]

[6] Grant, "The Overcapacity Crisis in the West European Petrochemicals Industry."

[7] Robert Balance, Janos Pogany, and Helmut Forstner, *The World's Pharmaceutical Industries, An International Perspective on Innovation, Competition and Policy* (Aldershot, Hants, England: Edward Elgar, 1992).

[8] OECD, 1985.

During the years 1952–68, the United States had the leadership of chemical exports, and from the middle of the 1880s, it was second after Germany (Tables A.13, A.14). Organic chemicals were the main goods exported by the United States, followed by artificial resins and plastics. The differences in the U.S. position in international production and trade of chemicals depended on the importance of the U.S. market. Before World War II, the large U.S. firms restricted themselves to the domestic market or Canada, whereas later they expanded to Latin America and then started to locate productive activities in Europe.[9] However, the U.S. specialization trade index showed a high value in fertilizers, followed by chemicals materials and inorganic chemicals.

From 1968, when it surpassed the United States, Germany has almost always been the world leader in exporting chemicals (Table A.15). Its share increased rapidly in the post-war years. Perfumes and cleaning products showed the fastest rate of growth over the entire period, followed by medicinal and pharmaceutical products (Graph A.4). If we look at the specialization index, dyes, tanning, and color products had the highest value, followed by artificial resins, plastic materials, and organic chemicals, which constitute the largest part of German chemicals export (Tables A.16).

Japan showed the highest rate of growth in chemicals in the group of countries considered. The best performance was in chemicals materials and products, followed by petroleum and rubber. If we look at the specialization index, fertilizers had the highest value, followed by rubber and resins.

The British position was very different. Except for the highly variable petroleum sector, the most dynamic branch of exports was organic chemicals, followed by fertilizers. The specialization index showed – apart from petroleum – perfumes and cleaning goods as the best performers, followed by dyes and color products.

France was the fourth world country in chemicals exports. Medicinal and pharmaceuticals were the goods with the best rate of growth in exports, followed by dyes. Its specialization index showed the highest value in perfumes and cleaning goods, followed by rubber, and inorganic chemicals.

Italian exports of chemicals grew quite rapidly, particularly before 1980. The branch with the highest rate of growth was petroleum, but the different branches increased with similar growth rates.

[9] Mussati, G., Soru, A. 1991. *"International Markets and Competitive Systems,"* in Alberto Martinelli, ed., *International Markets and Global Firms.*

Table A.16. German Specializations in the Chemicals Trade (1952–1992)

	Petroleum and Products	Organic Chemicals	Inorganic Chemicals	Dyes, Tanning, Color Product	Medicinal Pharmaceutical Product	Perfume, Cleaning Product	Fertilizers Manufactured	Artificial Resins and Plastics	Chemical Materials and Products, n.e.s	Rubber Manufactures n.e.s
1952	–	2.72	1.99	1.93	0.70	0.30	3.02		1.30	0.63
1962	0.35	1.48	1.32	1.97	0.92	–	1.48	1.41	0.93	–
1972	0.53	0.98	0.81	1.87	1.00	0.87	0.94	1.19	1.07	0.86
1982	0.37	1.31	0.94	2.42	1.14	1.07	0.82	1.50	1.23	0.96
1992	0.47	0.98	0.97	1.60	1.06	0.87	–	1.27	1.14	0.88

Source: Processed from *International Trade Statistics Yearbook*, Vol. I. Various years.

Table A.17. Japanese Export of Chemicals by Branch (Million U.S. Dollars 1990 and Percentage of the Japanese Export of Chemicals)

	Petroleum and Products	Organic Chemicals	Inorganic Chemicals	Dyes, Tanning, Color Product	Medicinal Pharmaceutical Product	Perfume, Cleaning Product	Fertilizers Manufactured	Artificial Resins and Plastics	Chemical Materials and Products, n.e.s	Rubber Manufactures n.e.s
1952	5	10	27	16	21	6	74	-	22	22
	2.67	*4.83*	*13.13*	*7.72*	*10.36*	*3.07*	*36.63*	*-*	*11.02*	*10.57*
1962	76	270	106	-	106	-	231	199	-	230
	6.22	*22.15*	*8.68*	*-*	*8.67*	*-*	*19.01*	*16.33*	*-*	*18.93*
1972	184	1,601	510	323	267	-	149	1,453	236	945
	3.25	*28.25*	*9.00*	*5.69*	*4.70*	*-*	*2.63*	*25.64*	*4.16*	*16.67*
1982	-	3,079	931	629	465	-	436	2,686	896	2,347
	-	*26.84*	*8.11*	*5.49*	*4.05*	*-*	*3.80*	*23.42*	*7.81*	*20.46*
1992	1,456	6,193	1,169	1,542	1,367	-	-	4,334	1,968	3,910
	6.64	*28.23*	*5.33*	*7.03*	*6.23*	*-*	*-*	*19.75*	*8.97*	*17.82*

Source: Processed from *International Trade Statistics Yearbook*, Vol. I. Various years.

Table A.18. Japanese Specializations in the Chemicals Trade (1952–1992)

	Petroleum and Products	Organic Chemicals	Inorganic Chemicals	Dyes, Tanning, Color Product	Medicinal Pharmaceutical Product	Perfume Cleaning Product	Fertilizers Manufactured	Artificial Resins and Plastics	Chemical Materials and Products, n.e.s	Rubber Manufactures n.e.s
1952	0.08	0.55	1.54	1.10	0.91	0.52	6.50[a]		1.11	1.15
1962	0.31	1.61	1.09	–	0.80	–	3.11	1.20	–	5.20
1972	0.23	1.64	1.01	0.76	0.48	–	1.03	1.52	0.42	2.01
1982	–	1.59	1.01	1.62	0.57	–	1.39	1.71	0.89	3.55
1992	0.50	1.47	0.83	1.16	0.52	–	–	1.14	0.87	2.37

[a] Index number partially distorted by the lack of fertilizers trade data for the other countries.

Source: Processed from *International Trade Statistics Yearbook*, Vol. I. Various years.

Table A.19. United Kingdom Export of Chemicals by Branch (Million U.S. Dollars 1990 and Percentage of the UK Export of Chemicals)

	Petroleum and Products	Organic Chemicals	Inorganic Chemicals	Dyes, Tanning, Color Product	Medicinal, Pharmaceutical product	Perfume Cleaning Product	Fertilizers Manufactured	Artificial Resins and Plastics	Chemical Materials and Products, n.e.s	Rubber Manufactures n.e.s
1952	782	84	402	324	414	267	165	-	512	493
	22.71	*2.43*	*11.68*	*9.40*	*12.03*	*7.76*	*4.80*	-	*14.87*	*14.31*
1962	1,615	517	277	533	653	385	-	627	639	-
	30.79	*9.86*	*5.27*	*10.15*	*12.45*	*7.33*	-	*11.96*	*12.18*	-
1972	1,698	1,142	607	820	1,272	600	27	1,208	1,092	-
	20.06	*13.49*	*7.16*	*9.69*	*15.02*	*7.08*	*0.32*	*14.27*	*12.90*	-
1982	26,309	4,253	1,667	1,143	2,421	1,293	177	2,304	2,235	921
	61.58	*9.95*	*3.90*	*2.67*	*5.67*	*3.03*	*0.41*	*5.39*	*5.23*	*2.15*
1992	11,767	6,129	1,762	2,251	5,254	2,579	-	3,381	3,451	1,278
	31.09	*16.19*	*4.65*	*5.95*	*13.88*	*6.81*	*0.00*	*8.93*	*9.12*	*3.38*

Source: Processed from *International Trade Statistics Yearbook*, Vol. I. Various years.

Table A.20. United Kingdom Specializations in the Chemicals Trade (1952–1992)

	Petroleum and Products	Organic Chemicals	Inorganic Chemicals	Dyes, Tanning, Color Product	Medicinal Pharmaceutical Product	Perfume Cleaning Product	Fertilizers Manufactured	Artificial Resins and Plastics	Chemical Materials and Products, n.e.s	Rubber Manufactures n.e.s
1952	0.68	0.27	1.37	1.34	1.06	1.31	0.85		1.50	1.56
1962	1.51	0.71	0.66	1.72	1.15	1.55	–	0.88	0.93	–
1972	1.45	0.78	0.80	1.30	1.54	1.40	0.13	0.85	1.29	–
1982	2.06	0.59	0.49	0.79	0.80	0.81	0.15	0.39	0.59	0.37

Source: Processed from *International Trade Statistics Yearbook*, Vol .I. Various years.

Table A.21. French Export of Chemicals by Branch (Million U.S. Dollars 1990 and Percentage of the French Export of Chemicals)

	Petroleum and Products	Organic Chemicals	Inorganic Chemicals	Dyes, Tanning, Color Product	Medicinal Pharmaceutical Product	Perfume, Cleaning Product	Fertilizers Manufactured	Artificial Resins and Plastics	Chemical Materials and Products, n.e.s	Rubber Manufactures n.e.s
1952	1,196	142	221	96	258	365	202	-	122	249
	41.97	*4.98*	*7.74*	*3.37*	*9.06*	*12.79*	*7.08*	*-*	*4.28*	*8.73*
1962	983	518	376	124	397	379	262	313	264	-
	27.18	*14.32*	*10.40*	*3.43*	*10.98*	*10.47*	*7.23*	*8.65*	*7.31*	*-*
1972	1,331	916	641	465	800	1,060	226	1,033	771	1,325
	15.53	*10.69*	*7.48*	*5.43*	*9.34*	*12.37*	*2.64*	*12.05*	*9.00*	*15.46*
1982	4,463	4,073	2,650	571	1,977	2,011	371	3,162	2,168	2,003
	19.03	*17.37*	*11.30*	*2.43*	*8.43*	*8.57*	*1.58*	*13.49*	*9.25*	*8.54*
1992	2,489	6,196	3,077	1,427	4,829	5,702	-	5,100	3,619	3,316
	6.96	*17.33*	*8.61*	*3.99*	*13.50*	*15.95*	*-*	*14.26*	*10.12*	*9.27*

Source: Processed from *International Trade Statistics Yearbook*, Vol. I. Various years.

Table A.22. French Specializations in the Chemicals Trade (1952–1992)

	Petroleum and Products	Organic Chemicals	Inorganic Chemicals	Dyes, Tanning, Color Product	Medicinal Pharmaceutical product	Perfume Cleaning Product	Fertilizers Manufactured	Artificial Resins and Plastics	Chemical Materials and Products, n.e.s	Rubber Manufactures n.e.s
1952	1.25	0.56	0.91	0.48	0.79	2.17	1.26		0.43	0.95
1962	1.33	1.04	1.31	0.58	1.02	2.21	1.18	0.64	0.56	–
1972	1.12	0.62	0.84	0.73	0.95	2.45	1.03	0.72	0.90	1.87
1982	0.64	1.03	1.41	0.72	1.18	2.29	0.58	0.99	1.05	1.48
1992	0.53	0.91	1.34	0.66	1.12	2.30	–	0.82	0.98	1.23

Source: Processed from *International Trade Statistics Yearbook*, Vol. I. Various years.

Italy's specialization index for the 1952–95 period showed an advantage in petroleum and rubber.

PRODUCTIVITY

The chemicals industry is a high value-added sector and also a capital intensive one. Its productivity was larger than that of total manufactures, globally and in every country considered here.

There were differences between the various branches: fine chemicals were characterized by a higher value added (drugs and medicines), while petroleum, rubber goods, and plastics had the lowest value.

Productivity of chemicals among countries did not converge. Together with Japan, the United States had the highest productivity level for industrial chemicals, basic chemicals, resins, other chemicals, drugs, and medicines. If we look at the value added (VA) in different countries, Japan showed the best performance in almost every branch, followed by the United States (Tables A.17, A.18).

Japanese productivity in the production of chemicals was more uniform, and was characterized by high levels for every branch except rubbers and plastics. Except for petroleum, which is the best performing branch everywhere, Japan had the absolute highest productivity in the other chemical branches.

The United Kingdom showed lower VA than other countries. Except for petroleum, the best VA was in basic chemicals (Tables A.19, A.20).

Germany had the best VA in petroleum. Despite its leadership in many branches and its growing market share, by comparison, Germany did not perform well (Tables A.21, A.22).

The contrast in productivity between oil and the other chemicals was evident also in France. In general, as was the case with Germany, the value added showed a correlation with the market share and the specialization index.

Over the whole period, Italy had the lowest VA level. This was particularly evident in the production of oil.

PLANT SIZE

Data on plants covers a smaller sample of countries (Germany, Italy, Japan, and the United Kingdom) (Tables A.23–A.31). Therefore, it is not possible to have an adequate comparison with information on production, trade, and VA. Generally, the employment per plant in the chemical

Table A.23. Italian Export of Chemicals by Branch (Million U.S. Dollars 1990 and Percentage of the Italian Export of Chemicals)

	Petroleum and Products	Organic Chemicals	Inorganic Chemicals	Dyes, Tanning, Color Product	Medicinal Pharmaceutical Product	Perfume Cleaning Product	Fertilizers Manufactured	Artificial Resins and Plastics	Chemical Materials and Products, n.e.s	Rubber Manufactures n.e.s
1952	-	54	130	49	48	41	74	-	20	62
	-	*11.27*	*27.23*	*10.21*	*10.05*	*8.68*	*15.44*	*-*	*4.22*	*12.91*
1962	1,059	338	184	-	185	73	269	309	82	167
	39.72	*12.68*	*6.91*	*-*	*6.94*	*2.73*	*10.08*	*11.58*	*3.08*	*6.28*
1972	2,451	721	304	-	600	-	129	946	273	748
	39.70	*11.69*	*4.93*	*-*	*9.72*	*-*	*2.09*	*15.33*	*4.42*	*12.12*
1982	7,050	1,491	1,010	353	970	265	349	2,040	683	1,049
	46.20	*9.77*	*6.62*	*2.32*	*6.36*	*1.74*	*2.29*	*13.36*	*4.48*	*6.87*
1992	3,882	2,230	550	790	2,472	962	-	3,397	1,335	1,887
	22.18	*12.74*	*3.14*	*4.51*	*14.12*	*5.50*	*-*	*19.41*	*7.63*	*10.78*

Source: Processed from *International Trade Statistics Yearbook*, Vol. I. Various years.

Table A.24. Italian Specializations in the Chemicals Trade (1952–1992)

	Petroleum and Products	Organic Chemicals	Inorganic Chemicals	Dyes, Tanning, Color Product	Medicinal Pharmaceutical Product	Perfume Cleaning Product	Fertilizers Manufactured	Artificial Resins and Plastics	Chemical Materials and Products, n.e.s	Rubber Manufactures n.e.s
1952	–	1.27	3.18	1.46	0.88	1.47	2.74		0.43	1.41
1962	1.95	0.92	0.87	–	0.64	0.58	1.65	0.85	0.23	1.73
1972	2.87	0.68	0.55	–	0.99	–	0.82	0.91	0.44	1.46
1982	1.55	0.58	0.82	0.68	0.89	0.46	0.84	0.98	0.51	1.19
1992	1.68	0.67	0.49	0.75	1.17	0.79	–	1.12	0.74	1.43

Source: Processed from *International Trade Statistics Yearbook*, Vol. I. Various years.

Table A.25. Value Added Per Worker in the Production of Chemicals by Branch (Thousand U.S. Dollars 1990. Annual Average, 1963–1991)[a]

	Industrial Chemicals	Basic Chemicals	Synthetic Resins	Other Chemicals	Drugs Medicines	Petroleum	Petroleum and Coal	Rubber	Plastic
1963–1970	42.24	25.18	22.02	20.23	28.61	74.72	15.99	15.24	12.30
1971–1980	61.62	39.48	29.68	39.72	38.68	187.25	28.90	31.39	25.81
1981–1990	92.17	65.55	61.85	73.86	73.18	325.94	56.73	45.56	44.29
1963–1991	**69.38**	**47.04**	**41.35**	**49.62**	**50.95**	**215.78**	**36.75**	**33.07**	**29.89**

[a] Average of value added per worker in France, Germany, Italy, Japan United Kingdom and United States.

Source: Processed from *Statistical Yearbook* and *Industrial Statistical Yearbook.* Various years.

Table A.26. U.S. Value Added Per Worker in the Production of Chemicals by Branch (1963–1991, Thousand U.S. Dollars 1990)

	Industrial Chemicals	Basic Chemicals	Synthetic Resins	Other Chemicals	Drugs Medicines	Petroleum	Petroleum and Coal	Rubber	Plastic
1963	93.12	106.17	81.91	83.54	115.83	107.59	72.27	49.34	41.94
1970	95.44	110.09	77.85	89.16	126.57	137.96	78.11	55.46	45.50
1980	126.86	135.27	99.94	119.06	121.24	347.34	90.25	53.58	48.10
1990	183.29	197.80	155.30	168.66	208.74	316.67	110.00	65.69	55.67

Source: Processed from *Statistical Yearbook* and *Industrial Statistical Yearbook*. Various years.

Table A.27. Japanese Value Added Per Worker in the Production of Chemicals by Branch (1963–1991, Thousand U.S. Dollars 1990)

	Industrial Chemicals	Basic Chemicals	Synthetic Resins	Other Chemicals	Drugs Medicines	Petroleum	Petroleum and Coal	Rubber	Plastic
1963	36.08					57.13		9.92	
1970	54.26	48.34	63.20	46.08	56.70	112.00	36.15	19.36	17.91
1980	113.12	125.60	97.41	132.36	153.84	396.43		76.67	47.57
1990	229.16	220.71	257.56	226.93	268.23	248.37	138.27	80.82	72.76

Source: Processed from *Statistical Yearbook* and *Industrial Statistical Yearbook. Various years.*

Table A.28. United Kingdom Value Added Per Worker in the Production of Chemicals by Branch (1963–1991, Thousand U.S. Dollars 1990)

	Industrial Chemicals	Basic Chemicals	Synthetic Resins	Other Chemicals	Drugs Medicines	Petroleum	Petroleum and Coal	Rubber	Plastic
1970	32.41	33.88	30.47	28.44	36.24	73.53	4.03	19.66	16.10
1980	63.93	69.08	39.04	64.63	72.52	422.46	71.73	38.95	38.22
1990	106.04	102.50	96.82	93.90	113.11	529.91	100.96	50.78	52.26

Source: Processed from *Statistical Yearbook* and *Industrial Statistical Yearbook.* Various years.

Table A.29. German Value Added Per Worker in the Production of Chemicals by Branch (1963–1991, Thousand U.S. Dollars 1990)

	Industrial Chemicals	Petroleum	Rubber	Plastic
1963	52.97			
1970	62.59			
1980	32.15	347.42	23.67	22.88
1990	64.88	926.76	70.95	65.64

Source: Processed from *Statistical Yearbook* and *Industrial Statistical Yearbook.* Various years

industry was higher than in total manufactures. Nonetheless, if we look at Table A.32 the average size of plants decreased in the last twenty years, both in chemicals and in manufacturing. Chemicals, like manufacturing, showed the greatest average size in 1980 when economies of scale were the main source of competitive advantage. Chemical plants, on average, decreased their size less than manufacturing ones in Germany, Japan and United Kingdom, whereas in Italy they had roughly the same rate of decrease. Petroleum, industrial chemicals, and resins had a greater average size of plants.

If we look at national experiences, we see that German chemicals had a larger absolute size in comparison with the other three countries, and showed a major difference in the size of plants in total manufacturing. Japan, in contrast had the smallest average size both in chemicals and in total manufacturing, due to its predominant form of organization based on networks of firms.

Different branches showed different sizes. The largest ones were those in oil production, while the smallest were producers of specialties

Table A.30. French Value Added Per Worker in the Production of Chemicals by Branch (1963–1991, Thousand U.S. Dollars 1990)

	Industrial Chemicals	Other Chemicals	Petroleum	Petroleum and Coal	Rubber	Plastic
1963	25.70		130.26			
1970	30.07		158.15			
1980	69.88	35.21	375.26	140.84	35.01	32.86
1990	87.40	78.42	730.99		36.73	58.81

Source: Processed from *Statistical Yearbook* and *Industrial Statistical Yearbook.* Various years.

Table A.31. Italian Value Added Per Worker in the Production
of Chemicals by Branch (1963–1991, Thousand U.S. Dollars
1990)

	Industrial Chemicals	Petroleum	Rubber	Plastic
1970	35.46	72.11	22.37	17.34
1980	57.15	100.43	40.96	28.34
1989	71.26	100.27	46.73	46.73

Sources: Processed from *Statistical Yearbook* and *Industrial Statistical
Yearbook.* Various years.

and plastics.There were no common patterns for plant size in different countries. Germany showed the largest average size of plants in any branch, except for petroleum production in which it lagged behind the United Kingdom; Japan showed the smallest one. For example, in petroleum production, Germany had an average size of around 370 employees; Japan had an average size that was one-third of the German. Moreover, except for plastics, German chemicals plants had roughly the same average size among the different branches, while in the other countries they varied significantly. In every branch of chemicals, the largest plants were in the 1980s. Except for Japan, petroleum production plants were the largest, followed by those for industrial chemicals. Only in the production of plastics did plant size decrease between 1970 and 1980.

Italian chemicals plants were also quite large. Industrial chemicals and petroleum plants were approximately the same size as German plants, whereas rubber and plastics were smaller than German plants but greater than Japanese and even British plants.

The United Kingdom showed a steady drop in the size of its chemicals plants in almost every branch. Almost all branches had already decreased their size in 1980, in contrast to other national experiences. Petroleum, for example, decreased from 720 employees to 354 between 1970 and 1980 and an even more drastic reduction of plant size continued in the following decade.

A relatively small average size characterized the Japanese chemicals plants, where plants for the production of plastics were about six times smaller than German ones. Synthetic resin plants were the largest, reaching around the same size as Western plants.

Table A.32. Employment by Plant in the Production of Chemicals by Branch and Total Manufactures (Number of Workers Per Plant. 1970–1980–1990)

	Industrial Chemicals	Basic Chemicals	Synthetic Resins	Other Chemicals	Drugs Medicines	Petroleum	Petroleum and Coal	Rubber	Plastic	Total Chemicals	Total Manufacture
1970	763.06	241.91	969.87	181.71	274.01	1233.49	648.76	796.21	210.34	520.77	344.71
1980	1035.97	239.55	660.73	161.31	290.99	1321.87	363.93	857.08	189.75	881.66	456.88
1990	557.37	202.15	307.72	161.27	296.48	759.97	75.45	499.13	178.95	328.69	226.46

Source: Processed from *Statistical Yearbook* and *Industrial Statistical Yearbook*. Various years.

Table A.33. Employment by Plant in the Total Production of Chemicals and Manufactures. Germany, Italy, Japan, and UK: 1970–1980–1990. Number of Workers Per Plant

	Germany		Italy		Japan		UK	
	Tot. Chemic.	Tot. Manuf.	Tot. Chemic.	Tot. Manuf.	Tot. Chemic.	Tot. Manuf.	Tot. Chemic.	Tot. Manuf.
1970	218.06	151.83			37.35	17.99	148.29	100.07
1980	455.68	162.48	182.61	134.02	29.00	15.42	93.78	66.41
1990	231.23	164.27	109.78	91.99	32.24	25.63	65.21	36.56

Source: Processed from *Statistical Yearbook* and *Industrial Statistical Yearbook. Various years.*

Table A.34. German Employment Per Plant in the Production
of Chemicals by Branch. (1970-1980-
1990. Number of Workers Per Plant)

	Industrial Chemicals	Petroleum	Rubber	Plastic
1970	284.50	309.73	385.67	96.25
1980	348.20	377.78	370.11	
1990	369.40	283.95	357.66	120.88

Source: Processed from *Statistical Yearbook* and *Industrial Statistical Yearbook*. Various years.

FIRMS

In 1990, among the world's top thirty chemicals companies by sales, there were seven companies from the United States, five from Germany and Japan, three from Britain and France, and one from Switzerland. There were no Italian companies, but in 1960, Montecatini was the eighth largest producer of chemicals in the world. In 1960, British ICI was the third chemicals company in the world, followed by Union Carbide and then by the "big three" of Germany. In 1980, Hoechst, Bayer, and BASF were the top three.

Du Pont (E.I.) de Nemours, the top producer of chemicals in 1960 and 1990, increased its sales, but not its profits. In 1960, they accounted for about 17% of total sales; in 1990, these fell to 5%. The other companies also showed the same trend. Imperial Chemicals Industries decreased its profits on total sales by 8% in 1960, to 5% in 1990. Monsanto decreased from 8% in 1960 to 6% in 1990.

Table A.35. Italian Employment Per Plant in the Production
of Chemicals by Branch. 1973-1980-1989. Number of Workers
Per Plant

	Industrial Chemicals	Petroleum	Petroleum and Coal	Rubber	Plastic
1973	232.60			253.47	81.11
1980	250.96	358.49		240.60	79.57
1989	270.11	379.31	82.19	139.88	63.09

Source: Processed from *Statistical Yearbook* and *Industrial Statistical Yearbook*. Various years.

Table A.36. United Kingdom Employment Per Plant in the Production of Chemicals by Branch (1970–1980–1990).
Number of Workers Per Plant)

	Industrial Chemicals	Basic Chemicals	Synthetic Resins	Other Chemicals	Drugs Medicines	Petroleum	Petroleum and Coal	Rubber	Plastic
1970	197.47	165.67	331.13	129.58	200.65	720.00	595.35	258.59	51.43
1980	142.56	153.94	129.41	111.99	211.59	354.17	89.39	143.71	48.26
1990	91.37	130.21	47.00	97.83	209.94	360.00	62.02	105.79	36.99

Source: Processed from *Statistical Yearbook* and *Industrial Statistical Yearbook.* Various years.

Table A.37. Japanese Employment Per Plant in the Production of Chemicals by Branch (1970-1980-1990. Number of Workers Per Plant)

	Industrial Chemicals	Basic Chemicals	Synthetic Resins	Other Chemicals	Drugs Medicines	Petroleum	Petroleum and Coal	Rubber	Plastic
1970	119.97	76.24	288.75	52.13	73.36	120.42	31.19	42.86	18.17
1980	113.49	85.61	334.76	49.32	79.39	164.77	253.12	19.81	13.43
1990	96.60	71.93	260.71	63.45	86.54	116.02	13.44	35.68	21.08

Sources: Processed from *Statistical Yearbook* and *Industrial Statistical Yearbook.* Various years.

Table A.38. Top Chemicals Companies Ranked by Sales and (by Profit) 1960-1970-1980-1990

Companies	COUNTRY	Rank in 1960	Rank in 1970 (Only in US)	Rank in 1980	Rank in 1990	Differences in Sales (Million U.S. Dollars) 1960-1990
E.I. DU PONT DE NEMOUR	US	15 (5)*	18	38	22 (9)	37,697
HOECHST	GERMANY	97		29	33 (52)	27,106
BAYER	GERMANY	68		30	39 (40)	25,272
IMPERIAL CHEMICAL INDUSTRIES	BRITAIN	24		40	44 (39)	21,783
DOW CHEMICAL	US	70 (21)*	51	55	53 (32)	19,224
CIBA-GEIGY	SWITZERLAND			96	80 (69)	
RHONE-POULENC	FRANCE	220		95	81 (163)	14,140
NORSK HYDRO	NORWAY				135 (120)	
MONSANTO	US	60 (29)	47	106	146 (103)	7,076
ASHAI GLASS	JAPAN			363	151 (174)	
UNION CARBIDE	US	27 (12)*	24	57	144 (180)	6,073

Rank by profit in 1960 refers only to U.S. companies.

CONCLUSIONS

There is much structure and little strategy in the picture of the chemicals industry provided above. With these limitations in mind, we can summarize the main conclusions in three points.

First, the evolution in the chemicals industry during the post–World War II period had three phases: the first one covered the rise of oil-based chemicals, with the United States as the leading country; the second one was the "Age of Petroleum," with a rapid convergence of all countries on the new technology in the 1960s and 1970s, accompanied by the astonishing rise of the Japanese chemicals industry; the third phase was the post–oil crisis, when the rate of growth of the chemicals industry dramatically decreased everywhere. During this phase, Germany emerged as the leading country. Indeed, both Germany and Japan were better able to recover from the oil crisis, and in this way acquired new shares of the chemicals markets for new products. The second point is that, in the long run, each country shows a path of specialization in different branches. If we look, for example, at the specialization index in the chemicals trade, we can see that the United States specialized in fertilizers, whereas Germany specialized in dyes, tanning, and color product. France and the United Kingdom specialized in perfumes and cleaning goods; Japan in fertilizers and rubber, as was also the case with Italy. The third point is that the chemicals industry generally shows decreasing employment by plant, and even, within the same branch, a heterogeneous average plant size in the different countries. If we disregard petroleum refineries, where economies of scale are very high, rubber, for example, shows an employment by plant of 499 in Germany in 1990, 63 in Italy (1989), 35.68 in Japan (1990), and 105 in the United Kingdom (1990) (Tables A.33–A.38).

ADDENDUM

Chemical trade data refers to Standard International Trade Classification revision 2 (SITC rev.2). During the period 1952-92, three revisions have been changed: SITC original classification, SITC revised, and SITC revised, 2. Equalization between them is made using UN statistical papers (1975, 1961, 1995).

SITC REV.2 code	Description	
51	Organic chemicals	Hydrocarbons, n.e.s., and their halogenated, sulphonated, nitrated or nitrosated derivatives; alcohols, phenols, phenol-alcohols, and their halogenated; carboxilic acids, and their anhydrides, halides, peroxides and peracids, and their halogenated; nitrogen-function compounds; organo-inorganic and heterocyclic compounds; other organic chemicals
52	Inorganic chemicals	Inorganic chemical elements, oxides and halogen salts; other inorganic chemicals, organic and inorganic compounds of precious metals; radioactive and associated materials
53	Dyes, tanning and coloring materials	Synthetic organic dyestuffs, natural indigo and color lakes; dyeing and tanning extracts, and synthetic tanning materials; pigments, paints, varnishes and related materials
54	Medicinal and pharmaceutical products	Medicinal and pharmaceutical products;
55	Essential oils and perfume materials, toilet, polishing and cleaning preparations	Essential oils, perfume and flavor materials; perfumery, cosmetics and toilet preparations, aqueous distillates and aqueous solutions of essential oils; soap, cleansing and polishing preparations
56	Fertilizers, manufactures	Fertilizers, manufactured
57	Explosives and pyrotechnic products	Explosives and pyrotechnic products
58	Artificial resins and plastic materials, and cellulose esters and ethers	Condensation, polycondensation and polyassition products, whether or not modified or polymerized, and whether or not linear (e.g., phenoplasts, aminoplasts, alkyds, polyallyl esters and other unsatured polyesters, silicones); polymerization and copolymerization products
59	Chemical materials and products, N.E.S.	Disinfectants, insecticides, fungicides, weed killers, antisprouting products, rat poison and similar products, put up in forms or packing for sail by retail or as preparations or as articles; starches, inulin and wheat gluten, albuminoidal substances, glues; miscellaneous chemical products, n.e.s.
62	Rubber products	Materials of rubber; rubber tires, tire cases, interchangeble tire treads, inner tubes and tire flaps, for wheels of all kinds; articles of rubber, n.e.s
33	Petroleum, petroleum products and related material	Petroleum oils, crude, and crude oils obtained from bituminous minerals; petroleum products, refined; residual petroleum products, n.e.s and related materials

Chemical production data refer to International Standard Industrial Classification revision 2 (ISIC rev.2). In the post–World War II era, there were three different revisions of the classification. The equalization of the revisions is made using the data from a UN annex table. All pictures and tables refer to the branches listed here.

ISIC Code	Description	
351	Industrial chemicals	Basic chemicals, fertilizers, synthetic resins, paints, drugs and medicines, soap, cleaning products, printers ink, explosives, photographic film, sensitized, in rolls, photographic paper
3511	Basic chemicals except fertilizers	Sulphur, styrene, acetylene, benzene, butylenes, ethylene, naphthalene, propylene, toluene, xylenes, trichloroethylene, methanol, butyl alcohol, ethanediol, glycerine, ethylene oxide, acetaldehyde, methanal, acetone, acetates, acetic acid, formic acid, phtalic anhydride, aniline, acrylonitrile, chlorine, hydrochloric acid, sulphuric acid, nitric acid, phosphoric acid, carbon bisulphite, zinc oxide, titanium oxides, lead oxides, ammonia, caustic soda, aluminium oxide, hydrated alumina, aluminium sulphate, copper sulphate, soda ash, sodium silicates, hidrogen peroxide, calcium carbide, dyestuffs, synthetic, vegetable tanning extracts, lithopone, activated carbon
3513	Synthetic resins	Rubber, synthetic, noncellulosic staple and tow, cellulosic staple and tow, alkyd resins, amino plastics, phenolic and cresylic plastics, polyethylene, polypropylene, polystyrene, polyvinychloride, regenerated cellulose, noncellulosic continuous filaments
352	Other chemical products	Paints-cellulose, paints-water, paints-other, mastics
3522	Drugs and medicines	Drugs and medicines
353	Petroleum refineries	Aviation gasolene, jet fuels, motor gasolene, naphthas, kerosene, white spirit, industrial spirit, gas-diesel oil, residual fuel oils, lubricants, petroleum wax, petroleum coke, bitumen, liquefied petroleum gas from natural plants, liquefied petroleum gas from petroleum refineries
354	Petroleum, coal products	hard coal briquettes, brown-coal briquettes, coke, coke-oven gas, tars
355	Rubber products	Inner tubes for motor vehicles, inner tubes for bicycles and motorcycles, tires for agricultural, bicycles, motorcycles and for road motor vehicles, rubber reclaimed, rubber-unhardened vulcanized plates, sheets, rubber-unhardened vulcanized piping and tubing, rubber hardened, rubber- transmission, conveyor, elevator belts, rubber footwear
356	Plastic products, etc.	Plastic footwear

REFERENCES

Aftalion, Fred, *A History of the International Chemical Industry.* Philadelphia. University of Pennsylvania Press, 1991.

Arora, Ashish, Ralph Landau, and Nathan Rosenberg, eds., *Chemicals and Long-Term Economic Growth.* New York. John Wiley & Sons, 1998.

Balance, Robert Janos Pogany, and Helmut Forstner, *The World's Pharmaceutical Industries, An International Perspective on Innovation, Competition and Policy.* Aldershot, Hants, England. Edward Elgar, 1992.

Burgess, T., B. Hwang, N. Shaw, C. De Mattos, *"Enhancing Value Stream Agility: The UK Speciality Chemical Industry."* European Management Journal, Vol. 20, no. 2, April 2002, pp. 199–212.

Grant, W., *"The Overcapacity Crisis in the West European Petrochemicals Industry,"* in Alberto Martinelli, ed., *International Markets and Global Firms, A Comparative Study of Organized Business in the Chemical industry.* Sage, London, Newbury Park, New Delhi., 1991.

Mussati, G., Soru, A. (1991), *"International Markets and Competitive Systems,"* in Alberto Martinelli, ed., *International Markets and Global Firms.*

OECD (1985), *The Pharmaceutical Industry, Trade Related Issues.* Paris.

United Nations (1961), *Standard International Trade Classification Revised,* in "Statistical Papers," Series M 34, New York, United Nations.

United Nations (1975), *Standard International Trade Classification Revision 2,* in "Statistical Papers," Series M 34/Rev 2, New York, United Nations.

United Nations (1995), *Standard International Trade Classification Revision 3,* in "Statistical Papers," Series M 34/Rev 3, New York, United Nations.

United Nations (various years), *Annual Bulletin of Trade in Chemical Products.* New York, United Nations.

United Nations (various years), *Annual Review of the Chemical Industry.* New York United Nations.

United Nations (various years), *Industrial Statistics Yearbook.* New York. United Nations.

United Nations (various years), *International Trade Statistics Yearbook.* New York, United Nations.

United Nations (various years), *International Yearbook of Industrial Statistics.* New York, United Nations.

United Nations (various years), *Statistics Yearbook.* New York, United Nations.

United Nations (various years), *The Chemical Industry Annual Review.* New York, United Nations.

United Nations (various years), *The Growth of World Industry.,* New York, United Nations.

United Nations (various years), *Yearbook of Industrial Statistics.* New York, United Nations.

United Nations (various years), *Yearbook of International Trade Statistics.* New York, United Nations.

United Nations (various years), *Yearbook of International Trade Statistics.* New York, United Nations.

United Nations (various years), *Yearbook of National Accounts Statistics.* New York, United Nations.

Selected Bibliography

Abelshauser, Werner, ed. *Die BASF. Eine Unternehmensgeschichte.* München: C.H. Beck, 2002.

Abescat, Bruno, La saga des Bettencourt. L'Oréal: une fortune française. Paris: Plon, 2002.

Achilladelis, Basil and Nicholas Antonakis. "The Dynamics of Technological Innovation: The Case of the Pharmaceutical Industry." *Research Policy* 30 (2001): 535–588.

Achilladelis, Basil, Albert Schwartzkopf, and Martin Cines, "A Study of Innovation in the Pesticide Industry: Analysis of the Innovation Record of an Industrial Sector." *Research Policy* 16 (August 1987): 175–212.

Aftalion, Fred. *History of the International Chemical Industry.* Philadelphia: University of Pennsylvania Press, 1989.

Allen, C. S. "Political Consequences of Change: The Chemical Industry." In Peter J. Katzenstein, ed. *Industry and Politics in West Germany.* Ithaca: Cornell University Press, 1989.

Amatori, Franco, and Bruno Bezza, eds. *Montecatini, 1888–1966: Capitoli di Storia di Una Grande Impresa.* Bologna: Il Mulino, 1991.

Berthoin, Antal. *The Transformation of Hoechst to Aventis: Case Study.* Berlin: WZB, 2001.

Aoki, Masahiko, and Hugh Patrick, eds. *The Japanese Main Bank System.* Oxford: Oxford University Press, 1994.

Arora, Ashish. "Patents, Licensing, and Market Structure in the Chemical Industry." *Research Policy* 26 (1997): 391–403.

Arora, Ashish, Ralph Landau, and Nathan Rosenberg, eds. *Chemicals and Long-Term Economic Growth: Insights from the Chemical Industry.* New York: John Wiley & Sons, 1998.

Arora, Ashish, Andrea Fosfuri, and Alfonso Gambardella. *Markets for Technology: The Economics of Innovation and Corporate Strategy.* Cambridge, MA: MIT Press, 2001.

Arora, Ashish, and Andrea Fosfuri. "Licensing the Market for Technology," *Journal of Economic Behavior and Organization* 52:2 (2003): 277–295.

Asso, Pier Francesco, and Marcello De Cecco. *Storia del Crediop: tra credito speciale e finanza pubblica, 1920–1960.* Roma: Editori Laterza, 1994.

Azzolini, Riccardo, Giorgio Dimalta, and Roberto Pastore. *L'industria chimica tra crisi e programmazione.* Roma: Editori Riuniti, 1979.

Backman, Jules. *The Economics of the Chemical Industry.* Washington, DC: Manufacturing Chemists Association, 1970.

Balance, Robert, János Pogány, and Helmut Forstner. *The World's Pharmaceutical Industries.* Aldershot: Edward Elgar, 1992.

Bamberg, James. *British Petroleum and Global Oil, 1950–1975: The Challenge of Nationalism.* Cambridge: Cambridge University Press, 2000.

Barnes, Pamela M., and Ian G. Barnes. *Environmental Policy in the European Union.* Cheltenham: Edward Elgar, 1999.

Barnett, Correlli. *The Audit of War: The Illusion and Reality of Britain as a Great Nation.* London: Macmillan, 1986.

Bathelt, Harald. "Global Competition, International Trade, and Regional Concentration: The Case of the German Chemical Industry during the 1980s." *Environment and Planning C: Government and Policy* 13 (1995): 411–12.

Bathelt, Harald. *Chemiestandort Deutschland: technologischer Wandel, Arbeitsteilung und geographische Strukturen in der chemischen Industrie.* Berlin: Sigma, 1997.

Bäumler, Ernst. *Ein Jahrhundert Chemie.* Düsseldorf: Econ, 1968.

Becker, Steffen and Thomas Sablowski, "Konzentration und Industrielle Organisation. Das Beispiel der Chemie- und Pharmaindustrie" in PROKLA, Zeitschrift für kritische Sozialwissenschaft 113, no. 4 (1988): 616–641.

Beer, John. *The Emergence of the German Dye Industry.* Urbana: University of Illinois Press, 1959.

Beltran, Alain and Sophie Chauveau. *Elf, des origines à 1989.* Paris: Fayard, 1999.

Benghozi, Pierre-Jean, Florence Charue-Duboc, and Christophe Midler, eds. *Innovation Based Competition and Design Systems Dynamics.* Paris: Harmattan, 2000.

Ben Mahmoud-Jouini, S., and Christophe Midler. "Compétition par l'innovation et dynamique des systèmes de conception dans les entreprises françaises. Réflexions à partir de la confrontation de trois secteurs." *Entreprises et Histoire*, 23 (1999): 36–62.

Bertini, G., P. Delmonte, and G. Rosa, "Ristrutturazione e cambiamento nell'industria chimica: l'esperienza degli anni Ottanta," *Rivista di Politica Economica*, 1987: 905–32.

Bibard, L. et al., "Recherche et développement et stratégie: Rhône-Poulenc Agrochimie et Rhône-Poulenc Santé. In *Stratégie technologique et avantage concurrentiel, rapport de recherche IREPD.* Grenoble: 1993.

Blair, Margaret, ed. *The Deal Decade*. Washington, DC: Brookings Institution, 1993.

Bonin, Hubert et al. *Transnational Companies, 19th-20th Centuries*. Paris: P.L.A.G.E., 2002.

Boswell, Jonathan, and James Peters. *Capitalism in Contention: Business Leaders and Political Economy in Modern Britain*. Cambridge and New York: Cambridge University Press, 1997.

Bozdogan, Kirkor. "The Transformation of the U.S. Chemical Industry." Working Paper of the MIT Commission on Industrial Productivity (1989).

Bram, Georges et al. *La chimie dans la société: son rôle, son image: actes du colloque interdisciplinaire du Comité national de la recherche scientifique*. Paris: Harmattan 1995.

Brandt, E. N. *Growth Company: Dow Chemical's First Century*. East Lansing: Michigan State University Press, 1997.

Brickman, Ronald, Sheila Jasanoff, and Thomas Ilgen. *Controlling Chemicals: The Politics of Regulation in Europe and the United States*. Ithaca, NY: Cornell University Press, 1985.

Brown, Shona L., and Kathleen M. Eisenhardt. "The Art of Continuous Change: Linking Complexity Theory and Time-Paced Evolution in Relentlessly Shifting Organizations." *Administrative Science Quarterly*, 42, no. 1 (Mar. 1997): 1-34.

Buchholz, Klaus. "Die gezielte Förderung und Entwicklung der Biotechnologie." In Wolfgang van den Daele, Wolfgang Krohn, and Peter Weingart, eds. *Geplante Forschung*. Frankfurt: Suhrkamp, 1979.

Burgess, T. et al. "Enhancing Value Stream Agility: The UK Speciality Chemical Industry." *European Management Journal*, 20, no. 2 (April 2002): 199-212.

Busset, Thomas, et al., eds. *Chemie in der Schweiz. Geschichte der Forschung und der Industrie* (Chemicals in Switzerland. History of Research and of the Industry). Basle: 1997.

Bustelo, Francisco. "Notas y comentarios sobre los orígenes de la industria española del nitrógeno." *Moneda y Crédito* 63 (1957).

Carosso, Vincent. *Investment Banking in America*. Cambridge: Cambridge University Press, 1970.

Catalán, Jordi. *La economía española y la segunda guerra mundial*. Barcelona: Editorial Ariel, 1995.

Cayez, Pierre. *Rhône-Poulenc, 1895-1975*. Paris: Colin et Masson, 1988.

Cesaroni, Fabrizio, Alfonso Gambardella and Walter Garcia-Fontes, eds. *R&D, Innovation, and Competitiveness in the European Chemical Industry*. Dordrecht: Kluwer, 2004.

Chandler, Alfred. *The Visible Hand*. Cambridge, MA: Belknap Press, 1977.

Chandler, Alfred and Stephen Salsbury. *Pierre S. Du Pont and the Making of the Modern Corporation*. New York: Harper & Row, 1971.

Chandler, Alfred D. *Scale and Scope: The Dynamics of Industrial Capitalism*. Cambridge, MA: Harvard University Press, 1990.

Chandler, Alfred D., Jr. *Shaping the Industrial Century: The Remarkable Story of the Modern Chemical and Pharmaceutical Industries*. Cambridge, MA: Harvard University Press, 2005.

Chapman, Keith. *The International Petrochemical Industry: Evolution and Location*. Oxford, UK; Cambridge, MA: Basil Blackwell, 1991.

Charue-Duboc, Florence. "Maîtrise d'oeuvre, maîtrise d'ouvrage et direction de projet, pour comprendre l'évolution des projets chez Rhône-Poulenc." *Gérer et Comprendre*, 49 (1997): 54–64.

Charue-Duboc, Florence, and Christophe Midler. "Le développement du management de projet chez Rhône-Poulenc." Rapport de recherche Rhône-Poulenc, 1994.

Charue-Duboc, Florence, and Christophe Midler. "Le développement du management de projet chez Rhône-Poulenc – II," Rapport de recherche Rhône-Poulenc, 1995.

Charue-Duboc, Florence, and Christophe Midler. "Le développement du management de projet chez Rhône-Poulenc – III." Rapport de recherche Rhône-Poulenc, 1998.

Clark, Kim B., and Takahiro Fujimoto. *Product Development Performance: Strategy, Organization, and Management in the World Auto Industry*. Boston, MA: Harvard Business School Press, 1991.

Cockburn, Iain, and Rebecca Henderson. "The Economics of Drug Discovery." In Ralph Landau, Basil Achilladelis, and Alexander Scriabine, eds. *Pharmaceutical Innovation*. Philadelphia: Chemical Heritage Press, 1999.

Cohendet, Patrick, ed., *La chimie en Europe: innovations, mutations et perspectives*. Paris: Economica, 1984.

Cohendet, Patrick, J. A. Herault, and M. Ledoux, "Quelle chimie pour l'an 2000?" *La Recherche*, 166 (1989): 1254–1257.

Colli, Andrea, and V. Maglia, "Medie e piccole imprese nella chimica italiana." *L'Industria*, XXVI, no. 2 (2005): 321–360.

Collins, Michael. *Banks and Industrial Finance in Britain, 1800–1939*. London: MacMillan, 1991.

Confalonieri, Antonio. *Banca e Industria in Italia, 1894–1906*. Milano: Banca Commerciale Italiana, 1976.

Confalonieri, Antonio. *Banca e Industria in Italia dalla crisi del 1907 all'agosto 1914*. Milano: Banca Commerciale Italiana, 1982.

Cottrell, Philip. *Industrial Finance 1830–1914*. London: Methuen, 1980.

Dalle, Francois. *L'Aventure l'Oréal*. Paris: Odile Jacob, 2001.

D'Antone, Leandra, ed. *Radici storiche ed esperienza dell'intervento straordinario nel Mezzogiorno*. Rome: Bibliopolis, 1996.

Da Rin, Marco. "German Kreditbanken 1850–1914: An Informational Approach." *Financial History Review*, 3, no. 2 (1996), 29–47.

Da Rin, Marco. "Finance and Technology in Emerging Industrial Economies: The Role of Economic Integration," *Research in Economics*, 51, no. 3 (Sept. 1997): 171–200.

Da Rin, Marco, and Thomas Hellmann. "Banks as Catalysts for Industrialization." *Journal of Financial Intermediation*, 11, no. 4 (2002): 366–97.

D'Attorre, Pier Paolo, and Vera Zamagni, eds. *Distretti, imprese e classe operaia. L'industrializzazione dell'Emilia-Romagna*. Milan: Angeli, 1992.

Davenport-Hines, R. P. T. *Dudley Docker: The Life and Times of a Trade Warrior.* Cambridge: Cambridge University Press, 1984.

Dell, Edmund. *A Strange Eventful History: Democratic Socialism in Britain.* London: HarperCollins, 2000.

Diego, Emilio de. *Historia de la industria en España. La química*. Madrid: 1996.

Dunning, John H. *Alliance Capitalism and Global Business*. London and New York: Routledge, 1997.

Dunning, John H. "Trade, Location of Economic Activity and the MNE: A Search for an Eclectic Approach," in Bertil Ohlin, Bertil Gotthard Ohlin, Per-Ove Hesselborn, and Per Magnus Wijkman, eds. *The International Allocation of Economic Activity: Proceedings of a Nobel Symposium Held at Stockholm*. London: Holmes & Meier, 1977.

Edgerton, David E. H., ed. *Industrial Research and Innovation in Business*. Cheltenham, UK, and Brookfield, VT: Edward Elgar, 1996.

Eisenhardt, Kathleen, and Behnam Tabrizi. "Accelerating Adaptive Processes: Product Innovation in the Global Computer Industry." *Administrative Science Quarterly*, 40, no. 1 (Mar. 1995): 84–110.

Elbaum, Bernard, and William Lazonick, eds. *The Decline of the British Economy*. Oxford: Clarendon Press, 1987.

Erni, Paul. *Die Basler Heirat, Geschichte der Fusion Ciba-Geigy* (The Marriage of Basle. History of the Merger Ciba-Geigy). Zurich: Buchverlag der Neuen Zürcher Zeitung, 1979.

European Environmental Agency. *Europe's Environment: The Second Assessment*. Luxembourg: Office for Official Publications of the European Communities, 1998.

Fauri, F. "The 'Economic Miracle' and Italy's Chemical Industry, 1950-1965: A Missed Opportunity." In *Enterprise and Society*, 1, no. 2 (June 2000): 279–314.

Fedor, Walter S. "Thermoplastics: Progress amid Problems." *Chemical and Engineering News*, 39 (May 29, 1961): 80-92.

Feldenkirchen, Wilfred. "Banking and Economic Growth: Banks and Industry in Germany in the Nineteenth Century and Their Changing Relationship during Industrialization." In Wang Lee, ed. *German Industry and German Industrialization*. London: Routledge, 1991.

Flechtner, H. J. *Carl Duisberg; vom Chemiker zum Wirtschaftsführer*. Düsseldorf, Econ Verlag, 1959.

Franke, J. F., and F. Wätzold, "Voluntary initiatives and public intervention – the regulation of eco-auditing." In Francois Lévêque, ed., *Environmental Policy in Europe*. Cheltenham: Edward Elgar, 1999, 175-199.

Freeman, Chris. *The Economics of Industrial Innovation*. London: Francis Pinter, 1982.

Freeman Chris. "Chemical Process Plant: Innovation and the World Market." *National Institute Economic Review*, 45 (Aug. 1968): 29-51.

Gaffard, J. L., et al. *Cohérence et diversité des systèmes d'innovation en Europe, rapport de synthèse du FAST*. Vol. 19, CEE, Bruxelles, 1993.

Galambos, Louis, with Jane Eliot Sewell. *Networks of Innovation: Vaccine Development at Merck, Sharp & Dohme, and Mulford, 1895-1995*. New York: Cambridge University Press, 1995.

García Delgado, J. L., ed. *Economía española, cultura y sociedad. Homenaje a Juan Velarde Fuertes*. Madrid, 1992.

Geilinger-Schnof, Ulrich. *175 Jahre Chemie Uetikon. Die Geschichte der Chemische Fabrik Uetikon von 1818 bis 1993* (175 Years Chemicals Uetikon. The History of the Chemical Works Uetikon from 1818 to 1993). Uetikon, 1993.

Goertz, Gary. *The World Chemical Industry Around 1910. A Comparative Analysis by Branch and Country.* Geneva: Centre of International Economic History, University of Geneva, 1990.

Goldsmith, Raymond. *Financial Intermediaries in the American Economy Since 1900.* Princeton, NJ: Princeton University Press, 1958.

Grabower, Rolf. *Die finanzielle Entwicklung der Aktiengesellschaften der deutschen chemischen Industrie.* Leipzig: Duncker & Humblot, 1910.

Grant, Wyn. *Government and Industry.* Aldershot: Edward Elgar, 1989.

Grant, Wyn. "Government-Industry Relationships in the British Chemical Industry." In Martin Chick, ed. *Governments, Industries, and Markets: Aspects of Government-Industry Relations in the UK, Japan, West Germany, and the USA since 1945.* Cheltenham: Edward Elgar, 1990, 142–56.

Grant, Wyn. "The Overcapacity Crisis in the West European Petrochemicals Industry," in Alberto Martinell, ed. *International Markets and Global Firms: A Comparative Study of Organized Business in the Chemical industry.* London and Newbury Park, CA: Sage Publications, 1991.

Grant, Wyn. *Pressure Groups and British Politics.* Basingstoke: Macmillan, 2000.

Grant, Wyn, Alberto Martinelli, and William Paterson. "Large Firms as Political Actors: A Comparative Analysis of the Chemical Industry in Britain, Italy and West Germany." *West European Studies*, 12, no. 2 (1989): 75–76.

Grant, Wyn, Duncan Matthews, and Peter Newell. *The Effectiveness of European Union Environmental Policy.* Basingstoke: Macmillan, 2000.

Grant, Wyn, William Paterson, and Colin Whitston. *Government and the Chemical Industry: A Comparative Study of Britain and West Germany.* Oxford: Clarendon Press, 1988.

Haber, L. F. *The Chemical Industry during the Nineteenth Century.* Oxford: Oxford University Press, 1958.

Haber, L. F. *The Chemical Industry, 1900–1930: International Growth and Technological Change.* Oxford: Clarendon Press, 1971.

Hall, Bronwyn. "The Impact of Corporate Restructuring on Industrial Research and Development." Brookings Papers on Economic Activity: Microeconomics. Washington, DC: Brookings Institution, 1990.

Hall, Peter A. *The Political Power of Economic Ideas: Keynesianism across Nations.* Princeton, NJ: Princeton University Press, 1989.

Hanisch, Tore Jørgen, and Gunnar Nerheim. *Norsk oljehistorie. Fra vantro til overmot*, 3 vols. Oslo: Leseselskapet, 1992–97.

Hansen, Kurt. "Die chemische Industrie von 1945 bis 2050." *Nachrichten aus Chemie, Technik und Laboratorium* 47 (1999): 1039.

Hansen, Povl A., and Görin Serin. *Plast. Fra galanterivarer til "high-tech." Om innovationsudviklingen i plastindustrien.* København, 1989.

Hardach, Gerd. "Banking and Industry in Germany in the Interwar Period 1919–39." *Journal of European Economic History*, 13 (1984): 203–34.

Hart, Oliver. *Firms, Contracts and Financial Structure*. Oxford: Oxford University Press, 1995.

Haynes, Williams. *American Chemical Industry*. 6 vols. New York: Van Nostrand, 1945-54.

Hilger, Susanne. *Die "Amerikanisierung" deutscher Unternehmen nach dem Zweiten Weltkrieg. Einflüsse auf Unternehmenspolitik und Wettbewerbsstrategien bei Henkel, Siemens und Daimler-Benz (1945-1975)*. Nürnberg-Erlangen: Habilitationsschrift Universität, 2002.

Hounshell, David A., and John Kenly Smith, Jr. *Science and Corporate Strategy: DuPont R&D, 1902-1980*. New York: Cambridge University Press, 1988.

Iijima, Takashi. *Nippon no Kagaku Gijyutsu: Kigyoshi ni miru sono Kozo*. Tokyo: Kogyo Chosakai, 1981.

Jemain, Alain. *Les conquérants de l'invisible, Air liquide 100 ans d'histoire*. Paris: Fayard, 2002.

Johnson, Jeffrey Allan. *The Kaiser's Chemists: Science and Modernization in Imperial Germany*. Chapel Hill: University of North Carolina Press, 1990.

Jones, Geoffrey. *Multinationals and Global Capitalism: From the Nineteenth to the Twenty-First Century*. Oxford and New York: Oxford University Press, 2005.

Kaplan, Steven. "Top Executive Rewards and Firm Performance: A Comparison of Japan and the United States." *Journal of Political Economy*, 102, no. 3 (1994): 510-46.

Karlsch, Rainer. "Capacity Losses, Reconstruction, and Unfinished Modernization: The Chemical Industry in the Soviet Zone of Occupation (SBZ)/GDR, 1945-1965." In John E. Lesch, ed., *The German Chemical Industry in the Twentieth Century*. Dordrecht: Kluwer, 2000, 375-92.

Karlsch, Rainer, and Raymond G. Stokes. *Faktor Öl. Mineralölwirtschaft in Deutschland 1859-1974*. München: C. H. Beck, 2003.

Keck, Otto. "The National System for Technical Innovation in Germany." In Richard R. Nelson, ed. *National Innovation Systems: A Comparative Analysis*. New York: Oxford University Press, 1993.

Keegan, William. *Mrs. Thatcher's Economic Experiment*. Harmondsworth: Penguin, 1984.

Kennedy, Carol. *ICI: The Company that Changed Our Lives*. London: Hutchinson, 1986.

Kennedy, William. *Industrial Structure, Capital Markets and the Origins of British Economic Decline*. Cambridge: Cambridge University Press, 1987.

Klein, Heribert. *Operation Amerika. Hoechst in den USA*. München: Piper, 1996.

Khoury, Sarkis. *The Deregulation of the World Financial Markets*. New York: Quorum Books, 1985.

Kudo, Akira, and Terushi Hara, eds. *International Cartels in Business History*. Tokyo: University of Tokyo Press, 1992.

Lamoreaux, Naomi R., Daniel M. G. Raff, and Peter Temin P., eds. *Learning by Doing in Firms, Markets and Countries*. Chicago: University of Chicago Press, 1999.

Lee, Wang, ed. *German Industry and German Industrialization: Essays in German Economic and Business History in the Nineteenth and Twentieth Centuries.* London: Routledge, 1991.

Lesch, John E., ed. *The German Chemical Industry in the Twentieth Century.* Dordrecht: Kluwer, 2000.

Liebermann, Marvin. "Exit from Declining Industries: 'Shakeout' or 'Stakeout'?" *Rand Journal*, 21, no. 4 (1990): 538–54.

Lobo, Félix. "El crecimiento de la industria farmacéutica en España durante los felices años del estado de bienestar." *Economía Industrial* 223, (1983): 121–133.

Loscertales, Javier. *Deutsche Investitionen in Spanien, 1880–1920.* Stuttgart: 2002.

Mansfield, Edwin, et al. *The Production and Application of New Industrial Technology.* New York: Norton, 1977.

Marchi, Alves, and Roberto Marchionatti. *Montedison, 1966–1989: l'evoluzione di una grande impresa al confine tra pubblico e private.* Milan: F. Angeli, 1992.

Martinelli, Alberto, ed. *International Markets and Global Firms: A Comparative Study of Organized Business in the Chemical Industry.* London and Newbury Park, CA: Sage Publications, 1991.

Mason, Mark. *American Multinationals and Japan: The Political Economy of Japanese Capital Controls, 1899–1980.* Cambridge, MA: Harvard University Press, 1992.

Mauskopf, Seymour H., ed. *Chemical Sciences in the Modern World.* Philadelphia: University of Pennsylvania Press, 1993.

McKelvey, Maureen. *Evolutionary Innovations: The Business of Biotechnology.* New York: Oxford University Press, 2000.

Midler, Christophe. "Modèles gestionnaires et régulations économiques de la conception," in Gilbert De Terssac et Ehrard Friedberg, eds. *Coopération et conception.* Toulouse: Octares, 1996.

Midler, Christophe, and Florence Charue-Duboc. "Beyond Advanced Project Management: Renewing Engineering Practices and Organizations." In Rolf A. Lundin and Christophe Midler, eds. *Projects as Arenas for Renewal and Learning Processes.* Dordrecht: Kluwer, 1998.

Mol, Arthur P. J. *The Refinement of Production: Ecological Modernization Theory and the Chemical Industry.* Utrecht: van Arkel, 1995.

Molony, Barbara. *Technology and Investment: The Prewar Japanese Chemical Industry.* Cambridge, MA: Harvard University Press, 1990.

Morandi, Luigi, and Giovanni Pantini. *Dialogo sull'industria chimica. Vie e modelli di sviluppo.* Milan: Etas libri, 1982.

Morikawa, Hidemasa. *Chiho Zaibatsu.* Tokyo: Toyo? Keizai Shuppansha, 1988.

Morikawa, Hidemasa. *Zaibatsu: The Rise and Fall of Family Enterprise Groups in Japan.* Tokyo: Tokyo University Press, 1992.

Mowery, David C., and Richard R. Nelson, eds. *Sources of Industrial Leadership. Studies of Seven Industries.* Cambridge: Cambridge University Press, 1999.

Müller-Fürstenberger, Georg. *Kuppelproduktion. Eine theoretische und empirische Analyse am Beispiel der chemischen Industrie.* Heidelberg: Physica, 1995.

Müller, Margrit. "Good Luck or Good Management? Multigenerational Family Control in Two Swiss Enterprises since the 19th Century." *Entreprises et Histoire*, 12 (1996): 19-47.

Murmann, Johann Peter. *Knowledge and Competitive Advantage: The Coevolution of Firms, Technology, and National Institutions.* Cambridge and New York: Cambridge University Press, 2003.

Nadal, Jordi. "La debilidad de la industria química española durante el siglo XIX. Un problema de demanda." In *Moneda y Crédito* 186 (1986): 33-70.

Nagel, Dieter. *Die ökonomische Bedeutung der Mineralöl-Pipelines.* Hamburg: Deutsche Shell AG, 1968.

Neukirchen, Heide. "Mühsamer Prozess. BASF: Ludwigshafen wird umgebaut." *Manager Magazin*, 32, no. 10 (1 October 2002): 46-49.

Nohria, Nitin, Davis Dyer, and Frederick Dalzell. *Changing Fortunes: Remaking the Industrial Corporation.* New York: Wiley, 2002.

Nouschi, André. *La France et le pétrole*, Paris: Picard, 2001.

Ohlin, Bertil, Per-Ove Hesselborn, and Per Magnus Wijkman, eds. *The International Allocation of Economic Activity.* London: Macmillan, 1977.

Pettigrew, Andrew M. *The Awakening Giant: Continuity and Change in ICI.* Oxford: Basil Blackwell, 1985.

Plumpe, Gottfried. *Die I.G. Farbenindustrie AG. Wirtschaft, Technik und Politik 1904-1945.* Berlin: Duncker & Humblot, 1990.

Porritt, Jonathon, and David Winner. *The Coming of the Greens.* London: Fontana, 1988.

Porter, Michael E. *Competitive Strategy: Techniques for Analyzing Industries and Competitors.* New York: The Free Press, 1980.

Pressnell, L. S., ed. *Money and Banking in Japan.* London: Macmillan, 1973.

Puig, Núria. "Una multinacional holandesa en España: La Seda de Barcelona." In *Revista de Historia Industrial*, 21 (2002): 123-58.

Puig, Núria. "The Search for Identity: Spanish Perfume in the International Market, 1901-2001." *Business History*, 45, no. 3 (2003): 90-118.

Puig, Núria and Javier Loscertales. "Las estrategias de crecimiento de la industria química alemana en España 1880-1936: exportación e inversión directa." *Revista de Historia Económica*, 19, no. 2 (2001): 345-87.

Puig, Núria. *Bayer, Cepsa, Puig, Repsol, Schering y La Seda: Constructores de la química española.* Madrid: Lid, 2003.

Radkau, Joachim. "Wirtschaftswunder" ohne technologische Innovation? Technische Modernität in den 50er Jahren." In Axel Schildt and Arnold Sywottek, eds. *Modernisierung im Wiederaufbau: Die westdeutsche Gesellschaft der 50er Jahre.* Bonn: Dietz, 1993.

Reader, W. J. *Imperial Chemical Industries, A History. Volume 1: The Forerunners, 1970-1926.* London: Oxford University Press, 1970.

Reader, W. J. *Imperial Chemical Industries, A History. Volume 2: The First Quarter Century, 1926-1952.* London: Oxford University Press, 1975.

Roe, Mark J. *Strong Managers, Weak Owners.* Princeton, NJ: Princeton University Press, 1994.

Rosenberg, Nathan, Ralph Landau, and David C. Mowery, eds. *Technology and the Wealth of Nations.* Stanford, CA: Stanford University Press, 1992.

Ruffat, Michèle. *175 ans d'industrie pharmaceutique française: Histoire de Synthélabo*. Paris: La Découverte, 1996.

Schildt, Axel, and Arnold Sywottek, eds. *Modernisierung im Wiederaufbau. Die westdeutsche Gesellschaft der 50er Jahre*. Bonn: Dietz, 1993.

Schröter, Harm G. "The International Dyestuff Cartel, 1927-39, with Special Reference to the Developing Areas of Europe and Japan." In Akira Kudo and Terushi Hara, eds., *International Cartels in Business History*. Tokyo: University of Tokyo Press, 1992.

Schröter, Harm G. "Strategic R&D as Answer to the Oil Crisis, West and East German Investment into Coal Refinement and the Chemical Industries, 1970-1990." *History and Technology*, 16 (Autumn 2000): 383-402.

Schröter, Harm G. "Die Auslandsinvestitionen der deutschen chemischen Industrie 1930 bis 1965." *Zeitschrift für Unternehmensgeschichte* 46 (2001): 186-9.

Schröter, Harm G. "Unternehmensleitung und Auslandsproduktion: Entscheidungsprozesse, Probleme und Konsequenzen in der schweizerischen Chemieindustrie vor 1914 (Governance of the firm and foreign production: decision-making processes, problems and consequences in the Swiss chemical industry before 1914)." *Schweizerische Zeitschrift für Geschichte* (Swiss Journal of History), 44 (1994): 14-53.

Seymour, Raymond B., and Tai Cheng, eds. *History of Polyolefins: The World's Most Widely Used Polymers*. Dordrecht: Kluwer, 1986.

Shimotani, Masahiro. *Nippon Kagaku Kogyoshi Ron*. Tokyo: Ochanomizu Shobo, 1992.

Smith, John Graham. *The Origins and Early Development of the Heavy Chemical Industry in France*. Oxford: Clarendon Press, 1979.

Smith, John Kenly, Jr. "World War II and the Transformation of the American Chemical Industry." In Everett Mendelsohn, Merritt Roe Smith, and Peter Weingart, eds. *Science, Technology and the Military*. Boston: Kluwer, 1988, 307-22.

Smith, John Kenly, Jr. "The End of the Chemical Century? Organizational Capabilities and Industry Evolution." *Business and Economic History*, 23, no. 1 (Fall 1994): 152-61.

Smith, John Kenly, Jr. "Patents, Public Policy, and Petrochemical Processes in the Post-World War II Era." In *Business and Economic History*, 27, no. 2 (Winter 1998): 413-19.

Spitz, Peter H. *Petrochemicals: The Rise of an Industry*. New York: Wiley, 1988.

Spitz, Peter, ed. *The Chemical Industry at the Millennium: Maturity, Restructuring, and Globalization*. Philadelphia: Chemical Heritage Press, 2003.

Stobaugh, Robert B. *Innovation and Competition: The Global Management of Petrochemical Products*. Boston: Harvard Business School Press, 1988.

Stobaugh, Robert B., and Louis T. Wells, Jr., eds. *Technology Crossing Borders: The Choice, Transfer and Management of International Technology Flows*. Boston: Harvard Business School Press, 1984.

Stokes, Raymond G. *Divide and Prosper: The Heirs of I.G. Farben under Allied Authority 1945-1951*. Berkeley: University of California Press, 1988.

Stokes, Raymond G. *Opting for Oil: The Political Economy of Technological Change in the West German Chemical Industry, 1945-1961*. Cambridge: Cambridge University Press, 1994.

Stokes, Raymond G. "Von der I.G. Farbenindustrie AG bis zur Neugründung der BASF (1925-1952)." In Werner Abelshauser, ed., *Die BASF. Eine nternehmensgeschichte*. München: C. H. Beck, 2002.

Streck, Wolf Rüdiger. *Chemische Industrie. Strukturwandlungen und Entwicklungsperspektiven*. Berlin: Duncker & Humblot, 1984.

Supple, Barry, ed. *Essays in British Business History*. Oxford: Clarendon Press, 1987.

Sylla, Richard. *The American Capital Market*. New York: Arno Press, 1975.

Teltschik, Walter. *Geschichte der deutschen Großchemie, Entwicklung und Einfluß in Staat und Gesellschaft*. Weinheim: VCH 1992.

Tilly, Richard. "Germany: 1815-70." In Rondo Cameron, ed. *Banking in the Early Stages of Industrialization*. New York: Oxford University Press, 1967.

Tokuhisa, Yoshio. *Kagaku Sangyo ni Miraiwa Aruka*. Tokyo: Nippon Keizai Shinbunsha, 1995.

Travis, Anthony S., et al., eds. *Determinants in the Evolution of the European Chemical Industry, 1900-1939*. The Netherlands: Kluwer, 1998.

Turner, John, ed. *Businessmen and Politics: Studies of Business Activity in British Politics, 1900-1945*. London: Heinemann. 1984.

Udagawa, Masaharu. *Shinko Zaibatsu*. Tokyo: Nippon Keizai, 1984.

Vogel, David. *National Styles of Regulation: Environmental Policy in Great Britain and the United States*. Ithaca, NY: Cornell University Press, 1986.

Wall, Bennett H. *Growth in a Changing Environment: A History of Standard Oil Company (New Jersey) 1950-1975*. New York: McGraw-Hill, 1988.

Whitehead, Don. *The Dow Story: The History of the Dow Chemical Company*. New York: McGraw-Hill, 1968.

Whittington, Richard, and Michael Mayer. *The European Corporation: Strategy, Structure, and Social Science*. London: Oxford University Press, 2000.

Yamaguchi, Takashi, and Ikue Nonaka. *Asahi Kasei and Mitsubishi Kasei: Sentan Gijutsu ni kakeru Kagaku*. Tokyo: Otsuki Shoten, 1991.

Zamagni, Vera. *An Economic History of Italy 1860-1990*. Oxford: Clarendon Press, 1993.

Zamagni, Vera. "L'ENI e la chimica." *Energia* 24 (2003): 16-24.

Index

AB Celloplast, 242
AB Nobel Plast, 245
Abbot Laboratories, USA, 162
ABCM (Association of British Chemical
  Manufacturers), 294
academia in the U.S. chemical
  network, 172
academic chemists, hired by DuPont,
  173
academic researchers, attracted to
  industrial technology, 172
accounting standards
  changing, 58
  for Ciba-Geigy, 217
acetylene
  liquefaction process, 275
  provider of, 145
acids
  Bayer integrating backward into,
    24
  large demand for, 23
acrylic polymers, production during
  World War II, 175
acrylonitrile, 150

action phase (mid-1980s to early
  1990s), 66–71
A.D. Little. *See* Arthur D. Little
added value products, firms searching
  for, 407
Addison, Christopher, 293
Age of Petroleum, 449
Agell, Josep, 378
AGFA
  Auguste Victoria coal mine interfirm
    agreement, 24
  diversification into photochemicals,
    23
  receiving crucial support from
    bankers, 88
  sold by Bayer, 80
  spin-off of by Bayer, 76
  starting as technology followers,
    23
aging industry, signs of, 57
AGIP, 350
agrarian sector, modernization of
  Spain's, 387
AgrEvo, 75

465

as a dynamic part of the economy,
220
employees, 198, 202
expansion and
internationalization of, 193–195
experts, 198
future prospects of, 222
overview of, 195–202
performance, 200
share in total industrial
production and exports after
World War I, 196
franc, overvaluation of, 210
Japanese exports and imports of
chemical products, 324
patents issued 1985–1995, 337
turnover and mean growth, 252
synergy benefits, achieved by Borealis,
249
Synthélabo
acquired by L'Oréal, 277
merger of Sanofi with, 280
synthetic dyes
Britain leading the way in, 84
lead time for developing, 88
synthetic fertilizers, Spain a net
importer of, 376
synthetic fiber industry
in Japan, 320
stabilization of the European, 157
synthetic fibers
consumption of, 224
driving Rhône-Poulenc's growth
after World War II, 255
remarkable development of at
Rhône-Poulenc, 259
synthetic materials
demand for a range of, 224
growth rate of after the Second
World War, 224
synthetic rubber
Allied ban on the production of,
142
prohibition on the manufacture of,
143
during World War II, 175
synthetics, Ciba production of, 203

Taiwan
Japanese exports and imports of
chemical products, 324
patents issued 1985–1995, 337
Takeda Pharmaceutical, 341
takeovers
environment supportive of in
Germany, 109
widening the scope for R&D
activities at Ciba-Geigy, 214
tar-dyes manufacture, Ciba and Geigy
roots in, 203
tariffs, establishment of, 377
Tarragona, 387
technical assistance services, 23
technical competencies in Japan, 331
technical education
deficient in Britain, 289
systematic required for the second
industrial revolution, 289
technical leadership, cost-leadership
related to, 56
technocratic culture in the German
chemical industry, 146–147
technocratic government in Spain, 385
technocratic management in Germany,
142
technological capabilities
of U.S. companies, 30
weakness of becoming significant in
Japan, 317
technological competencies
of Japanese chemical companies,
309
Japanese level of, 313
technological developments,
importing, 308
technological frontier, Japan
competitiveness conditioned
by, 313
technological implementation,
requiring steps beyond
scientific discovery, 84
technological revolution, international
chemical industry on the brink
of another, 367
technological strength of ICI, 295